Martin Wolfgang Krüger

Personalized Multi-Scale Modeling of the Atria

Heterogeneities, Fiber Architecture, Hemodialysis and
Ablation Therapy

Personalized Multi-Scale Modeling of the Atria

Heterogeneities, Fiber Architecture, Hemodialysis and
Ablation Therapy

Vol. 19
Karlsruhe Transactions on Biomedical Engineering

Editor:
Karlsruhe Institute of Technology (KIT)
Institute of Biomedical Engineering

Eine Übersicht über alle bisher in dieser Schriftenreihe erschienenen Bände finden Sie am Ende des Buchs.

Personalized Multi-Scale Modeling of the Atria

Heterogeneities, Fiber Architecture, Hemodialysis and Ablation Therapy

by
Martin Wolfgang Krüger

Dissertation, Karlsruher Institut für Technologie (KIT)
Fakultät für Elektrotechnik und Informationstechnik, 2012

Impressum

Karlsruher Institut für Technologie (KIT)
KIT Scientific Publishing
Straße am Forum 2
D-76131 Karlsruhe
www.ksp.kit.edu

KIT – Universität des Landes Baden-Württemberg und
nationales Forschungszentrum in der Helmholtz-Gemeinschaft

KIT Scientific Publishing 2013
Print on Demand

ISSN 1864-5933
ISBN 978-3-86644-948-0

Personalized Multi-Scale Modeling of the Atria: Heterogeneities, Fiber Architecture, Hemodialysis and Ablation Therapy

Zur Erlangung des akademischen Grades eines

DOKTOR-INGENIEURS

von der Fakultät für

Elektrotechnik und Informationstechnik

des Karlsruher Instituts für Technologie (KIT)

genehmigte

DISSERTATION

von

Dipl.-Ing. Martin Wolfgang Krüger

geb. in Meckenheim

Tag der mündlichen Prüfung: 08. November 2012
Referent: Prof. Dr. rer. nat. Olaf Dössel
Korreferenten: Dr. Kawal S. Rhode
Prof. Dr. rer. nat. Wilhelm Stork

Contents

Part IV Applications

Acknowledgments

This thesis was conducted at the Institute of Biomedical Engineering, Karlsruhe Institute of Technology (KIT). I sincerely thank all persons who contributed to the success of the research.

First of all, I am expressing my deepest thanks to Prof. Dr. Olaf Dössel for giving me the opportunity to work in a stimulating environment, for supporting my ideas and for trusting in me to work in the euHeart project. I thank Dr. Kawal Rhode for his continuous commitment in the clinical data acquisition and for refereeing my thesis. Furthermore, I would like to thank Prof. Dr. Wilhelm Stork for the interest in my work and his referee on the thesis.

I want to express my deep gratitude to Dr.-Ing. Gunnar Seemann for an excellent technical supervision as well as valuable discussions about cardiac modeling. I also would like to heartly thank Dr.-Ing. Frank Weber and Dr.-Ing. Christopher Schilling with whom I shared, besides frequent technical discussions, many nice moments in the "vierter Stock". Furthermore, I want to thank Walther Schulze for the interesting joint business travels and discussions and I would like to express my thanks to Thomas Fritz for the joint supervision of two student projects. Thanks also go to Dr.-Ing. David Keller with whom I overcame challenges in BSPM recording and image segmentation. I kindly thank all my former and current colleagues at the IBT for providing valuable feedback on my research activities and for critically proofreading this manuscript.

The people involved in euHeart, especially within work package 6, deserve my gratitude for an excellent collaboration. In particular, I would like to thank the people working at King's College London, who were impressively helpful in the patient selection and data acquisition. Futhermore, I want to thank Dr. Cristian Lorenz, Dr. Hannes Nickisch, Dr. Alexandra Groth and Dr. Hans Barschdorf from

Philips Research for providing cardiac segmentation software, an excellent support and the joint supervision of two Diploma theses. I also want to express my thanks to Prof. Dr. Pyotr Platonov, Dr. Fredrik Holmqvist and Dr. Jones Carlson from Lund University, with whom I worked together for two projects related to the atrial ECG. The hemodialysis project was conducted with Dr. Stefano Severi, University of Bologna, and his collaborators, whom I thank for the data collection and scientific exchange during the project. I thank Dr. Jichao Zhao and Prof. Dr. Bruce Smaill from the University of Auckland for providing a high resolution model of the sheep atria. I would like thank Prof. Dr. Dámien Sánchez-Quintana from the Universidad de Extramadura, Badajoz, for providing anatomical photographies and Prof. Dr. Javier Saiz and his co-workers at the University of Valencia for giving me the opportunity to temporarily work with them.

A number of students contributed through their own theses to the work described in this thesis: Lisa-Mareike Busch, Andreas Dorn, Wilfried Dzeakou, Dominik Kutra, Marieme Ly, Jan Richter, Viktor Schmidt, Max Sirkin, Arthur Teimourian and Bhawna Verma. I am thankful for their work and their dedication and passion to the project. Furthermore, I want to thank Bettina Schwab, Wilfried Dzeakou, Andreas Dorn and Arthur Teimourian for working for me outside of their theses.

At last, I want to thank my parents and family for supporting me throughout my studies and the course of this thesis. The greatest thanks go to my girlfriend Julia. Without her patience and support, especially in times when we did not see each other too often, this thesis could not have been realized.

The research leading to these results has received funding from the European Communitys Seventh Framework Programme (FP7/2007-2013) under grant agreement no 224495 (euHeart project).

Diese Arbeit entstand mit Unterstützung des
Karlsruher House of Young Scientists (KHYS).

List of Abbreviations

AF	atrial fibrillation
aHTD	atrial hybrid tension development (model)
AMDB	Anatomical Model Database
AP	action potential
APD$_{xx}$	action potential duration at xx% repolarization
APG	atrial appendage
AV	atrio-ventricular
AVR	atrio ventricular valve ring
BB	Bachmann's bundle
BCL	basic cycle length
bpm	beats per minute
BSPM	body surface potential map
CAM	common atrial myocardium
CAP	Cardiac Atlas Project
CC	correlation coefficient
Chaste	Cancer, Heart and Soft Tissue Environment (software)
CFAE	complex fractionated atrial electrogram
CGAL	Computational Geometry Algorithms Library
CLT	common left trunk
CRN	Courtemanche-Ramirez-Nattel (model)
CS	coronary sinus
CT	computed tomography
CT	crista terminalis
CV	(excitation) conduction velocity
DTI	diffusion tensor MRI
EAMS	electroanatomical mapping system
ECG	electrocardiogram
ECGI	ECG imaging
ERP	effective refractory period
ESRD	end stage renal disease
FO	fossa ovalis
GIMIAS	Graphical Interface for Medical Image Analysis and Simulation (software)
HD	hemodialysis

HTD	hybrid tension development (model)
IBT	Institute of Biomedical Engineering
II	inferior isthmus
ITK	Insight Segmentation and Registration Toolkit
IVC	inferior vena cava
KIT	Karlsruhe Institute of Technology
LA	left atrium
LAA	left atrial appendage
LAT	local activation time
LGE	late Gadolinium enhancement
LIPV	left inferior pulmonary vein
LPV	left pulmonary vein(s)
LSPV	left superior pulmonary vein
MAE	mean absolute error
MRI	magnetic resonance imaging
MVR	mitral valve ring
PCL	pacing cycle length
PM	pectinate muscle
PV	pulmonary vein
PVI	pulmonary vein isolation
PWd	P-wave duration
RA	right atrium
RAA	right atrial appendage
RF	radio frequency
RFA	readio frequency ablation
RIPV	right inferior pulmonary vein
RMPV	right middle pulmonary vein
RMSE	root mean squared error
RPV	right pulmonary vein(s)
RSb	beginning of starting repolarization
RSe	end of starting repolarization
RSPV	right superior pulmonary vein
SN	sinus node
SVC	superior vena cava
T2w	T2 weighted
TaWd	Ta-wave duration
TEE	transesophageal echocardiogram
TMV	transmembrane voltage
TVR	tricuspid valve ring
VCG	vectorcardiogram
VF	Visible Female
VM	Visible Man
VPH	Virtual Physiological Human
VTK	Visualization Toolkit
WL	wave length

1

Introduction

Atrial fibrillation (AF) is the most common cardiac arrhythmia [1]. It is mostly an age-related disease and thus patient numbers are constantly increasing in the developed world. Although AF itself is not lethal, it accounts for several secondary diseases and death-causes, e.g. formation of thrombi causing stroke. AF can be managed with anti-arrhythmic and anti-coagulation drugs, but the curative ablation therapy becomes more popular. Especially for younger patients, this therapy may be a better choice. However, the success rates reported for ablation interventions are still moderate and a major portion of patients need to undergo several interventions before the arrhythmia is cured [1]. Personalized computational models of the atria may contribute to understand the reasons for the moderate therapy success and improve the treatment in the future.

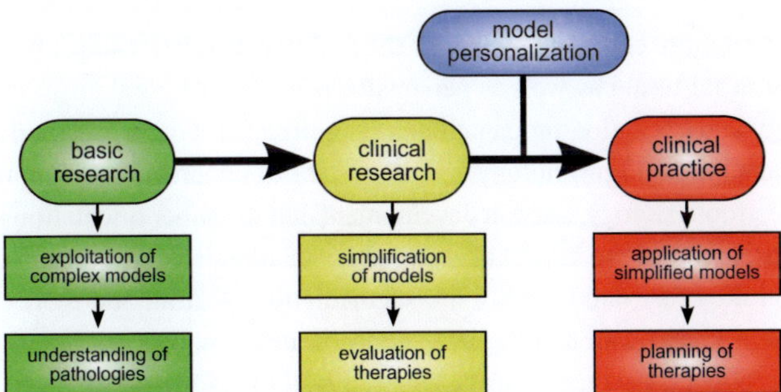

Fig. 1.1. Overview of the fields of usage of atrial models. Atrial models are currently exploited to understand pathologies in basic research. Atrial modeling is also in a transition phase to the application of atrial models in the clinical practice. This requires a personalization of the models to the specific patient. For examples to the fields of usage of atrial models see Figure 16.1 in the conclusions chapter.

Computational cardiac modeling started with modeling single cell ventricular electrophysiology over half a century ago. The interest in modeling the atria was comparable low for a long time, but atrial modeling has provided valuable insights into physiological and pathological mechanisms in basic science over the last decade or two. The studies most often focused on providing insights into mechanisms leading to AF [2]. As cardiac imaging and electrical measurements as well as computational resources refined to a great extent over the last decade, the use of cardiac modeling in clinical practice moved into the focus of interest [3, 4]. Application of cardiac modeling in the clinic requires a personalization of the models to reflect the patients anatomy and electrophysiology as well as to suit the clinicians needs and expectations. This is not an easy task to solve, especially for the atria, as their anatomy is more complex compared to the ventricles and atrial electrophysiology was studied in less detail in the past.

Atrial modeling is currently in a transition phase between the exploitation of simulations with well-established complex models to aid basic science and the introduction of simplified atrial models into the clinical practice [5] (Fig. 1.1). Both fields of applications, although often requiring contrary technical approaches, may contribute to the understanding of atrial diseases and therapy possibilities. Even more, knowledge gained from detailed simulations joined with insights from personalized macroscopic simulations may provide a more complete understanding of complex relationships between microscopic changes and macroscopic observations.

To generate a more complete picture of the multi-scale relationships, both detailed models of atrial tissue as well as personalized models of atrial anatomy, electrophysiology and excitation propagation are required [6]. For the detailed modeling of atrial anatomy and physiology it is important to understand and reproduce regional electrophysiology, tension development and myofiber orientation to e.g. aid the understanding of the impact of incomplete ablation lesions. Personalized atrial models on the other hand require the segmentation of atrial and thorax anatomy from patient image data and the adaptation of atrial conduction velocity and interatrial conduction sequence to fit electrophysiological measurements from the patient. Thereby, the fusion of electrical and anatomical data may benefit the personalization in both regards. Additionally, parameters of the complex cell electrophysiology models need to be identified which could allow a stable personalization. These methods need to be automated as much as possible and should require only minimal additional measurements to allow for an acceptance of the methods by the clinical personnel in the future. Otherwise the potential benefits in the disease diagnosis and therapy evaluation will be outruled by the prolongation of the patient treatment.

Once developed, detailed and personalized atrial models can be applied to various fields from basic science, clinical research and clinical practice. This will provide new insights to understand pathological and therapeutical mechanisms, ranging from AF initiation to moderate success rates of ablation therapy. In the long-run, personalized computational multi-scale atrial modeling can improve the outcome of ablation therapy, decrease actual treatment time through *in-silico* treatment planning and thus reduce both socioeconomic costs and patient burden.

Aims of the Thesis

The euHeart project, as part of the European Commisions seventh framework program, aims at creating integrated multi-scale patient-specific cardiac models for treatment planning and evaluation [7, 8]. Within the work of the seventeen European partners from industry, academia and clinical practice, the major goal of this thesis is the creation of personalized atrial models and the use of these for the model-based evaluation and planning of radio-frequency ablation. This aim is broken down into several tasks reaching from cellular modeling to ECG computation. In particular the following milestones are covered:

- Design of a new body surface potential mapping (BSPM) layout and establishment of an acquisition and post-processing workflow of BSPM and intracardiac data.
- Investigation of heterogeneous atrial electrophysiology and tension development. Adjustment of models of the atrial electrophysiology and tension development to the experimental findings. Study and realistic simulation of the atrial repolarization sequence regarding a realistic ECG.
- Development of methods and a framework for the personalization of models of atrial anatomy, electrophysiology and excitation conduction.
- Model-based examination of radio-frequency ablation from microscopic observations to multi-physics therapy evaluation.

- Understanding of the influence of hemodialysis therapy for renal disease patients on the increased incidence of AF in these patients.

Structure of Thesis

Part I provides an overview of the medical and modeling fundamentals underlying the work in this thesis.

- **Chapter 2** explains the general cardiac anatomy and electrophysiology and discusses AF and the measurement of cardiac electrical activity. The topics are covered only briefly, as these are part of the focus of the thesis and thus described in more detail in Chapter 3 and Part II. radio-frequency ablation as a curative therapy for AF is also introduced in this chapter.
- **Chapter 3** gives a comprehensive overview of the atrial anatomy and electrophysiology and introduces the basic methods of cardiac modeling in this regard.

Part II covers the description of the data acquisition and processing techniques as well as the atrial modeling methods.

- **Chapter 4** explains data acquisition and signal processing methods established as part of this thesis. In the end, section 4.5 summarizes the acquired clinical data.
- **Chapter 5** introduces and evaluates new techniques of atrial modeling. A method to augment geometrical atrial models with rule-based fiber orientation, tissue labels and generic ablation lesion patterns is presented. A well-established model of atrial electrophysiology is enhanced to reproduce electrophysiological properties from different regions in the atria. This model is also coupled with a model of atrial tension development which allows the realistic simulation of local atrial tension. Furthermore, a fast method to simulate atrial excitation is described.

Part III presents and evaluates techniques for the personalization of atrial models.

- **Chapter 6** introduces a framework for the personalization of multi-scale atrial models.

- **Chapter 7** describes the methods and results of the efforts of anatomical personalization. In the end of the chapter, an overview of the models created in this thesis and their use for projects related to this thesis and undertaken by external project partners is given.
- **Chapter 8** presents techniques to personalize atrial electrophysiological models.
- **Chapter 9** demonstrates methods to personalize the electrical excitation sequence in the atria. This spans from the individualization of interatrial conduction pathways to the estimation of global conduction velocity from ECG data.

Part IV gives examples of applications of atrial models spanning from basic research to clinical tasks.

- **Chapter 10** investigates the atrial repolarization sequence in a cohort of anatomical models.
- **Chapter 11** shows an example for the simulation of AF in a 3D personalized atrial model.
- **Chapter 12** examines the behavior of gaps in ablation lesions under different physiological environments.
- **Chapter 13** covers the results of patient-specific model-based evaluation of outcome prediction of radio-frequency ablation procedures based on late enhancement MRI data.
- **Chapter 14** displays the impact of ablation lesions on the cardiac function from electrical excitation to mechanical contraction and relaxation in a four chamber model.
- **Chapter 15** covers a study examining the impact of hemodialysis therapy on the atria from cell electrophysiology to ECG and relates the outcome to the increased incidence of AF in renal-disease patients.

At the end a summarizing conclusion for the thesis is drawn and an outlook for fields of further investigation is given.

Part I

Fundamentals

2

Clinical Fundamentals

2.1 Atrial Anatomy and Physiology

The heart is located posterior to the sternum between the left and right lung lobes. It is a hollow muscular organ of size comparable to a closed human fist. The heart is made up of four chambers. The right atrium collects venous blood from the main circulatory system via the inferior and superior vena cava (IVC/SVC) and also from the coronary sinus (CS). It passes the blood to the right ventricle through the triscuspid valve. From there, the blood is subsequently pumped to the lung. The left atrium (LA) collects oxygenated blood from the lung circulatory system, usually through four pulmonary veins. The LA passes the blood through the mitral valve into the left ventricle, which then pumps the blood into the body circulatory system. The heart itself is perfused by the coronary arteries which originate from the ascending part of the aortic arch. Blood from the coronary veins is collected in the CS, which runs posterior between the left ventricle and LA into the right atrium. The atria are electrically isolated from the ventricles by fibro-fatty tissues at the valve plane [9].

2.1.1 Anatomy

The right atrium can be divided into four components: The right atrial appendage (RAA), the smooth venous part of the posterior free wall, the rough part of the free wall and the septum [10]. The structure of the right atrium is mainly determined by its large muscle bundles: Crista terminalis (CT) [11, 12], 15-20 pectinate muscles (PM) [13], the intercaval bundle [14], tricuspid valve ring musculature and Bachmann's bundle [10]. The CT is the right atriums most prominent structure [11] and separates the smooth venous part and the rough free wall [10]. The CT leads from the right side of the orifice of the SVC on the posterior wall towards the right side of the IVC, encircling its orifice and smoothing out between the CS and IVC in the atrial septum. It diminishes in size between SVC and IVC [11]. The PMs lead from the posterior side over the right lateral and anterior portion of the right atrium towards the ring muscle around the orifice of the triscuspid valve. The first superior PM is called septum spurium and is thicker (4.2 mm vs. 1-3 mm) and less long than the other PMs [11]. The septum spurium forms the onset of the musculature of the RAA. The most inferior PM encircles the orifice of the CS. The sinus node (SN) is located at the superior beginning of the CT near the ostium of the SVC and the onset of the RAA [15]. The RAA is triangular-shaped and clasps the aortic arch [11].

The left atrium lies with its posterior wall next to the oesophagus and has a smooth endocardial surface [16]. The wall thickness of the smooth left atrial free wall reduces from the roof to the low posterior aspect (Fig. 2.1). The left atrial appendage (LAA) is located supero-anterior of the left superior pulmonary vein (LSPV). It is clearly delineated from the left atrial myocardium and has a rough endocardial surface. Its outer shape is tubular with multiple finger-like branches [17]. Under physiological conditions, the left atrium has a volume of 42 ml [18] and a diameter of 3.7 cm [18, 19]. Structural remodeling caused by AF commonly increases left atrial volume [20].

The region of the right atrium enclosed by the tricuspid valve and the orifice of the IVC forms an isthmus [9, 21]. In patients with atrial flutter, this region has a slower conduction velocity [22] and is therefore a substrate for macro-reentries and target of RFA [21]. In the left atrium, the region enclosed by the left inferior pulmonary vein (LIPV) and the mitral valve annulus is also an isthmus (mitral-isthmus) and may also be targeted by RFA in AF patients [23].

2.1.1.1 Septum and Interatrial Connections

The atria are separated by the atrial septum. The plane of the atrial septum is orientated posteriorly rightward oblique [9]. The septum contains an electrically isolating layer [25–28], but both sides of the septum are covered with regular myocardium from the right atrium and left atrium, respectively. The atria are electrically connected by various interatrial bridges and the limbus of the fossa ovalis

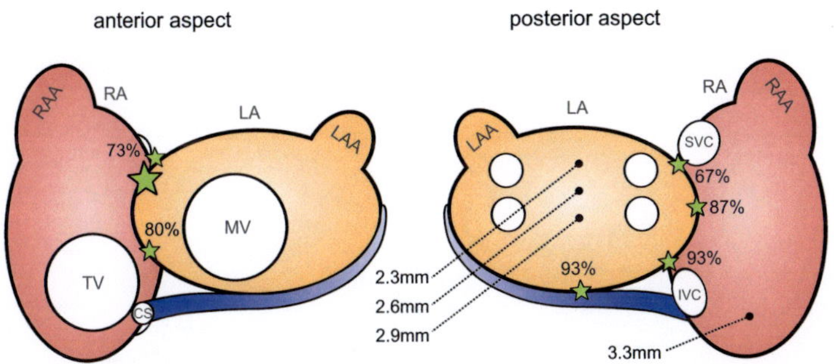

Fig. 2.1. Interatrial connections (green stars, with likelihood) and atrial wall thickness in the healthy population. Data from [10] and [24].

(FO) [29–33] (Fig. 2.1). The most prominent interatrial bridge is Bachmann's bundle (BB) on the supero-anterior side. BB extends between the two appendages, splitting at the beginning of the appendages and encircling these partly [10]. The inferior part of Bachmann's bundle connects with the atrio-ventricular valve ring musculature. Along its extent, BB bridges the interatrial grove [34, 35] and has a width of approximately 4.6 mm [36]. The CS is also covered with myocardium and therefore builds another electrical connection. The other interatrial connections can either bridge or tunnel the epicardial fat closely to the myocardium (bridge-type vs. path-type connection) [37]. The location and number of interatrial connections is highly variable between individuals [28, 37–39]. Figure 2.1 schematically shows the locations of the interatrial bridges and the probability of existence in the population.

2.1.1.2 Pulmonary Veins

Typically two left and two right pulmonary veins (PVs) drain blood from the five lung lobes into the left atrium [41]. The vessel of the right middle lung lobe joins the vessel of the right upper lobe prior to connecting to the left atrium [42]. The drainage patterns of the PVs vary between individuals. A multitude of anatomical studies addressed this issue [40–48]. The findings vary across studies, as neither the definition of the PV drainage pattern nor the focus of the studies are uniform. Marom et al. defined four different patterns of left PV ostia (1-2 PVs) and ten different patterns of right PV ostia (2-5 PVs) [40]. Figure 2.2 depicts schematically the observed patterns. The most common variations to the normal four PV pattern

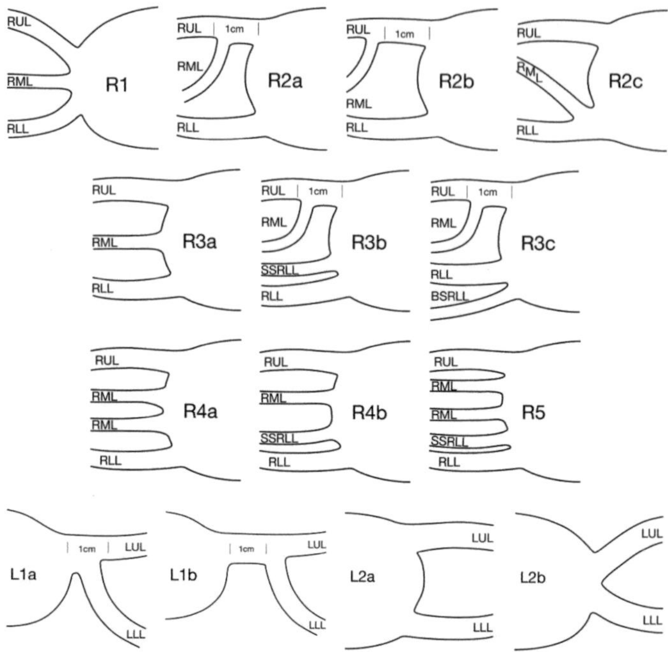

Fig. 2.2. Drainage patterns of right and left pulmonary veins as reported by [40]. Figure adapted from [40].

are a third middle right PV [42] and a joint trunk of the left PVs (common left trunk) [43, 45]. A common trunk is present, if the distance between the bifurcation of the PVs to the left atrial body is greater than 1.5 cm [45] or just outside the atrial body [43]. A short common left trunk may be present in more than half of all individuals [47] and a third right middle PV in up to 86% of individuals [42].

The wall of the PVs is approximately 0.5 mm thick [49]. The diameter of the PV ostia is larger in the upper PVs than in the lower PVs (11 vs. 8 mm) [50]. AF patients often show an increased diameter [50]. The PV onsets are covered with myocardial cells near their ostia [9, 51, 52]. The extent of these muscular sleeves are longer in the upper PVs than in the lower PVs [38, 49, 53–55]: LSPV 41 mm, RSPV 30 mm, LIPV 21 mm, RIPV 13 mm [53], 10.9-7.2 mm [49] and 14.8-6.0 mm [54]. This might concur with a higher rate of ectopic activity in the upper PVs [56]. Contradicting results were found whether the length of the sleeves varies between healthy and AF-patients [49, 54]. The fiber direction at the junction of the PV sleeves and the left atrial myocardium can show abrupt

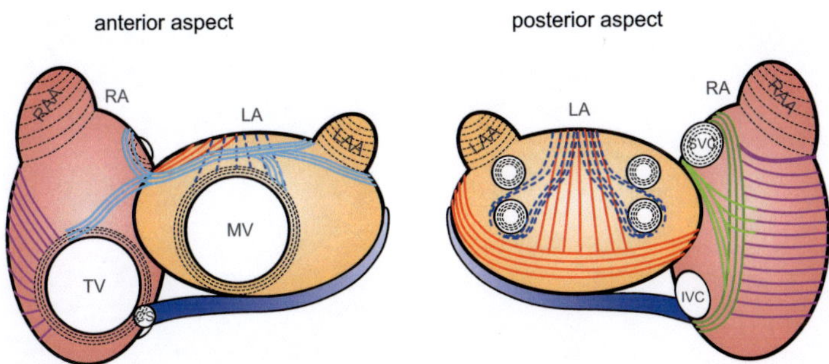

Fig. 2.3. Schematic representation of the major atrial fiber bundles and layers. Blue: septo-atrial bundle (subendocardium), red: septo-pulmonary bundle (subepicardium), light blue: Bachmann's bundle, green: crista terminalis, green: intercaval bundle, pink: pectinate muscles, black: circular fiber orientation around vessel orifices and valves as well in the appendages.

changes [57, 58]. Fibers are oriented circumferentially symmetric in the muscular sleeves of the PVs [52].

2.1.1.3 Fiber Architecture

The myofiber architecture in the human atria is characterized by non-transmural fiber layers and a number of prominent muscular bundles. Figure 2.3 schematically shows the major fiber bundles and layers in the right and left atrium. In the right atrium, myofibers encircle the tricuspid valve ring and SVC [59]. Fiber orientation is aligned along the longitudinal extension of the CT, PMs and BB. PM fibers end perpendicular in the CT and tricuspid ring muscle fibers (Fig. 2.4b). The myocardium between the pectinate muscles shows a fiber orientation similar to the pectinate muscles. The intracaval bundle is the main muscular structure of the venous component of the right atrium (Fig. 2.4e). It encircles the orifice of the SVC. One muscle bundle leaves this circular structure left lateral of the orifice and extends towards the right lateral side of the orifice of the IVC [14]. It connects with the CT on the posterior wall. There are no muscular extensions and thus no fiber orientation in the IVC [52].

The left atrium is comprised of two layers of myocardial fibers overlapping each other (Fig. 2.3). In contrast to ventricular fiber architecture, the transmural change in fiber orientation between those layers is abrupt. Fiber orientation is organized in two rather continuous layers compared to the right atrium with its multiple of

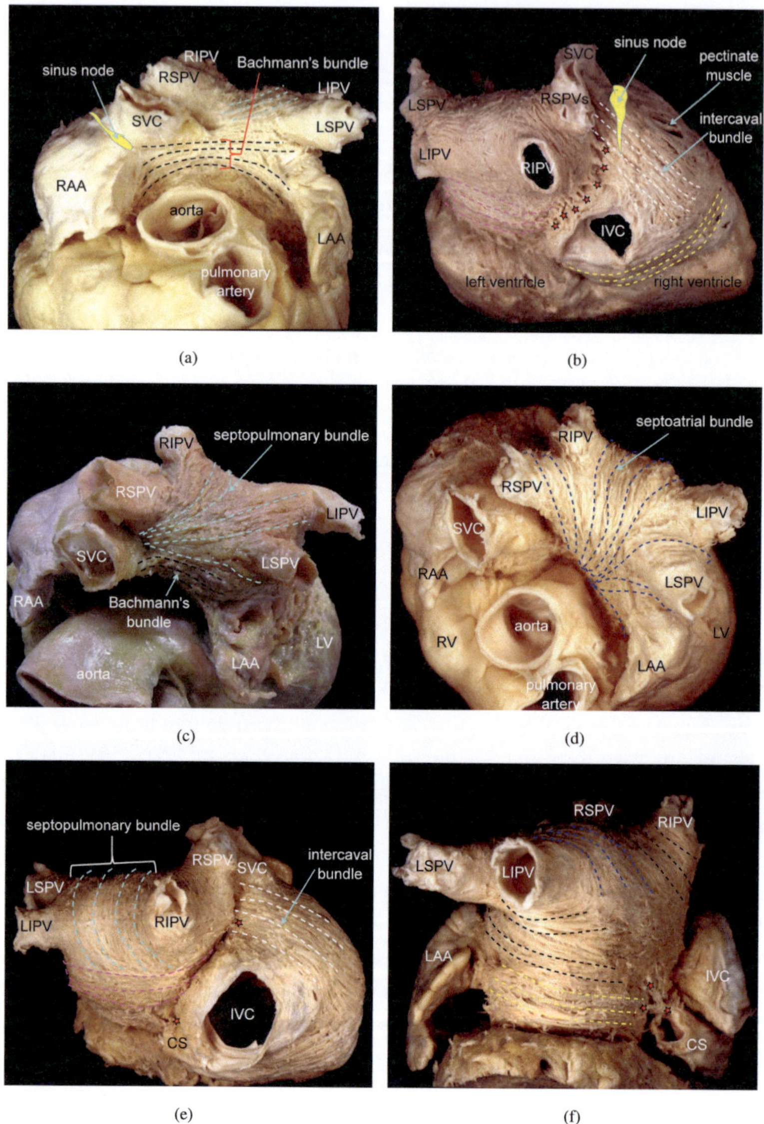

Fig. 2.4. Local fiber orientation (dashed lines) in various dissections of human atria. Black and red stars denote interatrial muscle bridges. Photographs provided by Damien Sanchez-Quintana and reproduced with permission. The photographs were partly also used in [26].

independent bundles. We find the septo-atrial bundle in the subendocardium and a septo-pulmonary bundle subepicardially [10] (Fig. 2.4c–e). The septo-atrial bundle circles the orifice of the mitral valve [59] and proceeds broad over the anterior wall towards the superior PVs [10] (Fig. 2.4f). It thereby incorporates the left atrial appendage (LAA) in its structure and continues to spread over the left atrial roof to encircle the orifices of the inferior pulmonary veins (Fig. 2.4f). Fibers encircle the sleeves of the PVs symmetrically from the ostium to the end of the sleeves [52, 59]. The fiber orientation in the atrial appendages may either be circularly organized or appear rather chaotic depending on the individual patient [10].

2.1.1.4 Wall Thickness

The myocardial wall of the atria is non-uniformly thick with values ranging between 1 mm and 3 mm [13, 16, 24, 60–65]. In the right atrium, the wall thickness is mainly determined by the thickness of the dominant muscular bundles (CT, PMs) [13], whereas in the left atrium also the smooth wall shows variation in thickness. The left atrial wall is thicker on the posterior and inferior side compared to the roof (2.9±1.3 mm vs. 2.3±1.0 mm) [24, 62]. Neher et al. reported that the left atrial wall is thinner than the right atrium wall (2.1 vs. 3.3 mm) [60], whereas Wang et al. found the opposite [13]. Females have a slightly thinner atrial wall compared to males [62]. AF patients usually have a thinner wall than the healthy population (2.1-2.5 mm vs. 2.3-2.9 mm) [24]. The regional wall thickness determines the RF energy needed to create transmural RFA lesions and is thus important to know for the individual patient [38, 61, 63, 66]. Table 2.1 summarizes the findings of the atrial wall thickness.

2.1.1.5 Implications for Modeling the Atrial Anatomy

Models of the human atrial anatomy need to be build purpose specific. In some cases it is suitable to create highly detailed models with a very fine resolution of specific parts of the atria, in order to understand regional conduction behavior [67–69]. In majority of cases however, especially if the model is to be applied in the field of clinical research or clinical practice, macroscopic bi-atrial models are needed. Such models should respect the specific anatomy of a patient, e. g. special attention should be laid on the PV drainage pattern. Additionally, the models

Table 2.1. Atrial wall thickness reported in literature.

study	species	condition	location	wall thickness
Neher et al. 2011 [60]	human (7)	control	RA (11 regions)	3.3 mm
			LA	2.1 mm
Otomo et al. 2010 [61]	porcine (13)		both atria	3.2 - 4.2 mm
Platonov et al. 2008 [24]	human (298)	control	LA SPV	2.3 mm
			LA CPV	2.5 mm
			LV IPV	2.9 mm
		AF	LA SPV	2.1 mm
			LA CPV	2.2 mm
			LV IPV	2.5 mm
Hall et al. 2006 [62]	human (34)	control	LA anterior	1.86 mm
			LA roof	1.06 mm
			LA posterior	1.4 mm
			LA isthmus	1.6 mm
			LA septum	2.2 mm
Deneke et al. 2005 [63]	human (7)	AF	LA isthmus	4-10 mm
			LA PV ostium	1-3 mm
			LA posterior	2-5 mm
			RA1	3-5 mm
			RA2	2-10 mm
Hassink et al. 2003 [49]	human (20)	AF & control	LA anterior	1.6 mm
			LA posterior	1.7 mm
Wang et al. 1995 [13]	human (9)	control	RA terminal grove	5-8 mm
			RA posterior vestibule	2 mm
			LA anterior	4-5 mm
			LA post/ant vestibule	3 mm
Coffey et al. 1981 [64]	human (n/a)	n/a	atrium	0.5 - 3.5 mm
			atrium	2 - 3 mm
Sunderman 1949 [65]	human (n/a)	n/a	atrium	2 mm

should include regionally varying atrial wall thickness to incorporate the multi-layer fiber architecture of the atria. Within the atrial wall, fast conducting bundles should be annotated and also other structures with altered electrophysiology or excitation behavior should be marked. Furthermore, the atrial septum should contain an isolating layer and several discrete interatrial bridges. If the computation of the atrial ECG signal is desired, also a thorax model needs to be created from image data of the same individual. In the thorax model, skeletal muscle, lungs, blood, intestine, fat and liver should be distinguished for simulations of the atrial ECG signal [70, 71].

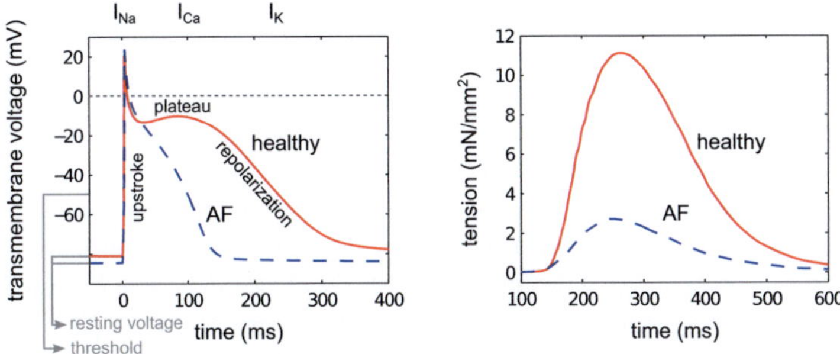

Fig. 2.5. Action potential and tension development in an healthy and a AF-remodeled (Sec. 3.2.2) human right atrial cell. Ionic currents mentioned above the action potential indicate the ionic currents most responsible for the different phases of the healthy action potential. Action potentials simulated with the CRN model [72, 73]. Tension data from [74].

2.1.2 Electrophysiology

The atrial electrophysiology is introduced in this section. Atrial electrophysiology and pathology is reviewed in more detail by Schotten et al. [75]. The myocardium consists of different cells. Cardiac myocytes form the main portion of these, fibroblasts a significantly smaller one. Myocytes are excitable and contracting cells. The cell membrane is selectively permeable for certain types of ions (mainly Na^+, K^+, Ca^{2+}) through ionic channels, pump and exchangers. Each pore is permeable for one or multiple ions. The channels may open and close depending on various factors, e.g. transmembrane voltage, ion concentrations or other transmitters. At rest, ionic concentrations inside and outside of the myocytes are in an equilibrium state and a transmembrane voltage of -80 to -90 mV forms.

If an external stimulus raises the potential difference between the extracellular and intracellular space above a certain threshold (approximately -50 to -60 mV), sodium channels open, allowing a fast influx of Na^+ ions. This is the so-called depolarization phase of the cardiac action potential (Fig. 2.5). After several milliseconds, the fast sodium channels are inactivated and calcium channels open. The slow Ca^{2+} influx prolongs the depolarization state of the cell, which is the plateau phase of the action potential. After 200 to 400 ms, repolarizing potassium channels open and the K^+ outflow begins. The sodium channels leave the inactivation state at the end of the repolarization phase. Thus, a new action potential cannot be triggered unless the previous one is nearly completed. This so-called refractory period prevents arrhythmias. Figure 2.5 depicts typical action potentials from the healthy and AF-remodeled human right atrium. AF remodeling is explained in section 3.2.2.

The intracellular spaces of adjacent myocytes are coupled via gap junctions. These are non-selective ion channels formed by two connexons, one in each cell membrane. A depolarization therefore spreads via intracellular currents through these gap junctions. So, the myocardium forms a syncytium: If the myocardium is activated at one place, a depolarization wave will spread over the whole myocardium. The conduction velocity of this wave is determined mainly by the gap junction density and the upstroke velocity of the action potential.

Some regions in the atria show a specialized electrophysiology. The sinus node (SN) and the atrio-ventricular (AV) node do not have a stable resting membrane potential, but constantly slowly depolarize through open sodium and / or calcium channels. As the SN depolarizes more quickly than the AV node, the frequency of induced action potentials is greater and the SN is thus overruling the pacemaker activity of the AV node.

Excitation Contraction Coupling

The Ca^{2+} influx during the plateau phase of the action potential triggers Ca^{2+} release from the junctional sarcoplasmic reticulum through the ryanodin-2-receptor-channels (Ca^{2+}-induced Ca^{2+} release, positive feedback loop). The cytosolic Ca^{2+} binds to troponin C. This removes the troponin complex from the myosin binding sites of actin. ATp attaches to the myosin and the myosin heads bind to the actin filaments [76]. Under ATP hydrolysis, the myosin heads pull the actin filaments towards the sarcomere center, which shortens the sarcomere. Afterwards, new ATP replaces the ADP at the myosin heads, which subsequently relax and can bind to a new position on the actin filament. The so-called cross bridge cycle is repeated until cytosolic Ca^{2+} concentration decreases again. The sarco-endoplasmic reticulum calcium-ATPase (SERCA) pumps calcium back into the sarcoplasmic reticulum and further ATPases pump Ca^{2+} out of the intracellular space into the extracellular matrix. Figure 2.5 shows the tension development measured in healthy and AF-remodeled atrial cells.

2.1.3 Sinus Rhythm Activity

The atrial excitation sequence starts in the SN. The SN is located in the right atrium between the orifice of the SVC and the onset of the RAA. The excitation wave travels faster along the CT, BB and within the PMs [85–87]. These structures have a higher conduction velocity and also an increased anisotropy compared to the common atrial myocardium (CAM) [88]. Bachmann's bundle usually activates the left atrium on the supero-anterior side [89]. A second wavefront activates the left atrium from the inferior right side. The latter wave is caused by fast downward activation of the right atrium via the CT and is carried over the septum with the musculature of the CS. The two waves merge below the LIPV approximately 100 ms after the depolarization of the SN. The conduction velocity in the atrium lies between 51 and 120 cm/s under physiological conditions (Tab. 2.2). Under pathological influences, the measured variation in conduction velocity increases [5]. The conduction velocity reduces with age [90]. Anisotropy ratios in fast conducting structures may lie between 4.76:1 and 8:1 [79, 88]. Ratios between 2:1 and 3:1 are commonly used for regular myocardium in simulation studies [91–93]. For AF conditions, this ratio is often increased in the simulations to 10:1 and 12:1 [94–96].

2.2 Atrial Fibrillation

Atrial fibrillation (AF) is the most common sustained cardiac arrhythmia [97]. It is characterized by chaotic excitation waves and fast, chaotic contraction of the atria [1]. Normally, the atria maintain a heart rate of 60–80 bpm in rest. The heart rate is controlled by the cardiac pacemaker function of the SN, which can adapt to the metabolic needs of the body. During AF, the rate is increased to 400-600 bpm. AF is not directly mortal but favors the formation of thrombi in the atria which can cause ischemic stroke. AF-caused strokes thereby form a significant portion of all ischemic strokes [98].

Mechanisms

AF is a self-begetting disease [99]. Paroxysmal AF (recurrent AF, spontaneous termination within 7 days) is most commonly caused by ectopic activity in the

Table 2.2. Conduction velocities measured in human atria. FW: free wall, H: healthy, SR: sinus rhythm, CAB: coronary artery bypass, Aflut: atrial flutter, OSA: obstructive sleep apnea, FAT: focal atrial tachycardia. Table adapted from [5].

study	subjects	region	long (cm/s)	BCL (ms)	condition
Feld et al. 1997 [22]	9	septum	58-61	300-600	H
		RA FW	58-64		
		isthmus	51-55		
Hansson et al. 1998 [77]	12	RA FW	88	SR	H
Lin et al. 1999 [78]	21	RA septum	96-98	200-500	H
		RA FW	94-100		
Spach et al. 1988 [79]	49	PM	70	800	H
Dimitri et al. 2011 [80]	20	RA	120	SR	H
	20	LA	120	SR	H
Kanagaratnam et al. 2002 [81]	16	RA FW	83	1006	CAB
			77	500	
Feld et al. 1997 [22]	9	septum	50-54	300-600	Aflut
		RA FW	53-57		
		isthmus	37-42		
Lin et al. 1999 [78]	25	RA septum	86-91	200-500	Aflut
		RA FW	91-98		
Schilling et al. 2001 [82]	8	isthmus	77	240	Aflut
		RA smooth	108		
		PM	133		
Shah et al. 1997 [83]	9	isthmus	70	243	Aflut
		RA sup	120		Aflut
		RA lat	110		Aflut
		RA septum	110		
Dimitri et al. 2011 [80]	20	RA	80	SR	OSA
		LA	90		
Weber et al. 2011 [84]	5	RA, LA post&roof	108	SR	AF / FAT
		RA, LA post&roof	86	500	
		RA, LA post&roof	77	300	

PVs [100] and develops to persistent AF (sustained AF > 7 days) upon non-treatment. Paroxysmal AF may then further progress to long-term persistent AF (continuous AF for > 12 month) and permanent AF (sinus rhythm cannot be restored by any means), when AF is not interrupted by regular sinus activity anymore (Fig. 2.6).

Epidemiology

In the European Union about 4.5 million patients suffer from AF [101]. Men are thereby slightly more prone to AF than women [101] and the prevalence of begetting AF is increasing with age (8% for individuals over 80 years) [102]. Caused

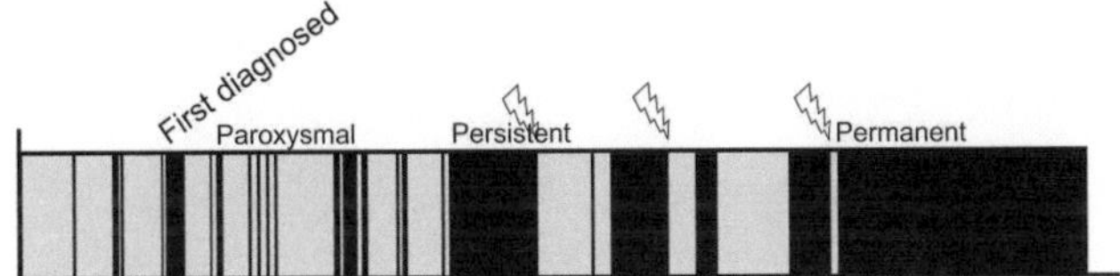

Fig. 2.6. Example for a natural time course of AF. Episodes of AF (black) and sinus rhythm (gray) alternate over time (x-axis) as the disease progresses from the undiagnosed state to paroxysmal, persistent to permanent AF. Flashes indicate cardioversions. Figure from [75].

by the aging of the western society and probably also because of environmental influences, the absolute number of AF patients will rise significantly in the future [103].

Treatment

The mechanisms leading to and sustaining AF are to-date not completely understood [1]. In patients with paroxysmal and persistent AF, the PVs are a common target for therapies, as ectopic triggers arising from the muscular sleeves of the PVs were shown to initiate AF [100]. Besides such triggers, pathological substrate in the atria is needed to sustain the arrhythmia. Both, the triggers and the substrate are altered under the presence of AF [104]. This so-called remodeling causes the cell electrophysiology and tissue properties to change in such way, that AF is easier initiated and/or perpetuated. Also, the left atrial autonomic ganglionic plexi are believed to play a role in the initiation and perpetuation of AF [1].

Therapies targeting AF can either manage the symptoms of the disease using drugs (palliative therapy) or try to eliminate the substrate and triggers causing AF (curative therapy). Catheter ablation of AF has been shown to result in better long-term success rates than antiarrhythmic drug therapy [105]. Nevertheless, the kind of treatment strongly depends on the state of AF of the individual patient, the expected improvement in quality of life and thirdly also in the health economical cost. The major cardiologist associations provide guidelines on how to treat the different atrial arrhythmias [1, 97].

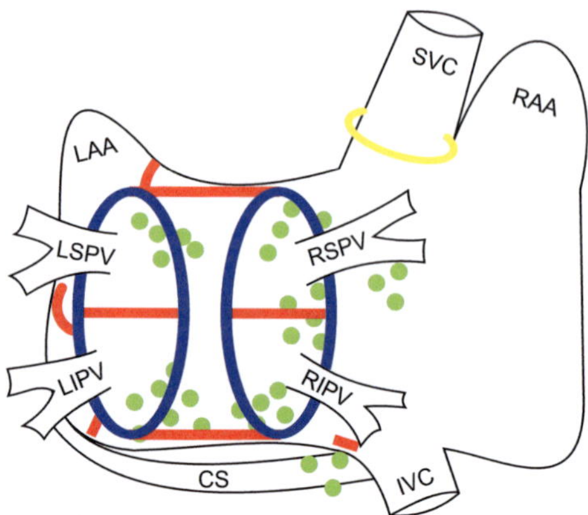

Fig. 2.7. Pulmonary vein isolation lesions (blue), linear lesions (red), circumferential SVC isolation (yellow) and additional local lesions (green). Figure based on figures from [1].

Economic Factors

Treatment of AF is very costly and the number of patients is constantly increasing. In the European Union, the overall economic burden caused by AF is estimated to about 13.5 billion EUR annually [97]. Pharmacological treatment of AF caused annual costs of about 780 to 1.500 EUR per patient (Germany, 2009) [106], although the clear delineation of costs between AF treatment and accompanying diseases is difficult [107]. Costs of curative catheter ablation treatment of AF vary between years of treatment. The initial ablation procedure causes costs around 40.000 EUR per patient [107] or at least 50.000 USD [1], but the annual follow-up costs are substantially lower. The time for equalization of costs between pharmacological treatment and curative ablation therapy strongly depends on the first-pass success rate of the catheter ablation and lies anywhere between 3 to 8 years [107].

2.3 Radio-Frequency Ablation

Radio-frequencey ablation (RFA) is a curative minimal-invasive therapy for paroxysmal and persistent AF. In Germany approximately 12.000 AF catheter ablations are performed annually (2011) [108]. A comprehensive review about the state-of-art in ablation of atrial fibrillation can be found in [1].

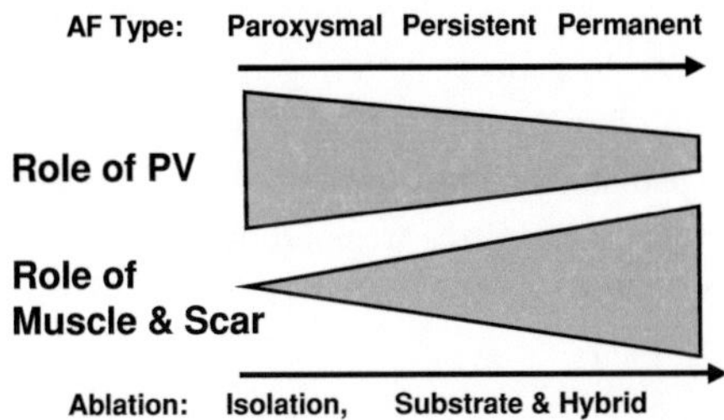

Fig. 2.8. Substrate evolution leads to changes in ablation technique. Figure from [109].

During RFA intervention, various catheters are introduced into the right heart via the femoral veins (Sec. 2.4.2). Transseptal puncture in the atria allows access to the left heart. Besides the ablation catheter, mapping catheters and a coronary sinus catheter may be introduced (Fig. 2.11). By applying radio-frequency currents via the ablation catheter, the clinician tries to electrically isolate tissue regions, mostly in the left atrium, which are thought to be cause and/or substrate of the abnormal activation. The myocardium is thereby heated with the radio-frequency current until it is necrotic and thus electrically isolating. In a first step, the PVs are usually isolated circumferentially from the left atrial myocardium to isolate ectopic triggers in the PV sleeves (pulmonary vein isolation, PVI) [1, 100]. Afterwards, linear lesions may be added to form a second barrier for abnormal excitation (Fig. 2.7). Ablation lines are realized as a set of point lesion. Regions in which complex fractionated electrograms (CFAEs) were measured or regions with pathological dominant frequency may also be target of ablation (electrogram-guided ablation). These are believed to act as a substrate for the maintenance of AF. From an engineering perspective, the process of RF-ablation of AF can be simplified as a three-step approach (Fig. 2.7):

1. Isolation of the pulmonary veins, to isolate ectopic foci as initiators of AF
2. Ablation of regions with strong fractionation in electrograms or abnormal dominant frequency, to eliminate substrate for maintenance of AF
3. Placement of additional linear lesions, to reduce anatomical substrate for excitation of rotors, which maintain AF or atrial flutter

Depending on the type of AF, not all of these steps will be conducted [1] (Fig. 2.8). Ablation of paroxysmal AF will only include step 1 and sometimes additionally step 2. For the treatment of persistent or long-term persistent AF all steps might be performed. Common linear ablation lesions include a roof-line and a mitral-isthmus line (Fig. 2.7). In patients suffering from atrial flutter, atrial isthmuses need to be ablated to eliminate the substrate for flutter. This may also be done prospectively in AF patients to reduce the risk of development of atrial flutter after RFA therapy. As the AF mechanisms are not fully understood, it is not yet possible to find the most appropriate ablation strategy for each individual AF cause and each individual patient [1].

The success rates of RFA of AF reported in literature show a large variation [1]. This is mainly due to variations in the definition of the success of therapies. Quite commonly, patients need to undergo a second RFA procedure after atrial arrhythmias reoccurred, which is or is not accounted for in the success rates. For the termination of paroxysmal AF, circumferential PVI has been reported to have the best success rates compared to other ablation approaches [109], but it leaves room for improvement [110]. A 3-month cure rate of 65% was reported by Fisher et al. [109], but 24% of the patients needed a redo-procedure to achieve this number. During redo-procedures, gaps in ablation lesions are targeted. In circumferential PVI, these can occur nearly anywhere along the lesion although lesions around the left PVs seems to be slightly more prone to gaps than lesions around the right PVs [111]. The isthmus area between the LIPV and mitral valve is especially prone to gaps [111].

2.3.1 Process of Scar Formation

In RFA procedures, RF alternating current is administered via a catheter to the atrial myocardium with a continuous sinusoidal unmodulated waveform of 300 - 1000 kHz. The most important mechanism of myocardial necrosis induction is based on the conversion of electrical energy into heat within the myocardial tissue (hyperthermia). The myocardium needs to be heated above 50°C in order to become necrotic [112, 113]. Recently Wood et al. found this threshold to be 60°C [114]. If new RFA lesions are placed in areas of previous scar, an increased impedance between catheter tip and myocardium as well as decreased tissue temperatures can be observed [115]. Nevertheless, lesion formation is not affected by underlying scar as long as electrode size, tissue contact and temperature are controlled [116].

These factors need to be managed in such a way that transmural lesions are formed. As the thickness of the atrial myocardial wall varies regionally [117], the parameters also need to be adjusted during an intervention. Overheating may occur if too much energy is distributed to the myocardium. Tissue overheating may lead to coagulum formation and charring on the electrode tip. The electrical impedance between the catheter tip and myocardium then rapidly increases, leading to a loss in effective myocardium heating. To prevent such complications, temperature-controlled energy application systems have been developed. A thermistor or thermocouple embedded in the tip of the ablation catheter allows temperature monitoring at the electrode-tissue interface during energy application. Maximal RF energy (usually 50 W) is delivered until the preselected target temperature has been reached and thereafter automatically titrated down to maintain the target temperature. Additionally, recent RFA systems cool the tip of the catheter with water to prevent non-uniform heating at tissue directly exposed to the catheter tip [118] (irrigated tip catheter, Fig. 2.11(c)). This approach also allows for an application of higher RF currents [119].

The extent of the lesion formed by RFA is influenced by many parameters such as delivered RF power, electrode length, electrode orientation, blood flow and tissue contact [120]. In particular, the following relationships have been found experimentally:

- Linear dependency of lesion depth on
    - myocardial temperature [113, 118, 121].
    - the applied RF current [118].
    - the tip-electrode radius [119, 121].
    - the applied power [118, 121, 122].
    - the delivered energy [118].
- Logarithmic dependency of lesion depth & width on the application time (steady state after approximately 45-60 seconds) [113, 122, 123].
- Linear dependency of lesion width on the contact force [123, 124].

During RFA, various heat transfer processes take place in the region of the catheter tip and myocardium. The desired effect is that the catheter tip exchanges heat conductively with the myocardium. Resistive heating of myocardial tissue is only effective for distances less than 2 mm from the catheter tip [119]. Additionally, a resistive heating of blood and tissue near to the catheter tip is taking place. The catheter as well as the endocardial myocardium is cooled by convective heat loss to the circulating blood pool. Additional heat loss is observed in areas with large coronary arteries [121, 125].

The electrophysiological properties of myocardial cells change under hyperthermia. These effects are reversible with falling temperature and are thus not subject to modeling of the long-term behavior of RFA scars. The processes are described here for the matter of completeness. During the process of heating, tissue conductance is increased by 2% per degree Celsius [125]. Wood and Fuller reported an increase in conduction velocity during hyperthermia, most likely caused by cell shrinkage and / or changes in cell-to-cell coupling [126]. Haines reports an increased automaticity in cardiac cells which are heated around 50°C in contrast to cells at 45°C [118]. Nath et al. revealed a sigmoidal relationship between tissue temperature and resting membrane voltage [112, 113]. They also report changes in action potential morphology and duration [112]. Additionally, three stages of excitability of myocardial cells during hyperthermia were shown in this study. The median temperature associated with normal excitability (44.0°C) is significantly lower than the median temperature associated with reversible loss of excitability (48.0°C) and irreversible loss of excitability and tissue injury (50.5°C) [112, 113]. Acutely after application of RF current, a significant reduction in action potential duration APD_{50} (-41%), APD_{90} (-19%), APD_{max} (-16%), and conduction time (16±3 ms vs. 13±4 ms) can be observed in the border zone surrounding the necrotic core of the ablation lesion [126]. These effects seem to resolve after approximately 3 weeks. Other groups found similar effects in a radius of up to 8 mm around acute lesions [127, 128].

Necrotic regions caused by RFA show a loss of cell definition, separation of the fibers by edema, extravascular red blood cells, and a loss of nuclei and cross-striations [129]. In the border zone around the necrosis, the cross striations and sarcolemma are partially preserved. Cells in this zone might recover and thus the electric isolation might be undone. Tissue in RFA lesions loses its normal fiber orientation [119]. In the border zone these changes are similar to peri-infarct regions, but the tissue will have a texture of diffuse fibrosis instead of patchy fibrosis. Macroscopically seen, diffuse fibrosis has rather isotropic conduction properties [130] and is thus also reflecting a loss in myofiber orientation.

More detailed information about lesion formation and behavior can be found in various review articles. Ndrepepa and Estner provide a comprehensive review of

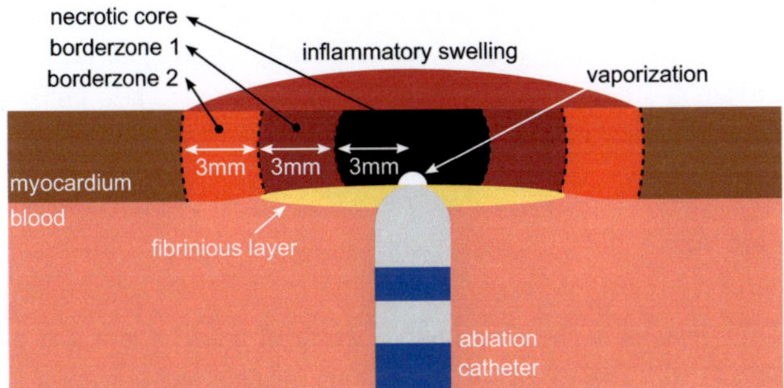

Fig. 2.9. Schematic representation of an acute radio-frequency ablation lesion in the atria.

biophysics of ablation scars with a special focus on RFA in the atria [119]. Similar reviews are provided by Haines [118] and by Haemmerich [131]. Additional information on the biophysics of ablation scars ca be found in [132]. The thermodynamical effects of RFA are summarized in reviews by Nath et al. [113] and Berjano [133].

2.3.2 Lesion Structure

In the acute phase after RFA, lesions have two border zones (3 mm radius each) [113, 132, 134]. These have a reduced blood flow due to microvascular injury [134]. In contrast, others studies describe only two zones: A necrotic core with spots of hemorrhage and a hemorrhagic border zone [119, 135]. The latter is caused by disruption of endothelia cells and erythrocyte passage.

The shape of the lesions is determined by the thermal processes during RFA. Circulating blood in the chamber cools the endocardial surface and thus prevents a great endocardial extent of the hyperthermia. This cooling process has only very limited transmural extent and thus, the lesion extent at the endocardial surface is less than in the midmyocardial region. Epicardially, the scar is narrower than in the center, because the resistive heating from RFA is also decreasing with distance. Additionally, coronary vessels at the epicardial surface form another heat sink.

The acute lesion is covered endocardially with a fibrinious layer [119]. Volume loss in ablated tissue may lead to an impressed endocardial scar surface [119].

In excised tissue, lesions have a pale white color, due to the local denaturation of myoglobin [119]. Overheating of tissue directly underneath the electrode tip may lead to small areas of vaporization of tissue. There are also hints, that if the myocardial wall is very thin, an epicardial inflammatory swelling might occur.

2.3.3 Lesion Size

Regular RFA lesions in the atria have a width of 5.3 ± 1.4 mm and a depth of 2.2 ± 1.3 mm [136]. The extent of the lesions changes post-ablation [113, 126, 136, 137]. Saul et al. found an increased lesion width one month post-ablation (8.7 ± 4.3 mm), but a decrease in lesion depth (1.2 ± 0.4 mm) [136]. The decrease in lesion depth may contradict findings from other groups which found post-ablation lesion growth [113, 137] and also reported an increase in lesion transmurality over time [137]. A more recent MRI-based study also found a change in lesion extent within the first three month after ablation [138]. The previous study did not find any further atrial lesion size changes between month 1 and 3 post-ablation [136].

A decrease in lesion depth may be linked to an inflammatory swelling in the atrial tissue which resolves over time. A long-term increase in lesion transmurality is plausible as AF may reoccur shortly after an RFA procedure, but 60% of these patients do not have AF in the long-run [1]. Deneke et al. reported a chronic lesion width of 6-13 mm (mean 10 mm) and chronic atrial lesion depth of up to 5.5 mm [63]. These results should be regarded carefully, as a lesion depth of over 5 mm is greater than the thickness of the atrial myocardium (<3 mm) [24].

2.3.4 Implications for Modeling Ablation lesions

To model RFA lesions it is important to distinguish between modeling the process of scar formation and modeling the results of an RFA procedure. If a model of the process of scar formation is desired, it is important to take acute changes in tissue properties into account (temperature, electrophysiology). Usually, the heat transfer processes are modeled in this regard [133]. Simulations incorporating the results of an RFA procedure need to be separated in those investigating the acute effects after RFA and those investigating the long-term effects of RFA. For modeling of the acute effect, a slowed conduction and shorter APD in the border zone can be chosen. For modeling of the chronic effects, a healthy action potential morphology along with a model of diffuse fibrosis can be set in the border zone, as fibrinious layer and inflammatory swelling around a lesion can be neglected. Depending on the complexity of the simulations, it can also be advisable to reduce the complexity of the anatomical effects surrounding the RFA lesion. E. g. in simplified models of cardiac excitation, it may not be necessary to include fibroblast-models in the border zone, but instead reduce excitation velocity and anisotropy in this region, as this will not change the simulation outcome significantly. The change in fiber orientation in and around RFA lesions will also impact the active and passive mechanical properties of the myocardium. Therefore, existing models of cardiac elastomechanics may need to be adapted to reflect this behaviour. See Section 3.1.3 for an overview of existing models of RFA in the atria.

2.4 Measurement of Cardiac Electrical Activity

2.4.1 Electrocardiogram and Body Surface Potential Map

Electrical wavefronts in the myocardium cause an electrical field in the body and on its surface. The electrical signals on the body surface can be measured as electrocardiogram (ECG). Routinely a 12 lead (nine electrodes) ECG is recorded clinically. The system contains bipolar (Einthoven I-III between the extremities) and unipolar leads (Goldberger aVR, aVL, aVF and Wilson V1-V6). Long-term ECG recordings and ECG recordings during interventions have a reduced lead system using one to three channels. The vector cardiogram (VCG) is the spatial representation of the temporal changes in the field integral vector. The projection of the field integral vector on the frontal, sagittal and transversal plane results in three orthogonal VCG leads V_x, V_y und V_z. Compared to the standard ECG, the VCG contains less information, as it has fewer leads. On the other hand, its interpretation is sometimes easier, as events in the VCG signals can be better correlated to geometric structures in the heart. More detailed information about the potential distribution on the body surface and thus about the cardiac activity can be gained using a multi-electrode body surface potential mapping (BSPM) systems (Sec. 4.2).

The electrical activity in the atria causes a distinct signal in the ECG (Fig. 2.10). The depolarization wave can be observed in the ECG signal as the P-wave. The P-wave lasts approximately 100 ms and usually ends with the beginning of the QRS

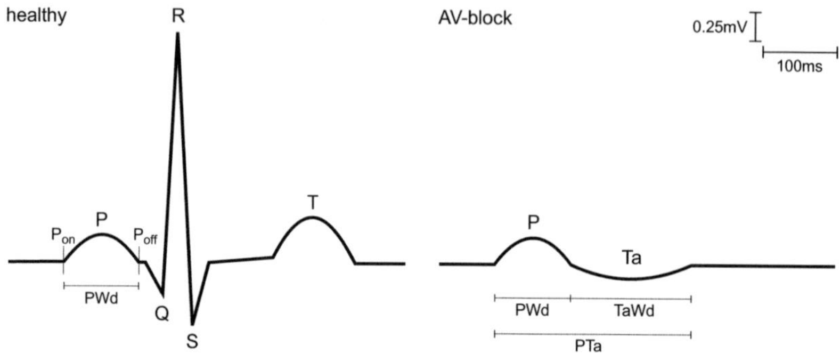

Fig. 2.10. Schematic drawing of the ECG from a healthy person and an AV-block patient. Beginning and end of the P-wave are defined as the crossing with the isoelectric line. PWd: P-wave duration, TaWd: Ta-wave duration, PTa: PTa interval.

complex. The P-wave is a smooth, monophasic positive, Gaussian-like curve in the Einthoven II lead. The signal is usually strongest in this lead with an amplitude of approximately 0.25 mV. In lead aVR, it may be inverted and the P-wave is often biphasic in V1 [139]. Enlargement of the atria, e.g. caused by persistent AF, may result in a bifid P-wave (left atrial enlargement, P mitrale) or in a peaked P-wave (right atrial enlargement, P pulmonale). Ectopic triggers or multifocal arrhythmias can lead to inverted or variable morphologies [139].

The P-wave duration (PWd) corresponds to the time of complete atrial depolarization. Paroxysmal AF patients, or persistent AF patients between AF episodes, show a prolonged P-wave duration [140, 141]. The diagnosis of AF purely based on the PWd is possible [142], but has only moderate specificity and sensitivity [97, 143]. Variations in the dominant interatrial conduction pathway (Sec. 2.1.1.1), which may vary between individuals, may be detected in the ECG and VCG signals [117].

Integrated BSPM over the duration of the P-wave show a dipolar pattern on the thorax front. The maximum lies near the left hip, the minimum near the right shoulder [144–146]. Ectopic foci or a shift in sinus rhythm origin change this pattern [144].

The signal of the atrial repolarization is occluded by the QRS complex in healthy persons. In patients with third degree AV block, the so-called atrial T-wave (Ta-wave) is observable [147–151]. Its polarity is inverted compared to the P-wave and its amplitude is also smaller (Fig. 2.10). The Ta-wave duration is approximately 300 ms [150–152].

2.4.1.1 Calculation of the VCG from 12 lead ECG

Dower et al. [153] introduced a possibility to calculate the standard 12 lead ECG signal from the Frank VCG leads. For this, a matrix D transforms the signals of the VCG leads S_{VCG} into the signals of the eight ECG electrodes (V_1-V_6,I,II) S_{ECG}:

$$S_{ECG} = DS_{VCG}. \tag{2.1}$$

$$D = \begin{pmatrix} -0.515 & 0.157 & -0.917 \\ 0.044 & 0.164 & -1.387 \\ 0.882 & 0.098 & -1.277 \\ 1.213 & 0.127 & -0.601 \\ 1.125 & 0.127 & -0,.086 \\ 0.831 & 0.076 & 0.230 \\ 0.632 & -0.235 & 0.059 \\ 0.235 & 1.066 & -0.132 \end{pmatrix} \tag{2.2}$$

If the opposite operation is desired, e.g. to retrieve the VCG leads from the 12 lead ECG, the Dower matrix would need to be inverted. As the matrix is not quadratic, a pseudo inverse can be used. Carlson et al. suggested to use the Moore-Penrose inverse for this purpose [154]:

$$D^\dagger = (D^T D)^{-1} D^T. \tag{2.3}$$

With this, the VCG leads can be calculated as

$$\hat{S}_{VCG} = D^\dagger S_{ECG} = D^\dagger D S_{VCG} = (D^T D)^{-1} D^T D S_{VCG} = S_{VCG}, \tag{2.4}$$

with the inverse Dower matrix

$$D^\dagger = \begin{pmatrix} -0.172 & -0.074 & 0.122 & 0.231 & 0.239 & 0.194 & 0.156 & -0.010 \\ 0.057 & -0.019 & -0.106 & -0.022 & 0.041 & 0,048 & -0.227 & 0.887 \\ -0.229 & -0.310 & -0.246 & -0.063 & 0.055 & 0.108 & 0.022 & 0.102 \end{pmatrix}. \tag{2.5}$$

In this work, the VCG is used in a study of the interatrial conduction (Sec. 9.1.2). The ECG leads used for the VCG calculation were extracted from the recorded BSPM leads (Figs. 7.7–7.14) to receive signals which are directly comparable to signals measured and post-processed in the same way. By this, no additional error was introduced by differently placed VCG electrodes.

2.4.2 Local Activation Time Maps

Electrophysiological studies can be performed to record local electrical activity in the atria as intracardiac electrograms using electroanatomical mapping systems (EAMS). Multiple catheters are introduced into the right heart via the femoral veins (Fig. 2.11). Usually a near stationary catheter is placed in the CS and at least one mapping catheter and one ablation catheter are put into the atria. The left atrium is accessed by puncturing the fossa ovalis. The endocardial surface of the atria is determined by moving a localized catheter along the atrial wall.

The electrodes on the catheters record the extracellular potential in the proximity to the catheter. Signals can be recorded in two ways. During bipolar measurements, the potential difference between two (adjacent) electrodes is recorded. Unipolar recordings measure the potential difference between one intracardiac electrode and a reference electrode, e.g. on the body surface. A regular, single wavefront passing the electrode results in a biphasic signal. In AF patients, fractionated signals (complex fractionated atrial electrograms, CFAEs) can be observed [1]. The cause for the appearance of CFAEs is under debate.

Regular intracardiac signals in roving catheters from the endocardial wall can be analyzed by the eletrophysiologist and the time of activation per electrode can be manually marked for each measurement. The local activation times (LAT) are thereby determined as the difference to a reference signal, most commonly measured in one of the CS catheter electrode pairs. A collection of activation times over the whole atria and over several sinus rhythm or pacing cycles produces data for a LAT map. Depending on the density of the data collection, the recording of a complete right or left atrial LAT map takes at least five minutes. Afterwards, the LATs are interpolated by the EAMS in regions with no or sparse data.

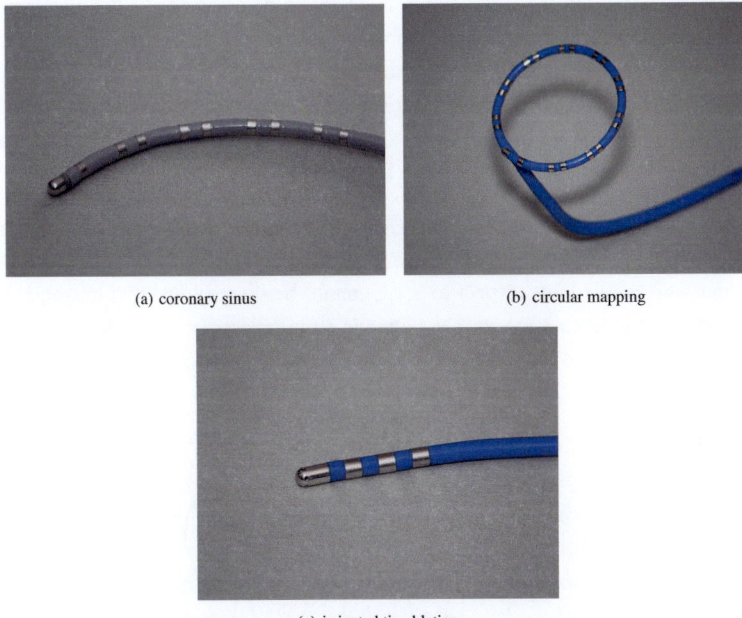

(a) coronary sinus   (b) circular mapping

(c) irrigated tip ablation

Fig. 2.11. Distal tips of various catheters. a) Example of a coronary sinus catheter (Livewire™, St. Jude Medical, St. Paul, MN, USA). b) Circular mapping catheter with ten electrode pairs (Inquiry™ Optima™, St. Jude Medical). c) Irrigated tip ablation catheter (St. Jude Medical).

2.4.3 Conduction Velocity Vector Estimation

The spread of excitation conduction can be determined from LATs [84, 155–160]. If no interpolation has been performed before, sparse activation time data, e.g. from mapping catheters, can be interpolated using radial basis functions [158, 159] or by fitting a cosine function to circularly distributed data [84, 160].

LATs at discrete locations in a 3D object can be expressed in 3D+t Cartesian coordinates x_i, y_i, z_i, t_i. To estimate the conduction velocity, a smooth polynomial surface can be fitted to the active site and its neighboring sites. Barnette et al. introduced this method and chose a quadratic polynomial to account for the wavefront curvature [156]:

$$t_{fit} = ax^2 + by^2 + cz^2 + dxy + exz + fyz + gx + hy + iz + j \qquad (2.6)$$

To solve the fit, the ten coefficients need to be determined. In the ideal case, ten coordinate–time nodes need to be involved in the fit. As linear dependencies between nodes in the fit were commonly observed for activation time data, more nodes needed to be considered for the determination of the coefficients. The overdetermined system of equations could then be solved using a singular value decomposition least squares approach [156].

To determine the direction of the propagation of the excitation wavefront, the gradient can be calculated in closed form from the polynomial fit. Thereby, the partial derivatives of t with respect to x, y, z cannot be directly inverted to obtain the conduction velocity vector, as x, y, z change simultaneously. Barnette et al. [156] provide a solution to calculate the conduction velocity vector components from the gradient in activation times:

$$
\frac{dx}{dt} = \frac{\partial x}{\partial t} + \left(\frac{\partial x}{\partial y}\right)\left(\frac{dy}{dt}\right) + \left(\frac{\partial x}{\partial z}\right)\left(\frac{dz}{dt}\right),
$$
$$
\frac{dy}{dt} = \frac{\partial y}{\partial t} + \left(\frac{\partial y}{\partial x}\right)\left(\frac{dx}{dt}\right) + \left(\frac{\partial y}{\partial z}\right)\left(\frac{dz}{dt}\right),
$$
$$
\frac{dz}{dt} = \frac{\partial z}{\partial t} + \left(\frac{\partial z}{\partial x}\right)\left(\frac{dx}{dt}\right) + \left(\frac{\partial z}{\partial y}\right)\left(\frac{dy}{dt}\right). \tag{2.7}
$$

The excitation propagation direction is assumed to be normal to the wavefront to solve equation 2.7. The partial differentiation of t with respect to x, y, z were defined as

$$
t_x = \frac{\partial t}{\partial x},
$$
$$
t_y = \frac{\partial t}{\partial y},
$$
$$
t_z = \frac{\partial t}{\partial z}. \tag{2.8}
$$

For equation 2.6, these were

$$
t_x = 2ax + dy + ez + g,
$$
$$
t_y = 2by + dx + fz + h,
$$
$$
t_z = 2cz + ex + fy + i. \tag{2.9}
$$

Due to the orthogonality of the wavefront surface and the propagation direction

$$CV_y = \frac{t_y}{t_x} CV_x,$$

$$CV_z = \frac{t_z}{t_x} CV_x \tag{2.10}$$

held true. The combination of equations 2.7 and 2.10 then led to the formulation of the conduction velocity vector components for the fit.

$$CV_x = \frac{dx}{dt} = \frac{t_x}{t_x^2 + t_y^2 + t_z^2}$$

$$CV_y = \frac{dy}{dt} = \frac{t_y}{t_x^2 + t_y^2 + t_z^2}$$

$$CV_z = \frac{dz}{dt} = \frac{t_z}{t_x^2 + t_y^2 + t_z^2} \tag{2.11}$$

Each node i was used in a number of fits N_i. Each of these fits was centered around a reference point j. To achieve a spatial smoothing of the velocity vectors, the weighted sum of velocity vectors calculated per discrete node x_i, y_i, z_i, t_i could be calculated from all fits in which the node i was involved [156].

$$CV_x^i = \sum_{j=1}^{N_i} CV_x^j * w_{ji}$$

$$CV_y^i = \sum_{j=1}^{N_i} CV_y^j * w_{ji}$$

$$CV_z^i = \sum_{j=1}^{N_i} CV_z^j * w_{ji}$$

$$CV^i = \sqrt{(CV_x^i)^2 + (CV_y^i)^2 + (CV_z^i)^2} \tag{2.12}$$

The weight is given as

$$w_{ji} = \frac{1}{LRMSE * D_{ji}} + \frac{3}{RMSE_j}. \tag{2.13}$$

RMSE is the root mean squared error of the fit, LRMSE the RMSE of the linear part of the fit and D_{ji} is the distance between active sites of the fit which also takes the spatio-temporal window size $(x_{win}, y_{win}, z_{win}, t_{win})$ of fit j into account

$$D_{ji} = \sqrt{(x_i - x_j)^2 + (y_i - y_j)^2 + (z_i - z_j)^2 + (t_i - t_j)^2}$$
$$+0.5 * \sqrt{x_{win}^2 + y_{win}^2 + z_{win}^2 + t_{win}^2}. \qquad (2.14)$$

The RMSE and LRMSE provided a measure between the measured time $t_{ji,meas}$ and the calculated times $t_{ji,calc}$, $t_{ji,calc,lin}$. The latter is the time calculated using only the linear part of the fit equation 2.6.

$$t_{ji,calc} = ax^2 + by^2 + cz^2 + dxy + exz + fyz + gx + hy + iz + j$$
$$t_{ji,calc,lin} = gx + hy + iz + j$$

$$RMSE = \sqrt{mean((t_{ji,meas} - t_{ji,calc})^2)}$$
$$LRMSE = \sqrt{mean((t_{ji,meas} - t_{ji,calc,lin})^2)} \qquad (2.15)$$

Figure 2.12 provides an example of conduction velocity vectors calculated from a LAT map recorded with an EAMS. The vectors help to identify the direction of wave propagation and may help to identify zones of altered conduction or conduction block.

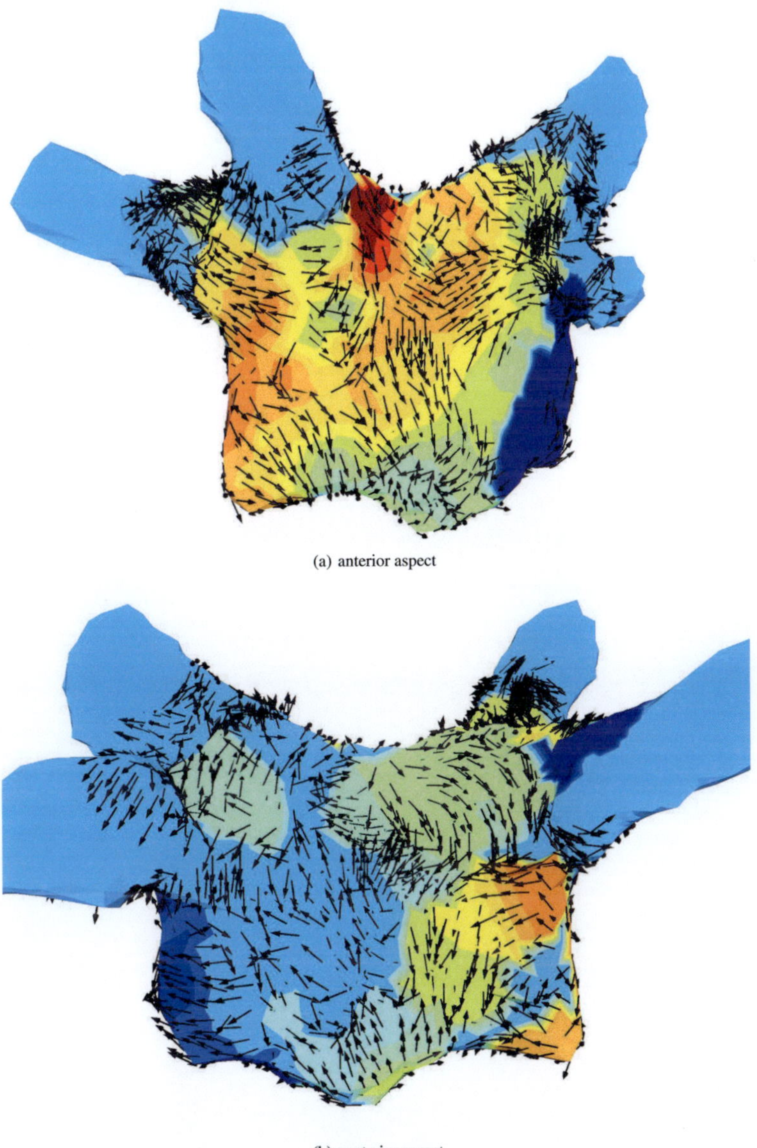

(a) anterior aspect

(b) posterior aspect

Fig. 2.12. Example of a left atrial LAT map with overlay of conduction velocity vectors. The LAT was collected after successful pulmonary vein isolation in a patient suffering from paroxysmal AF (subject 12, Tab. 4.1). Therefore there's no activation in the PVs (constant activation time).

3

Introduction to Modeling of the Atria

3.1 Fundamentals of Anatomical Modeling

3.1.1 Existing Models of Atrial Anatomy

The first models of human atrial anatomy have been established in the 1990s (Tab. 3.1). Models with different complexity were used for the simulations. Either rather simplistic shapes (peanut) and surface models were used with complex propagation models, or larger datasets, e. g. 4 chamber anatomies, were simulated with rule-based propagation models (cellular automata). As computational power increased, detailed anatomical models could also be used with complex electrophysiology and propagation models (monodomain). The creation of the anatomical models was generally a manual process. Thereby, image data from volunteers was manually segmented and further post-processed to build the detailed anatomical models. These models were then often used over a long period of time for various *in-silico* studies. Annotations of different atrial structures allowed the incorporation of few heterogeneities of conduction velocity, anisotropy and action potential. Some models were also extended with generic ablation lesions. Only few groups extended their models to be able to compute the body surface ECG. In most such cases, the body model was not derived from the same volunteer, from whom the atrial anatomy was acquired from. A significant number of the published atrial models, especially models with a finite wall thickness, did not incorporate an isolating layer in the atrial septum. In geometries which only represent the endocardial surfaces, both atria are naturally separated and needed to be connected manually. Table 3.1 provides an overview of published 3D atrial models.

3.1.2 Fiber Architecture in Atrial Models

Almost from the beginning of the use 3D atrial anatomy models, the importance of fast conducting structures and atrial fiber orientation for the atrial excitation sequence was recognized. Already the first complex models of human atrial anatomy [91, 161, 162] incorporated parts of the atrial fiber architecture. Over the last decade, a significant number of new models of atrial anatomy were presented [5]. Some of these models included atrial fiber architecture [92–94, 163–165]. All of these models represented singular anatomies derived from volunteer image data which was manually segmented and annotated. The majority of these models were 3D surface models [93, 163–165] which cannot reflect the complex, multi-layer fiber architecture of the human atria (Sec. 2.1.1.3). Existing volumetric models [92, 94] took fiber architecture to be transmurally constant, thus neglecting intersecting and overlaying bundles and layers. In contrast to this, myofiber architecture is commonly set automatically in ventricular model utilizing statistical [166] or rule-based (data from [167]) methods. For the atria, such methods were not available but initial, incomplete approaches were presented in 2008 and 2009 [94, 168, 169]. Most recently, a high resolution model of sheep atria containing detailed information about the sheep fiber architecture has been presented [96]. Table 3.1 lists the fiber orientation properties of atrial models described in the literature.

3.1.3 Models of Radio-Frequency Ablation

Computational studies of RFA split in three different types of physics. Thereby, the majority of studies focused on modeling the thermodynamical effects of RFA to gain a deeper understanding of the processes leading to the scar formation [115, 125, 170–182]. The main achievements are summarized in a review by Berjano [133].

Electrophysiological modeling of the atria focused for a long time on aiding the understanding and prevention of AF [2]. Only very few groups have tried to model the electrophysiological effects of RFA on cardiac tissue. Some groups tried to examine the temporal electrical behavior of ablation lesions [125, 179, 183]. None of these investigated or included changes in atrial electrophysiology after RFA. Two recent studies model the effects of RFA on intracardiac electrograms to examine the role of late-potentials around lesions [183] and to evaluate the transmurality of lesions from electrograms [61].

Whole atria simulation studies concentrated on investigating the effects of different ablation lesion and Maze procedure patterns on the atrial activation sequence and the success to terminate AF. In all of these studies lesion tissue conductivity was set to zero [93, 184–187]. In a first approach, Ellis et al. modeled ablation lesions in a bi-atrial model in 1997 [184]. The Lausanne Heart Group compared several years later, as computational resources had increased, different catheter ablation patterns to the former state-of-the-art treatment, the Maze procedure [185]. Within this study, they also evaluated the impact of gaps (1.3, 1.7, 2.4, 3.0 mm) in the ablation lines on the outcome of the procedure. As did Rotter et al. (also from the Lausanne group) two years later to evaluate the feasible use of biophysical models to develop new ablation patterns and anticipate RFA success rates [186]. Reumann et al. modeled different ablation patterns in a computer model based on an adaptive cellular automaton [187]. They tested various ablation lesion patterns reported in the clinical literature for the ability to prevent AF onset from ectopic firing in the pulmonary veins. Additionally, the effects of non-transmural lesions were evaluated.

3.1.4 Thorax Models

From the beginning of atrial modeling only few group investigated the simulation of atrial body surface ECG signals [188, 228]. For this purpose, models of the human thorax were needed besides the atrial model. These thorax models represented either very detailed anatomy derived from the Visible Man Project [70, 71, 162, 214] or generically generated thorax shapes including simple models of the lungs were used [217–219, 229–231].

Table 3.1. 3D geometrical atrial models. Table edited and extended from [5].

Study	Basis	Species	Compartments	Geometry	Mesh Type	EPM	Prop.	Fibers	Sep.	Heterogeneities		Ext.
Lorange et al. 1993 [188]		human (CT)	4 chamber	surf-2D	voxel	n/p	eikonal	-	c	-		ECG
Ellis et al. 1997 [184]		generic	biatrial	surf-2D	icosahedral	n/p	n/p	-	n/p	-		Abl
Gray et al. 1998 [189]		generic	2 spheres	surf-3D	cubic	[190]	[189]	-	c	-		
Freudenberg et al. 2000 [191]		Visible Female	biatrial	vol	voxel	n/p	CA	-	c	CT	(G/-/-)	
Harrild et al. 2000 [91]		human (MRI)	biatrial	vol	hexahedral	Nyg	MD	B	i	BB, CT, PM, IS, FO	(G/CV/-)	
Werner et al. 2000 [162]		Visible Man	4 chamber	vol	voxel	n/p	CA	V	c	-		ECG
Vigmond et al. 2001 [163]		generic	2 spheres	surf-3D	cables	[192]	[163]	ML	i	BB, CT , PM, CS, FO	(G/-/-)	
Zemlin et al. 2001 [164]		Visible Female	biatrial	surf-3D	triangular	CRN	[164]	ML	c	CT, PM	(G/CV/-)	
Blanc et al. 2001 [193]		generic	2 spheres	surf	voxel	[194, 195]	MD	-	c	-		
Virag et al. 2002 [196]		human (1,MRI)	biatrial	surf-3D	triangular	CRN	MD	-	i	BB, CT, FO	(G/CV/-)	
van Dam et al. 2003 [197]		human (1,MRI)	biatrial	surf-3D	triangular	n/a	FM	-	c	-		
Jacquemet et al. 2003 [165, 185, 186, 198–201]	[196]	human (1,MRI)	biatrial	surf-3D	cubic	CRN	MD	ML	i	BB, CT, FO	(G/-/-)	Abl, ECG
Seemann et al. 2006 [92]		Visible Female	biatrial	vol	voxel	CRN	MD	B	c	BB, CT, PM, AVR, APG, SN	(G/CV/EP)	
Ridler et al. 2006 [202]		human (1,CT)	biatrial	3D	cables	[192]	MD	yes	i	CS, FO, CS	(G/-/-)	
Dokos et al. 2007 [203]		generic	2 spheres	surf-2D	triangular	[204]	n/p	-	c	-		
Kuijpers et al. 2007 [205]	[197]	human (1,MRI)	biatrial	surf-3D	triangular	CRN	BD	-	c	-		
Gong et al. 2007 [206]	[91]	human (averaged)	biatrial	vol	hexahedral	CRN	n/p	B	i	BB, CT, PM, CS, FO, IS	(G/CV/-)	
Plank et al. 2008 [94]	[92]	Visible Female	left atrium	vol	voxel	CRN	MD	V	n/a	-		
Tobon et al. 2008 [93, 207–209]	[91]	human (1,MRI)	biatrial	surf-3D	triangular	Nyg	MD	ML	i	BB, CT, PM, CS, IS, FO	(G/CV/-)	Abl
Wieser et al. 2008 [210, 211]	[212]	human (1,MRI)	biatrial	surf-3D	triangular	[192]	MD	-	i	BB, CT, CS, FO	(G/CV/-)	
Reumann et al. 2008 [187]	[92]	Visible Female	biatrial	vol	voxel	CRN	CA	B	c	BB, CT, PM	(G/-/-)	Abl
Kharche et al. 2008 [213]	[92]	Visible Female	biatrial	vol	voxel	CRN	CA	B	c	BB, CT, PM, AVR, APG, SNs	(G/CV/EP)	
Krueger et al. 2009 [169]	[92, 162]	Vis. Man, Vis. Female	biatrial	vol	voxel	CRN	MD	B	c	BB, CT, PM	(G/CV/EP)	
Krueger et al. 2010 [8, 214]		human (2,MRI)	biatrial	vol	voxel	CRN	MD	B	c	BB, CT, PM, AVR, APG	(G/CV/EP)	Abl, ECG
Abed et al. 2010 [215]		Visible Man	biatrial	vol	voxel	[215]	MD	-	n/p	BB, CT, CS, APG	(G/-/-)	
Krueger et al. 2011 [216]		human (15,CT,MRI)	biatrial	vol	voxel	CRN	MD	VL	c	BB, CT, PM, IS, AVR, APG	(G/CV/EP)	
Lu et al. 2011 [217]		human (1,MRI)	4 chambers	vol	spheroidal	Nyg	CA	-	n/p	BB, CT, PM, IS, FO	(G/-/-)	ECG
Aslanidi et al. 2011 [95, 218, 219]	[92]	Visible Female	biatrial	vol	voxel	CRN	[218]	BS	c	BB, CT, PM, SN	(G/CV/EP)	ECG
Cristoforetti et al. 2011 [220]		human (1,CT)	left atrium	surf	triangular	CRN	n/p	-	n/a	-		
Ridler et al. 2011 [221]	[202]	human (1,CT) (dog size)	biatrial	3D	cables	[222]	MD	-	i	CT, CS, FO	(G/-/-)	
Burdumy et al. 2012 [223]		human (4,CT)	biatrial	vol	voxel	CRN	CA	-	c	-		
Zhao et al. 2012 [96]		Sheep (1, microscopy)	biatrial	vol	voxel	[224]	MD	V	c	BB, CT, PM, PV	(G/-/-)	
Krueger et al. 2012 [225]		human (9,CT,MRI)	biatrial	vol	voxel	CRN	MD	V	i	BB, CT, PM, IS, AVR, APG	(G/CV/EP)	Abl, ECG
Aslanidi et al. 2012 [226]		canine (1,μCT)	biatrial	vol	voxel	n/a	n/a	V	c	BB, PV	(G/-/-)	

EPM: electrophysiological model, CRN: model of Courtemanche et al. [72], Nyg: model of Nygren et al. [227], comp.: compartments, prop.: propagation model, sep.: septum, B: bundles, V: volumetric, ML: mono-layer, VL: various layers, BS: bundles and SN, ext.: Model extension, Abl: Ablation lesion model, ECG: ECG is computed, surf: surface model, surf-3D: 3D surface model, vol: 3D volumetric model, CA: cellular automaton, MD: monodomain, BD: bidomain, i: isolating, c: connective, n/a: not applicable, n/p: not specified, IS: right atrial isthmus, AVR: atrio-ventricular ring, APG: appendages, G: structures in geometry, CV: structures have different conductivity, EP: structures have different electrophysiology.

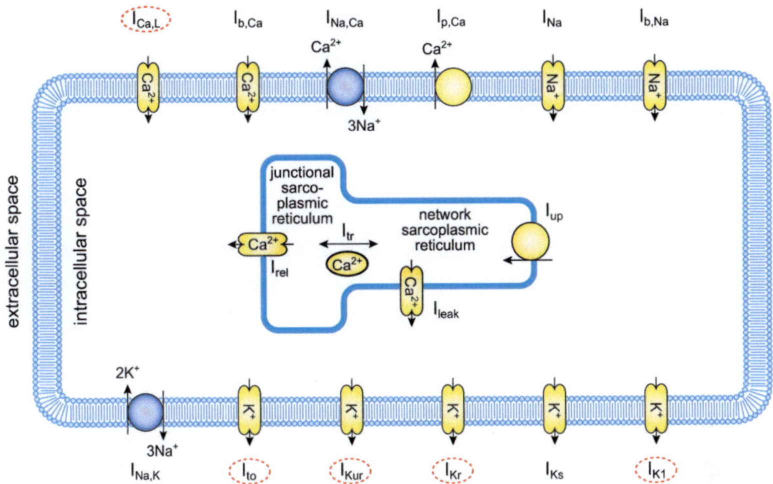

Fig. 3.1. Schematic structure of the Courtemanche-Ramirez-Nattel model [72] of atrial electrophysiology. Ionic currents which are altered during electrical remodeling are encircled in dashed red. Figure modified from www.cellml.org.

3.2 Fundamentals of Electrophysiological Modeling

The first model of the electrophysiology of a cell was presented by A.L. Hodgkin and A.F. Huxley in 1952 [232]. The model contained a mathematical description of the ionic mechanisms underlying electrophysiological phenomena, e.g. the action potential. Their model was based on voltage-clamp experiment data on the axons of a giant squid. In 1962, D. Noble published the first mathematical model of the electrophysiology of a cardiac cell, which also included pace-maker activity [233]. It took another 35 years, until the first models of human atrial electrophysiology were published [72, 227].

3.2.1 Existing Models of Atrial Electrophysiology

The Courtemanche-Ramirez-Nattel (CRN) model [72] provides a mathematical description of the action potential based on ionic current data obtained from human atrial cells (Fig. 3.1). Nonlinear-coupled ordinary differential equations describe the electrical behavior of a human atrial myocyte by calculating ion concentrations, ionic currents, bindings to intracellular structures, and the transmembrane voltage. The change in transmembrane voltage V_m is thereby described by

$$\frac{dV_m}{dt} = \frac{-(I_{ion} + I_{st})}{C_m}, \tag{3.1}$$

where I_{st} is the externally applied stimulus current and C_m the cell membrane capacitance. The net current across the cell membrane I_{ion} is given in the CRN model by the sum of twelve ionic currents

$$I_{ion} = I_{Na} + I_{K1} + I_{to} + I_{Kur} + I_{Kr} + I_{Ks} + I_{Ca,L}$$
$$+ I_{p,Ca} + I_{NaK} + I_{NaCa} + I_{b,Na} + I_{b,Ca}. \tag{3.2}$$

Each current I_x is defined following Ohm's law as the product of the specific ion channel conductivity g_x and an ion specific voltage weighted with a product of gating variables γ_i

$$I_x = g_x \prod_i \gamma_i (V_m - E_{Nernst,x}), \tag{3.3}$$

where $E_{Nernst,x}$ is the Nernst voltage of the specific ion. The conductivity g_x scales the amplitude of the current and the gating variables γ_i describe the kinetic behavior of the ion channels permeable to ion x. More specific, they describe the open probability of the channels. For channels which have only two states (opened, closed) the change in the number of opened channels is described by

$$\frac{d\gamma_i}{dt} = \alpha_i (1 - \gamma_i) - \beta_i \gamma_i. \tag{3.4}$$

The voltage-dependent rate constants α_i and β_i describe the transition from opened (γ_i) to closed ($1 - \gamma_i$) state (and vice versa) of the ion channel gates.

Atrial electrophysiology shows variations in the APD and morphology as well as in its dynamic behavior. Besides the sinus node, which depolarizes spontaneously, also the other atrial structures (Sec. 5.2) show differences in the cell electrophysiology [234, 235]. A detailed review about the atrial electrophysiology and pathologies is given in [75].

Currently, five mathematical models describing human atrial electrophysiology are available [72, 227, 236–238]. Although they are partly based on the same data,

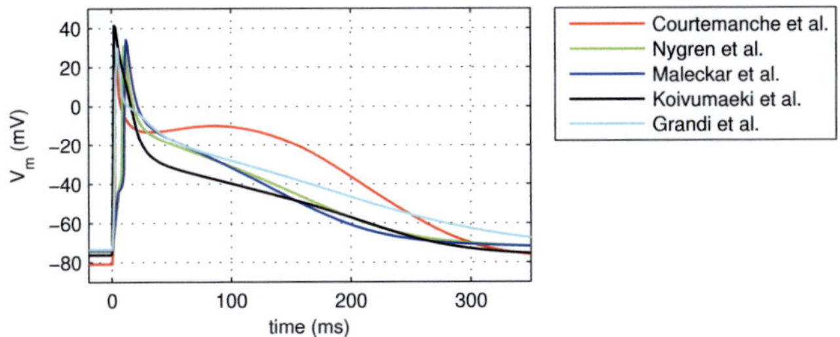

Fig. 3.2. Simulated single cell human atrial action potentials.

their action potential morphology and restitution behavior differs significantly
(Fig. 3.2). The electrophysiology models can be adapted to reproduce the au-
torhythmicity of the sinus node [92, 239]. Less complex phenomenological models can be parameterized to reproduce the action potential and restitution behavior
of physiological models with only few currents across the cell membrane [240].
Figure 3.1 exemplary depicts the schematic structure of the CRN model [72].

3.2.2 Electrical remodeling

Models of atrial electrophysiology can be adapted to reflect electrophysiological changes caused by persistent AF. This so called electrical remodeling alters
ion channel behavior [241] and may also reduce the excitation conduction velocity [242, 243]. Based on data from [241, 244, 245], the maximum ion conductances
of $I_{Ca,L}$ (-65%), I_{to} (-65%) and I_{K1} (+110%) in the CRN model were altered to reproduce the experimental findings [73, 246]. Recently, additional alterations of the
sustained outward K$^+$ current I_{sus}/I_{Kur} and the slow delayed rectifier K$^+$ current
I_{Ks} under persistent AF conditions have been reported [247]. The electrical remodeling has different impacts depending on the progression of AF (paroxysmal,
persistent, permanent) and eventual accompanying diseases, e.g. congestive heart
failure [241, 248–250]. Due to inconsistencies in the experimental data, various
other implementations of persistent AF electrical remodeling have been developed [68, 201, 208, 251] (Tab. 3.2). Thereby, besides changes in the maximum ion
channel conductances, also the mathematical formulations of the activation and
inactivation curves and channel kinetics may be altered [208, 251]. The general
effects (shorter APD, slower conduction velocity) are similar in all remodeling
models (Fig. 3.3).

Table 3.2. Different implementations of persistent AF electrical remodeling. Simulated action potentials are shown in Figure 3.3. Zhang et al. [251] and Tobon et al. [208] additionally altered the mathematical formulation of channel kinetics.

Study	EP Model	I_{K1}	$I_{Ca,L}$	I_{to}	I_{Kr}	I_{IKur}
Zhang et al. 2005 [95, 251]	[72, 227]	+235%	-74%	-85%	-	-
		+90%	-64%	-65%	-	-
Zhao et al. 2009 [68]		-	-45%	-63%	-68%	-
Seemann et al. 2010 [73]	[72]	+110%	-65%	-65%	-	-
Tobon et al. 2010 [208]	[227]	+250%	-74%	-85%	-	-
Uldry et al. 2012 [201]	[72]	-	-30%	-80%	+50%	-90%

3.3 Fundamentals of Modeling the Excitation Propagation

The atrial muscle mainly contains myocyte cells within the extracellular matrix. The intracellular spaces of the myocytes are electrically interconnected via gap junctions. A change in cell current or transmembrane voltage in one cell is therefore distributed to the neighboring cells. So, a depolarization wave can spread over the myocardium. The conduction behaviors in the extracellular and intracellular space are anisptropic, having greater conductivity along the myocyte orientation. The anisotropy is caused by the cylindrical cell shape and because the majority of gap junctions is located at the short sides of the myocytes.

3.3.1 Mono-/Bidomain

Excitation propagation in atrial models is commonly modeled with a macroscopic approach, in which the atrial tissue is described as a continuum [5]. One node in the model represents several hundred myocytes. Depending on the problem, the resolution of such models varies between $50\,\mu$m and several millimeters.

The excitation sequence can be described mathematically by detailed reaction-diffusion models. The bidomain model computes the electrical currents in the extracellular and intracellular space using two Poisson equations [252, 253]

$$\nabla \cdot (\sigma_e \nabla \Phi_e) = -\beta I_m - I_{se}, \tag{3.5}$$

$$\nabla \cdot (\sigma_i \nabla \Phi_i) = \beta I_m - I_{si}. \tag{3.6}$$

Where Φ_i and Φ_e are the intracellular and extracellular potentials, σ_i, σ_e are the conductivity tensors and β is the cell surface-to-volume ratio. $I_{si,e}$ are externally

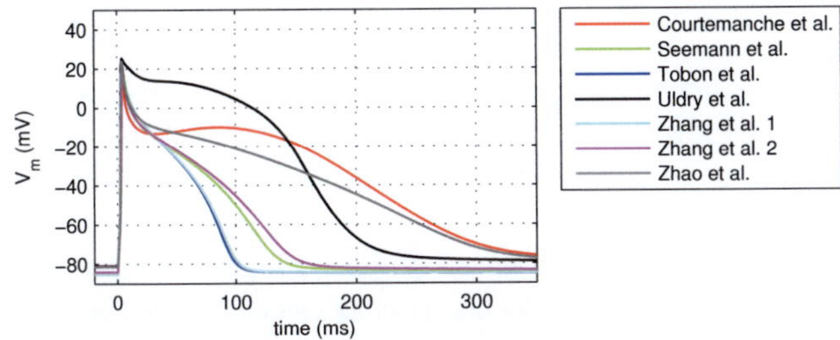

Fig. 3.3. Simulated single cell human atrial action potentials with different implementations of AF remodeling in the CRN model [72]. Maximum ion channel conductances were implemented as listed in Tab. 3.2. Modifications of the mathematical formulations of the current kinetics as described by [208, 251] were neglected.

applied current sources in both domains. The transmembrane current density I_m is the sum of all membrane currents

$$I_m = C_m \frac{dV_m}{dt} + I_{ion}. \tag{3.7}$$

C_m is the cell membrane capacitance and I_{ion} is the sum of the ionic currents across the cell membrane (Eqn. 3.2). The bidomain model can handle differences in conductivity tensors between the two domains. Further domains can be added to the formulations to represent other myocardial spaces [254]. If the anisotropy ratio (longitudinal/transversal) is considered to be equal in both domains, the bidomain model can be reduced to one domain. The monodomain model calculates the current in the intracellular space and through gap junctions

$$\nabla \cdot (\sigma \nabla V_m) = \beta \left(C_m \frac{dV_m}{dt} + I_{ion} \right), \tag{3.8}$$

with

$$\sigma = \frac{\sigma_e \cdot \sigma_i}{\sigma_e + \sigma_i}. \tag{3.9}$$

The monodomain and bidomain models are coupled with models of the cardiac electrophysiology (Sec. 3.2) and can therefore reflect the complex reaction-diffusion processes in the human heart. A more detailed description of the bidomain model and the simplification to the monodomain model can be found in [253]. With regard to atrial modeling, the monodomain model is commonly used (Tab. 3.1). The bidomain model plays only a minor role.

3.3.2 Cellular Automaton

The detailed reaction-diffusion models are computationally very expensive. Rule-based simulation systems were therefore also commonly used to compute the electrical activation in the atria. Such automata enabled atrial simulations as early as 1964 [255]. The electrical excitation is modeled by a rule-based extension of the activated region. Thereby, tissue anisotropy [162, 256], wavefront curvature [256] and rate-adaptive APD and conduction velocity can be included in the rules. Automata are not suitable to reflect the diffusion processes in the atria, although they may imitate the cells action potential morphology [257].

3.3.3 Eikonal / Fast-Marching

Cardiac excitation propagation can also be computed by adapting shortest paths algorithms, which are commonly used in image processing [258]. The fast-marching method is based on the Eikonal equation [259]

$$\| \nabla T \| F(\mathbf{x}) = 1, \tag{3.10}$$

where T is the arrival time at point $\mathbf{x}$ and F is the speed of the wavefront expansion. For the implementation of the fast-marching level set method, nodes in a model are divided into three sets: 1) active nodes, 2) nodes on the border of the activation wave front and 3) inactive nodes. Each node i can initially be assigned to one of these groups and has an initial arrival time T_i. Let s be the given seed points.

$$\Omega_{inactive} = \{i \mid i \neq s\}$$
$$\Omega_{activated} = \{i \mid i = s\}$$
$$\Omega_{trial} = \{i \mid i \text{ neighbor of } s\}$$
$$T_i = \inf \forall i \in \Omega_{inactive}$$
$$T_i = 0 \forall i \in \Omega_{activated}$$
$$T_i = \text{distance}(i,s) \forall i \in \Omega_{trial}$$

The state of each node can then be updated iteratively until the set of inactive nodes is empty.

```
while (Ω_inactive ≠ ∅) do
   find {x | x ∈ Ω_trial | T_x = min(T)}
   Ω_activated ← x
   for all (n neighbor of x) do
      if (n ∉ Ω_activated) then
         Ω_trial ← n
         if (T_n > distance(n,x)) then
            T_n = distance(n,x)
         end if
      end if
   end for
end while
```

The distance function corresponds to the speed of the wavefront F and may be given as the Euclidean distance between two neighboring nodes. Other distance functions can also be used, as well as weighted distances (see Sec. 5.7).

The fast-marching level set approach can be extended to reflect diffusion [260, 261] and can also be coupled with models of the cell electrophysiology to reproduce the behavior of the detailed reaction-diffusion models [262]. Such systems0 become computationally more intensive again (yet remaining faster than detailed reaction-diffusion systems), but can be used to reliably model reentries and other arrhythmic patterns [263, 264]. Recently, a graph-based method to model cardiac excitation was presented [265] which might allow a further increase in computational speed in trade-off of further detail.

3.4 Forward Calculation of the ECG

The electrical signals at the body surface which are caused by the excitation of cardiac tissue can be measured in the ECG (Sec. 2.4.1). To compute the ECG signals from simulated transmembrane voltages, the forward problem needs to the solved. In a first step, the impressed current sources I_{imp} can be calculated from the transmembrane voltages

$$I_{imp} = -\nabla \cdot (\sigma_i \nabla V_m). \tag{3.11}$$

In a second step, the parabolic part of the bidomain formulation [252, 266], a Poisson equation, is solved

$$\nabla \cdot [(\sigma_i + \sigma_e)\nabla\Phi_e] = I_{imp}. \tag{3.12}$$

For the reference potential and the thorax-air boundary, Dirichlet and Neumann boundary conditions are set. Equation 3.12 can then be discretized to a linear system of equations using either finite elements or boundary elements [5]. The finite element method, although computationally more expensive, has the advantage that it can handle conductivity anisotropy within the tissue. The forward-calculated potentials form body surface potential maps (BSPM) at each time instance. The integration of the surface node potentials over time results in a static integral BSPM [144].

Part II

Developed Methods

We actually made a map of the country,
on the scale of a mile to the mile.

Lewis Carroll

Sylvie and Bruno Vol. II

4

Clinical Data Acquisition and Processing

Clinical data were acquired at three medical centers:

1. Guy's Hospital, Guy's and St. Thomas' NHS Foundation Trust, London, UK
2. Pädiatrische Radiologie, Heidelberg University Hospital / Radiologie, German Cancer Research Center (DKFZ), Heidelberg, Germany
3. Städtisches Klinikum Karlsruhe, Germany

Some datasets were collected as part of the PhD work of Dr.-Ing. David Keller [267], Dr.-Ing. Frank Weber [268] and Dipl.-Phys. Thomas Fritz. Table 4.1 on page 75 summarizes the data collection.

4.1 Image Data Acquisition

Image data were acquired for two purposes. First, cardiac images were needed to create patient specific atrial models. Second, thorax image data were needed for those patients from whom BSPM data were recorded. Image acquisition in London was performed using a 1.5 T Achieva MRI scanner (Philips Healthcare, Best, The Netherlands), in Heidelberg using a 1.5 T Magnetom Avanto MRI scanner (Siemens AG, Munich, Germany) and in Karlsruhe using a Sensation 64 CT scanner (Siemens AG, Munich, Germany).

To image the heart, a respiratory-navigated and ECG-gated steady-state free precession cardiac MRI sequence was run. The resolution varied between $1.0 \times 1.0 \times 1.0\,\mathrm{mm}^3$ (Heidelberg) and $1.25 \times 1.25 \times 1.37\,\mathrm{mm}^3$ (London). CT images had a resolution of $0.4 \times 0.4 \times 3\,\mathrm{mm}^3$ (Karlsruhe).

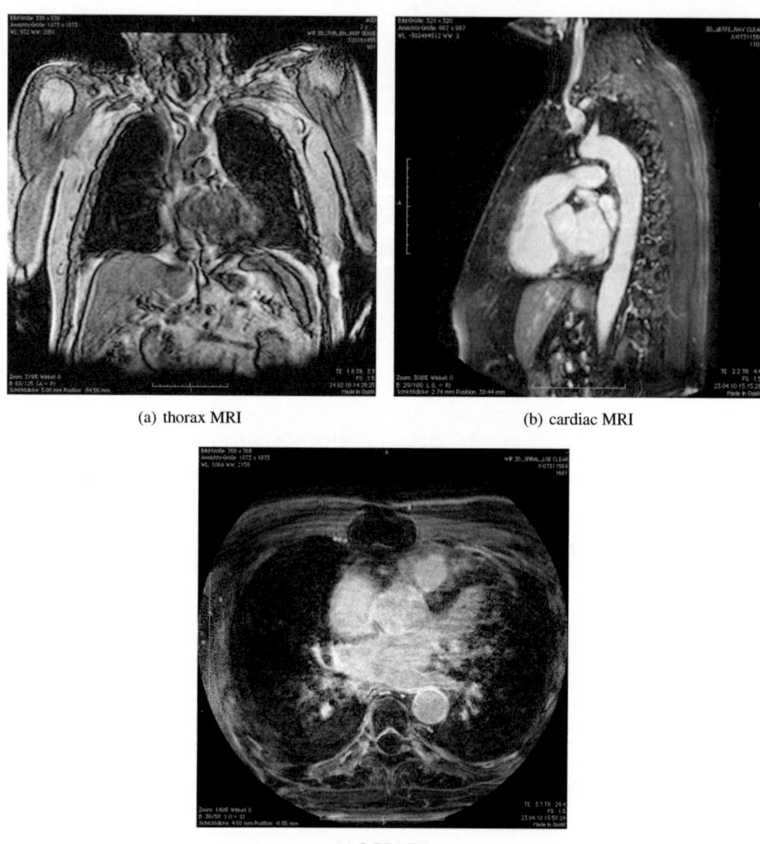

(a) thorax MRI

(b) cardiac MRI

(c) LGE MRI

Fig. 4.1. Example images acquired at Guy's Hospital London, UK, with a Philips Achieva 1.5 T MRI scanner (subject 8 in Tab. 4.1). a) Coronal slice of thorax MRI. b) Sagittal slice of SSFP MRI. c) Transverse slice of LGE-MRI of the left atrium.

Thorax images were only acquired on MR scanners. A breath-hold ultra fast gradient echo sequence was run using the same receiver coils as for the cardiac scans. This way the patient did not need to be moved between the scans and the image stacks were co-registered in the MRI scanner coordinate system. Thorax images had a resolution between $0.98 \times 0.98 \times 2.0\,\mathrm{mm}^3$ (Heidelberg) and $1.48 \times 1.48 \times 2.0\,\mathrm{mm}^3$ (London).

In AF patients, also late Gadolinium enhancement (LGE) MRI was performed. Myocardial regions with reduced perfusion have a slower contrast agent wash-out than regular myocardium. An image acquisition 15–30 minutes after infusion of a contrast agent has strong image intensity in these regions. Thereby, scar tissue from RFA [269–271] and fibrotic tissue [272] can be imaged. Imaging of RFA scar tissue acutely after the intervention (imaged $< 24\,h$ after RFA) may not provide a reliable prediction of the chronic scar delineation. Within the first three to six months after an RF intervention, damaged tissue can regenerate and scar outline will change [138, 273]. For three patients, chronic LGE-MRI scans were acquired.

4.1.1 Discussion

Cardiac CT images had a resolution and image quality superior to the MRI data. Nevertheless, CT acquisition is based on ionizing radiation. For patients awaiting RF intervention, the additional radiation dose caused by the CT scan might be negligible low compared to the radiation dose from fluoroscopy. Nevertheless, MRI provided additional functional information of the myocardium and could also be performed on volunteers. The latter was useful to test and establish imaging and post-processing workflows, which then provided more reliability when working with patient data.

MRI data collected at the Heidelberg centers had better resolution compared to the data collected in London. The resolution and image quality from the data collected in London was sufficient to segment and post-process the images for the model generation. None of the MR images could show the atrial wall. MRI resolution was in the range of half the atrial wall thickness (Sec. 2.1.1.4). Thus, partial volume effects overlap at the border between the endocardial blood pool, myocardium and surrounding tissue. Additionally, MRI does not provide a good contrast between myocardium and surrounding tissue anyway.

During the first image acquisition sessions in London, no MRI sequence to image the thorax using the cardiac coils with sufficient image quality was available. Therefore, a number of datasets were not selected for post-processing (see Sec. 4.5.1). Philips Research provided later on a thorax MRI sequence which was then used for the subsequent data acquisition sessions.

Care needed to be taken to image the heart and the thorax in the same breathing state. This allowed a good alignment of both image stacks. If similar breathing states cannot be achieved during the acquisition, the heart position might need to be manually corrected along the sagittal axis.

Image intensity in LGE-MR images cannot be used as a quantitative measure. The image intensity depends on various factors, such as the amount of contrast-agent given, time after contrast-agent injection and accumulation of contrast-agent in tissue surrounding the myocardium. This also contributes to difficulties during the segmentation of scar of fibrotic tissue from LGE data [274].

4.2 BSPM Acquisition

Several companies offer body surface potential mapping (BSPM) systems. The products most commonly cover not only the ECG recording system, but also solve the inverse of problem of ECG (ECG imaging, ECGI). Among the offered products are Prime ECG®(80 leads, Heartscape Technologies Inc., Bothell, WA, USA), AMYCARD-01C ($\leq$256 leads, Amycard, Moscow, Russia) and ECVUE™(252 leads, CardioInsight™Technologies Inc., Cleveland, OH, USA). These products have in common, that the raw ECG data cannot be accessed directly, as the system setup is proprietary. Additionally, all but the Amycard system use (expensive) disposable electrode vests and do not allow a flexible adjustment of the electrodes to new schemes (see below). BioSemi (Amsterdam, The Netherlands) offers a multi-purpose biosignal recording system which allows for flexible electrode setups, multi-channel ECG recordings and full data access (Active Two). Furthermore, the modular setup of the hardware allows the storage and transport of the system in a regular hand baggage case. The system could therefore be used in different hospitals in Europe. The other systems mentioned above a more integrated and designed for use in a single clinical site, as they cannot be transported between hospitals due to the size of the overall casing.

4.2.1 BioSemi System

The Active Two system is comprised of different hardware modules, as well as a signal recording software (Fig. 4.2(a)). The active electrodes are organized in

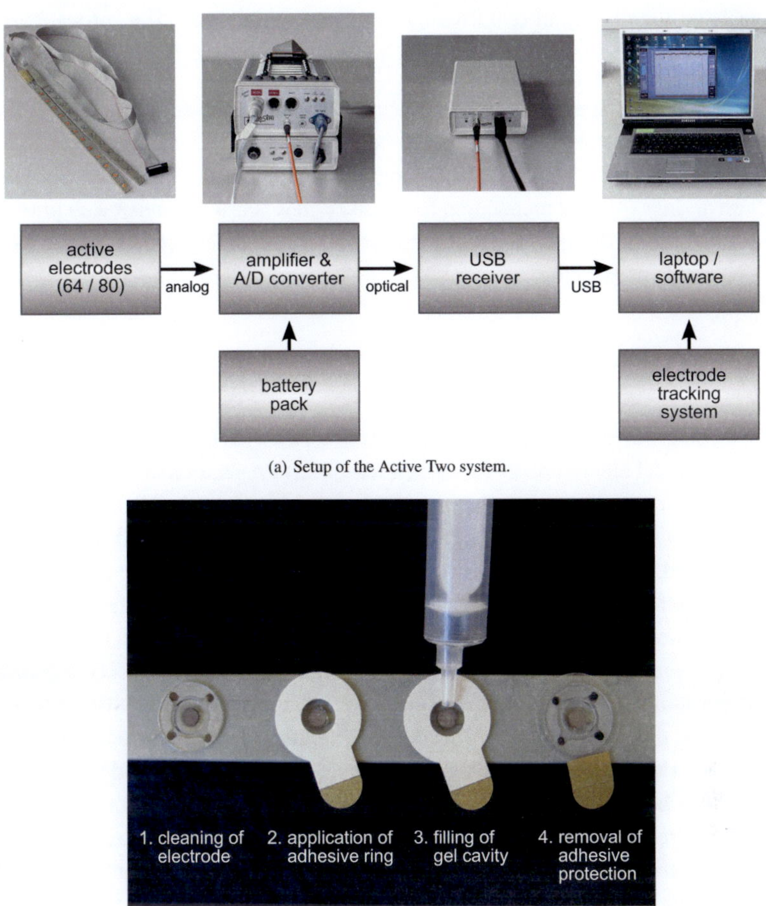

(a) Setup of the Active Two system.

(b) Preparation of active electrodes.

Fig. 4.2. Hardware setup of the BioSemi Active Two system.

flexible strips of eight or twelve electrodes and connected to the analog/digital (A/D) converter with flat cables. The A/D converter is powered by a battery pack and includes a 50 Hz hardware realized band-stop filter. The digital signals are sampled with 2048 Hz and transmitted via a fiber optical cable to the USB receiver (galvanic separation). This connects to a laptop computer, which visualizes the signals on the display in real time by a software (ActiView). The software also provides a low and high pass filter, which were turned off for the ECG recordings.

Before sticking the electrodes on the skin, the electrodes need to be cleaned and disinfected (1), adhesive rings need to be applied (2), electrode gel needs to be filled in the gel cavity (3) and at last the protection needs to be removed from the adhesive rings (4). The preparation for all electrodes is usually undertaken prior to the actual use of the system to not lengthen procedure time. Figure 4.2(b) depicts the electrode preparation.

4.2.1.1 80 Leads Electrode Setup

The 16 single electrodes of the first IBT Active Two system proved to be impractical in clinical environments for several reasons. i) Preparation of multiple single electrodes for application to the patient often led to electrodes sticking together and gel covering the adhesive rings. ii) Localization of the electrodes was more complicated, as the electrodes did not underlie a fixed interelectrode distance. iii) Single electrodes were very light and often did not stick as well to the skin as the strips. iv) Cables connecting to the distal end of the single electrodes often broke, as these were rather thin and not flexible enough. Altogether this caused a greater time consumption in the setup of the system and the application to the patients. This led to delays in the clinical workflow. Therefore, a new electrode setup was designed. The major requirements were to create a setup which could more easily and timely be used in clinical environments and which would be more robust for electrode tracking (Sec. 4.2.2). These requirements were realized by using only electrode strips. A combination of four strips of twelve electrodes and four strips of eight electrodes increased the electrode count to 80 (Figs. 4.4(a) and 4.4(b)). The new setup additionally offered the optional use of three single electrodes to record a standard Einthoven ECG if desired and the workflow allowed.

Due to a constant electrode spacing within each strips, the electrode localization was more robust. The strips themselves could be easily prepared prior to a measurement in the clinical environment, as the strips could be laid out next to each other on a tray without moving. Nevertheless, the strips with two different electrode spacings allowed a flexible adjustment of the electrode placement on the body surface, e.g. if additional ECG leads or navigation system reference electrodes needed to be placed on the chest (Fig. 4.5). The strips also allowed for a

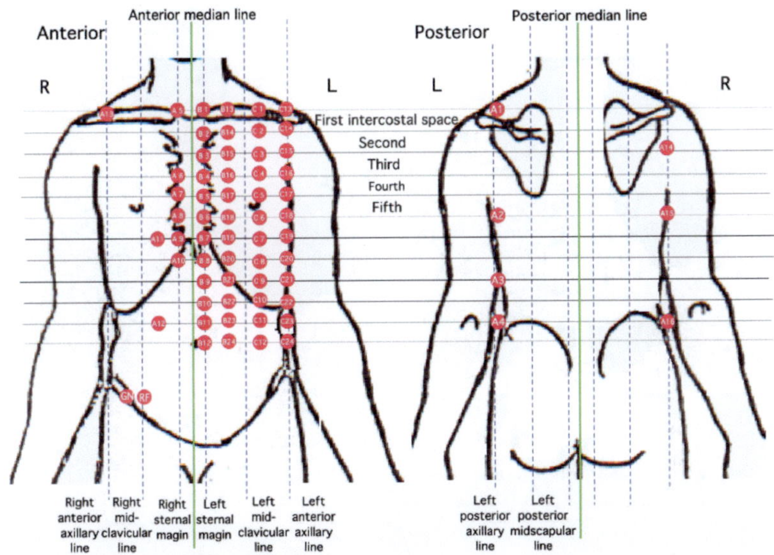

(a) Schematic drawing of 64 electrode BSPM setup. Figure from [275].

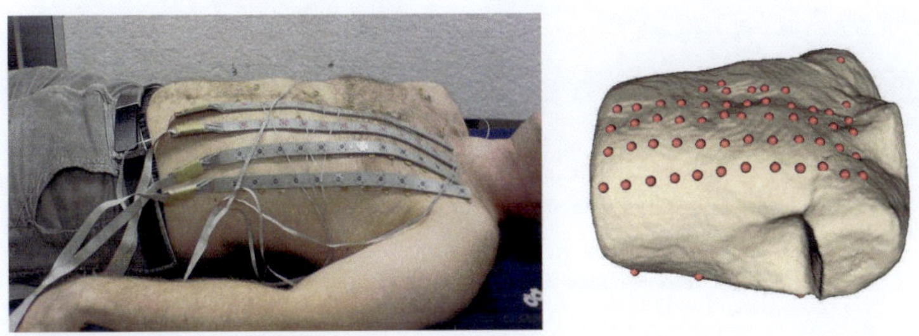

(b) Photography of 64 electrode BSPM setup (subject 3, Tab. 4.1).

(c) Model of 64 electrode BSPM setup (model of subject 3).

Fig. 4.3. BSPM electrode setups. Classic 64 channel setup.

better electrode to chest contact, as they weighed more than the single electrodes and thus the elastic forces of the cable did not counteract the adhesive connection. Overall the preparation time for the hardware setup for a patient was reduced from 15–30 minutes to 5–10 minutes. Figures 4.3 and 4.4 show both setups as schemata, applied to volunteers and as electrode positions mapped to thorax models of the volunteers.

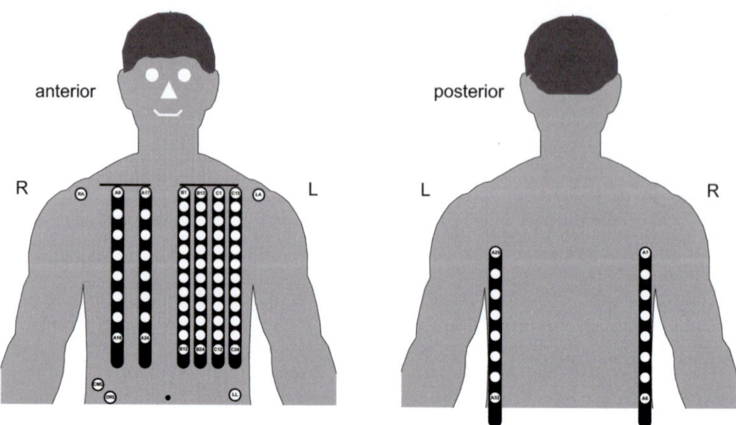

(a) Schematic drawing of 80 electrode BSPM setup.

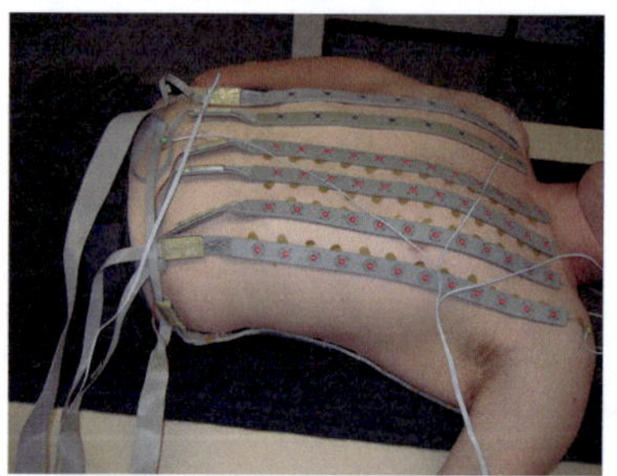

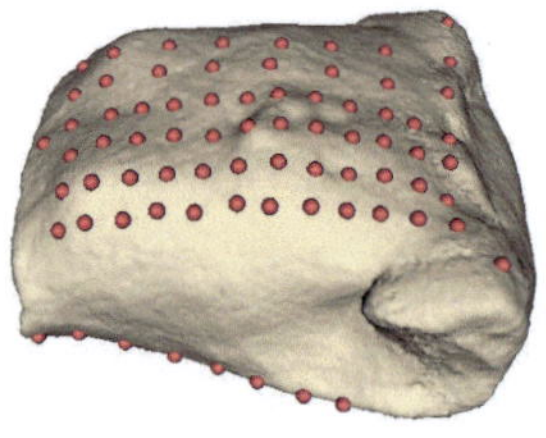

(b) Photography of 80 electrode BSPM setup (subject 1, Tab. 4.1).

(c) Model of 80 electrode BSPM setup (model of subject 1).

Fig. 4.4. BSPM electrode setups. Newly developed 80 channel setup.

4.2.2 Electrode Tracking

The positions of the electrodes on the patients' thorax need to be precisely located and registered to the thorax MRI data, to allow for a reliable comparison of simulated and measured ECG signals in these positions. Tracking of electrodes can be performed with various systems. In this section, two systems used during the data acquisition in this work are presented. Additionally, a localization approach using X-ray images is presented.

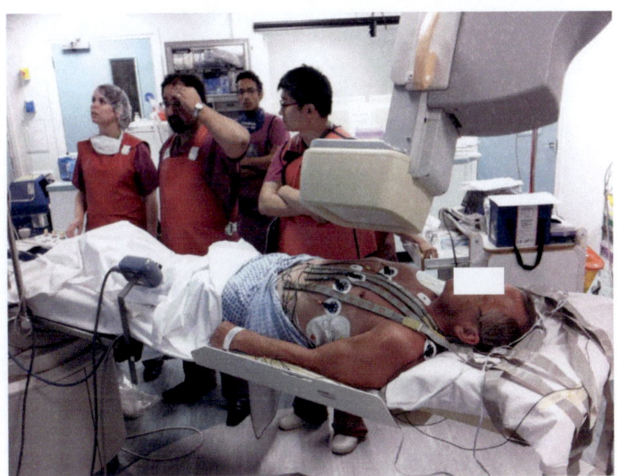

Fig. 4.5. Photography of 80 electrode BSPM setup on patient in cath lab during a VT stim procedure. Photography provided by Walther H. W. Schulze, Institute of Biomedical Engineering, Karlsruhe Institute of Technology (KIT).

Electrode locations were tracked either magnetically using Fastrak®(Polhemus Inc., Colchester, Vermont, USA) or optically using Optotrak (Northern Digital Inc., Waterloo, Ontario, Canada). Both systems had in common that each electrode position needed to be tracked individually with a pen marker. The Fastrak®system registers the positions to a local coordinate system. Later on, this coordinate systems was manually fitted to the MRI coordinate system. One of the X-MRI EP laboratories at Guy's Hospital, London, UK, in which some of the measurements were conducted, had an Optotrak system installed. This system registers a marker pen using three fixed infrared cameras. The system was coupled to the MRI and X-ray systems and thus the captured electrode positions were tracked within the MRI coordinate system [276].

In cases where both information could not be acquired, electrode positions were manually marked on the thorax model surface of the patient following a detailed protocol. This was only done for the 80 channel setup using electrode strips. The electrode strips were placed along the median line on the thorax front beginning with their distal electrode on the clavicle bone. The strips were horizontally equally spaced on the clavicle bones. Additionally, the constant electrode spacing within the electrode strips could be used to ensure a best possible match between model and measurement.

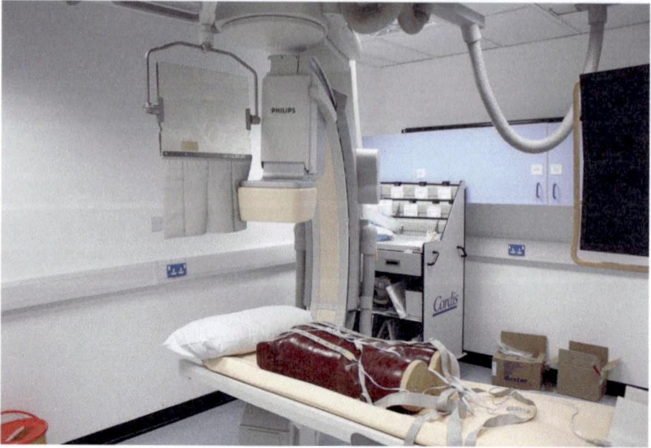

(a) phantom in cath-lab

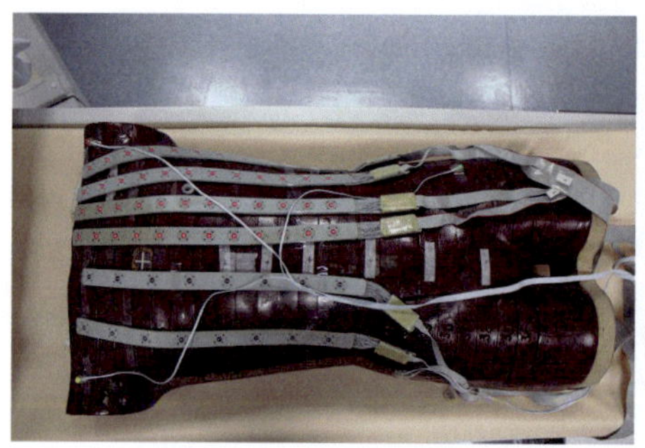

(b) active electrode strips on phantom

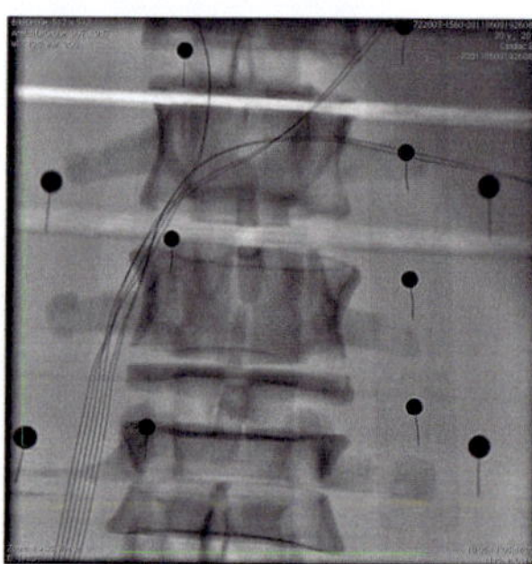

(c) example x-ray image of phantom with electrodes

Fig. 4.6. Test setup of acquisition of electrode positions using an X-ray system. For this test a RANDO®Man phantom (The Phantom Laboratory, NY, USA) was used.

To further reduce the clinical procedural time of acquiring BSPMs, BSPM electrode locations might also be tracked using conventional low radiation dose X-ray images. As AF patients are admitted to RFA, and all BSPM measurements and tracking was done prior to the EP intervention, this approach promises significant time reduction as well as a direct registration of the electrode locations to the coordinate system of an appended electroanatomical mapping system (Sec. 2.4.2) and thus the atrial geometry. In an initial experiment, a thorax dummy (RANDO®Man, The Phantom Laboratory, NY, USA) was placed on the patient table in an EP laboratory (Fig. 4.6(a)). The BSPM electrodes were placed on the dummy according to the regular 80 channel BSPM scheme (Fig. 4.6(b)). 44 pairs of biplane X-ray images were acquired. The collaborators at KCL reconstructed the electrode positions from these data with an accuracy of 2.0 ± 1.3 mm [277] using the epipolar constraint [278] and also found that 12 pairs of X-ray images were sufficient for the reconstruction of the electrode positions.

4.2.3 Discussion

The new 80 electrode BioSemi BSPM system enabled a smoother clinical workflow with less preparation time with the patient compared to the previous 64 electrode setup. Additionally, the electrode tracking became more robust and the signal quality was better overall, as the single electrodes tended to loose contact during the measurement. Having less cables around the patients while still being able to flexibly position the electrode strips also allowed for a use of the system during EP interventions [277]. Using an electrode vest, as done by some ECGI system manufacturers, would further ease the handling of the system in the clinical environment but would also restrict the use of the system to ECGI purposes.

The optical tracking system installed in the London XMR suites was easier to use and more robust in the tracking process. With the magnetic tracking system, field distortions due to metal in the patient bench may result in unusable tracking results. No problems were observed while using the magnetic tracking system in the same room as the MRI system outside the five Gauss line. Recent developments in the localization of BSPM electrodes during the RF intervention may in the future provide a precise localization of the electrodes while decreasing procedural time for the trade-off of negligible additional radiation dose.

Using the optical tracking system, traced electrode positions were directly registered with the MRI scanner coordinate system. The electrode positions from the magnetic tracking system needed to be manually registered with the thorax model. Other systems with passive electrodes allow the imaging and registration of electrode positions directly with the acquisition of the thorax geometry [279, 280]. This is, to this point, only possible using CT and not MRI for the acquisition of the thorax geometry. Thus, if also functional information about the cardiac muscle is desired (Sec. 4.1), two expensive and time consuming image acquisition sessions (CT & MRI) need to be carried out [279]. This is usually not feasible neither for the hospital nor for the patient.

4.3 ECG Signal Processing

In the course of this work, ECG signals from different modalities were recorded and processed. This comprises ECG signals from multi-channel ECG systems, electroanatomical mapping systems and VCG signals calculated from standard 12 leads ECG. The ECG signal processing workflow is shown in Figure 4.7.

4.3.1 ECG Signal Processing

ECG signals were post-processed to retrieve a representative average ECG signal of the atria (P-wave). The complete signal processing workflow was realized in Matlab (The MathWork Inc., Natick, MA, USA). For this purpose an I/O interface for the BioSemi file format (BDF) was created in Matlab based on the BioSig toolbox (http://biosig.sourceforge.net).

In a first step, a signal of at least 60 seconds length with least motion artifacts, was manually chosen from the complete ECG data which was usually several minutes long. Then baseline wander was removed using the wavelet-based technique presented in [281]. Subsequently, high frequency noise was removed in a similar manner. A further increase in signal-to-noise ratio could be achieved through common noise removal as described in [282]. As a consequence, one signal channel was removed from the data. This work step was optional. Afterwards remaining 50 Hz noise was removed by a 3rd order FIR Notch filter.

A template heart beat was created from the data by channel-wise signal averaging. This eliminated singular events (e. g. extra systoles), uncorrelated noise was suppressed and thus SNR was increased by a factor of $\sqrt{N}$, where N is the number of superimposed cycles. The template heart beat represented the mean ECG morphology and amplitude of the subject. This was especially important for the use of the data for model personalization, as the models behave deterministically. The modeling error could therefore increase if the model was fitted to a single ECG complex, instead of an averaged signal.

For the creation of the template heart beat, first the R-peaks in one low-noise, high amplitude channel were delineated using a Hilbert-transform approach [283]. ECG signals in all channels were cut 350 ms before and 650 ms after the detected R-peak times. ECG complexes belonging to RR-intervals which were shorter than

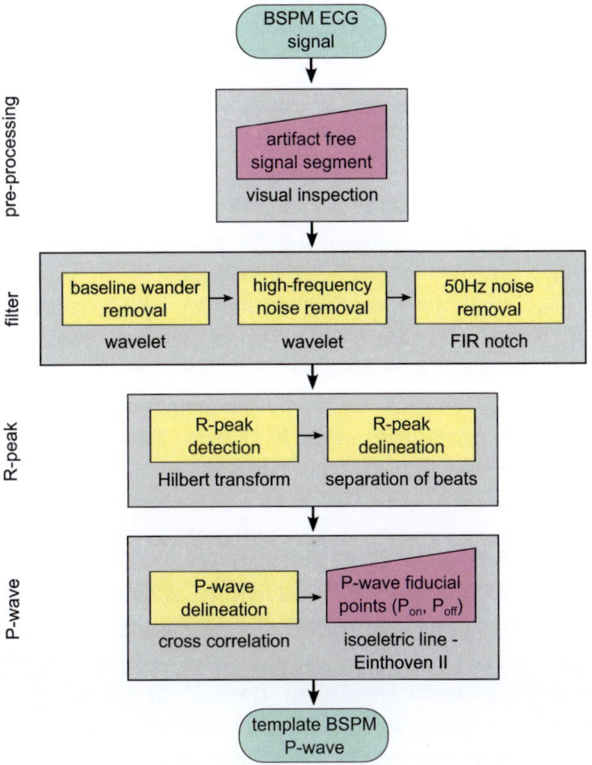

Fig. 4.7. ECG signal processing workflow.

0.1 times the average RR interval or greater than 1.9 times the average RR interval were considered to be falsely detected and thus removed from the data. As small differences in P-R time may be present in the signals, the averaged P-wave might be broadened during the averaging process. To remove this artifact while preserving the signal-to-noise ratio improvement, the P-waves preceding the R-peaks were shifted to have the greatest possible cross-correlation. Thus, the R-peaks were not precisely aligned, but the averaged signal of the P-wave was best-possibly aligned.

The P-wave duration was determined as the difference of P-onset and P-offset. The fiducial points were defined as the crossing of the P-wave with the isoelectric line in the Einthoven II signal [284] (Fig. 2.10). The points were marked manually in each template heart beat.

4.3.2 Spatial BSPM Interpolation

To visualize the measured multi-lead ECG signals as BSPMs on the patient thorax, the averaged ECG signal was interpolated on the surface of the finite element thorax model using a variant of the Laplacian interpolation [285]. As the size of the problem was small (typically about 5000 unknowns), the interpolation could be performed fast. Simulation results were written do disk every every millisecond. Thus, prior to the ECG signal interpolation, the signals were sampled down from 2048 Hz to 1000 Hz. ECG signals usually contain frequencies of up to 120 Hz in the QRS complex. The atrial parts of the ECG signal contain frequencies much lower. The down sampling to 1000 Hz does therefore not result in a loss of information.

To create integral BSPM maps, the ECG signals were normalized and integrated over the P-wave duration using a trapezoidal function. Afterwards the integral values were interpolated on the thorax surface. For simulated BSPMs, signals at the electrode locations were virtually recorded and treated in the same manner to create integral maps (integration, interpolation). This reduced uncertainty between the simulated and measured integral maps, as potential errors from the interpolation process were common to both measured and simulated integral maps.

4.3.3 Discussion

The ECG signal processing pipeline described above was set up to create a representative template ECG signal of the electrical activity of the atria for each patient. There could be variations in morphology and duration between subsequent P-wave in a patient. Multi-scale cardiac and thorax models are not able to recreate such behavior as they are deterministic. Thus, simulation results should be compared to an average ECG signal from the patient. ECG signals will need to be adapted to other investigations. E.g. the inverse problem of ECG might be better solved using continuous ECG signals instead of the averaged signal.

The signal averaging techniques were originally engineered to create a ventricular template ECG complex. The techniques were refined and further adapted to the atria by aligning the P-waves through signal cross correlation. The heart rate was assumed to be constant (60–80 bpm) over the averaging interval. There seems to be no relationship between PWd and heart rate [150] and thus no broadening of the P-wave is to be expected even if the heart rate would change over the averaging interval.

The spatial interpolation of BSPM signals between electrodes on the thorax surface can provide a visual impression of the potential distribution on the body surface. Nevertheless, the information content remains the same as in the ECG signals. In areas with few electrodes, e.g. the thorax back or the right side of the thorax when using the 64 channel setup, the interpolation may produce unrealistic potential distributions. On the other hand, the visual impression of the potential distribution might highlight regions of interest better than the signals could do.

The vector cardiogram is calculated from the standard ECG lead signals (Sec. 2.4.1.1). This describes a reduction of information content. Nevertheless, the three VCG signals may provide an easier understanding of electrical processes in the atria. For the future, BSPMs could be used to find another reduced number of electrode positions, which may under certain pathologies provide better insights into the electrical dysfunction of the heart, e.g. ECG-silent cardiac ischemia [286].

4.4 Adaptation of the Conduction Velocity Vector Estimation

For five patients (8–12, Tab. 4.1), LAT maps (Sec. 2.4.2) from the left atrium were collected using electroanatomical mapping systems (CARTO, BiosenseWebster, Haifa, Israel and NavX, St. Jude Medical, St. Paul, MN, USA). From the LAT maps, local conduction velocity was estimated based on the algorithm from Barnette et al. [156], which is described in Section 2.4.3. In this section, two extension to the algorithm are described. The algorithm is evaluated for the use with LAT map data and the results are discussed.

4.4.1 Adaptation to Measured and Simulated LAT maps

The original algorithm proposed to fit a second order polynomial to the LAT data. Under certain conditions it could be useful to apply a linear fit instead of a second order polynomial. This could e.g. be the case if a planar wavefront is propagating over a patch of tissue. For such linear fit, only four parameters are needed to solve the system of equations. This allows to retrieve a higher spatial resolution of conduction vectors. The linear fit was realized by exchanging equation 2.6 with

$$T = ax + by + cz + d, \tag{4.1}$$

and equation 2.9 with

$$\begin{aligned} t_x &= a, \\ t_y &= b, \\ t_z &= c. \end{aligned} \tag{4.2}$$

It is worthwhile to note that in the linear case the linear RMSE (LRMSE) and RMSE are the same.

On the other hand, complex activation patterns in the atria with a low spatial resolution, e.g. as observed in clinically measured LATs (Sec. 2.4.2), may require higher order polynomials to depict the wavefront curvature. Equation 2.6 could therefore be replaced by a third order polynomial with 20 unknowns

$$\begin{aligned} T = {}& ax^3 + by^3 + cz^3 + dx^2y + ex^2z + fy^2x + gy^2z + hz^2x + iz^2y \\ & + jxyz + kx^2 + ly^2 + mz^2 + nxy + oxz + pyz \\ & + qx + ry + sz + u, \end{aligned} \tag{4.3}$$

with its corresponding derivatives (Eqn. 2.9)

$$\begin{aligned} t_x &= 3ax^2 + fy^2 + hz^2 + 2dxy + 2exz + jyz + 2kx + ny + oz + q, \\ t_y &= 3by^2 + dx^2 + iz^2 + 2fyx + 2gyz + jxz + 2ly + nx + pz + r, \\ t_z &= 3cx^2 + ex^2 + gy^2 + 2hzx + 2izy + jxy + 2mz + ox + py + s. \end{aligned} \tag{4.4}$$

Activation times retrieved from electrophysiological simulations usually had a very high spatial resolution (e.g. 0.33 mm). Using all computational nodes M

for the calculation of local conduction velocity vectors would outrun computer's memory capacity in most cases, as an $M \times M$ distance matrix needed to be calculated. A spatial sampling of 1.32–2.00 mm was sufficient to ensure a good conduction velocity estimation in most cases, especially in whole atria simulations. The introduction of randomized variation for the spatial sampling of the nodes on the regular simulation grid reduced linear dependencies between adjacent nodes involved per fit.

For the visualization of the conduction velocity vectors in the 3D geometry, fewer vectors than calculated were visualized to avoid an overlay of adjacent vectors. Additionally, the endocardial surface of the atria was included in the visualization to avoid an overlay of vectors from overlapping tissue regions (e.g. anterior and posterior side). If only surface data were used in the calculation of the conduction velocity vectors (e.g. using clinically measured LAT maps), the conduction velocity vectors were geometrically projected into the plane orthogonal to the surface normals. This constraint corrected vectors pointing inside or outside the endocardial surface.

4.4.2 Evaluation of Artificial Setups

The algorithm was tested on two artificial environments (Fig. 4.8). Activation times were computed with an adaptive cellular automaton [162, 187, 223]. First, a patch with heterogeneous conduction velocity and second an arc with homogeneous conduction velocity was simulated and analyzed. Additionally, the arc setup was recreated in Matlab using an analytical calculation of the activation times.

$$t = 0 \ldots \pi$$
$$x = r \cdot \cos t$$
$$y = r \cdot \sin t$$
$$z = constant$$
$$AT = r \cdot t \cdot \frac{1}{CV} \tag{4.5}$$

where r was the (constant) radius of the arc (geometry and activation times not shown). The simulated activation time distributions are shown in Figures 4.8a,c. Figures 4.8b,d show histogram plots of conduction velocity magnitude calculated using the cubic fit. The conduction velocity in the flat patch was determined pre-

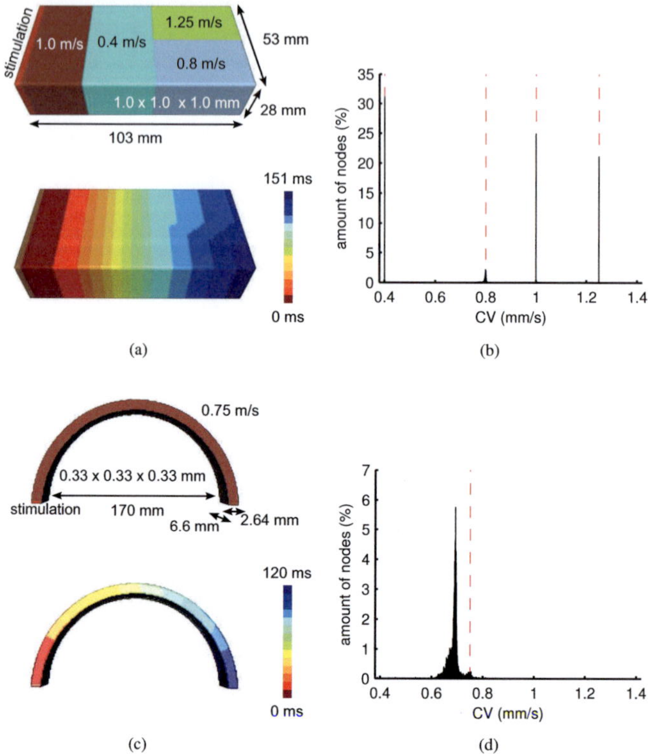

Fig. 4.8. Conduction velocity values calculated from simulated activation times in two test environments. Conduction velocity histogram plots were generated using the cubic fit method. Dashed red vertical lines indicate the conduction velocities which were used as input for the cellular automaton.

cisely with all three fit methods (linear, quadratic, cubic). Using the linear fit resulted in more detailed conduction velocity vector directions at the borders of conduction velocity regions, but also produced erroneous conduction velocity vectors within the regions. There was little difference between the solution using the quadratic and cubic fit. The conduction velocity which was set in the automaton for the curved geometry seemed to be underestimated by all fit methods (peak at 0.68 m/s instead of 0.75 m/s). The linear fit also calculated a variation in conduction velocity magnitudes. For the analytically determined activation times in the arc, the conduction velocity was determined precisely with all methods and for all tested input conduction velocities. Histogram plots are not shown for these results, as the calculated conduction velocity did not show a variation and thus the histograms were single vertical lines overlaying another.

4.4.3 Discussion

In this section a method to calculate local conduction velocity magnitude and direction from simulated and measured activation time data was presented. The method was originally designed to analyze data from *in-vitro* experiments in canine ventricles. It was extended by a linear and cubic fit option and adapted to clinical data collected *in-vivo* in human atria.

The algorithm produced exact results for the flat test environment. The conduction velocity was determined precisely from the analytical arc setup realized in Matlab with all fit methods. The conduction velocity set in the automaton was not the resulting conduction velocity observed in the automaton arc simulation. This was most likely caused by a longer pathway for the excitation wavefront in the discretized arc geometry. This is the explanation for the discrepancy between the conduction velocity which was input into the automaton and the conduction velocity determined with the fit.

The linear fit method showed better performance at edges of regions with different conduction velocity but also produced more errors. This was due to the smaller number of neighboring coordinate–time tuples needed for the linear fit compared to the quadratic or cubic fit. The cubic fit produced more realistic conduction velocity vectors at wavefront positions with multiple bends in the wavefront curvature but required more neighbor nodes than the quadratic fit. Depending on the given activation wavefront curvature, the most suitable fit method should be used.

The method to determine local conduction velocity used LATs as input. From simulations such data may be easily accessed in high quality. Intracardially measured LAT maps on the other hand are prone to several errors. First, the endocardial surface acquired with the electroanatomical mapping systems (EAMS) may not correlate to the real atrial morphology, as e.g. derived from MRI or CT scans. This will influence the localization of the activation time measurements and will thus impact on the calculated conduction velocity. Second, the time of activation is determined by the electrophysiologist manually from the catheter signals, which may introduce a user-specific error. Third, if the LAT measurements are not dense, the activation times are interpolated between the measurement positions by the EAMS. This could occlude activation pattern variations between the measurement positions. Fourth, catheter signals recorded at the left atrial septal wall may overlay with the far field signals from the right atrial septal wall. This could lead to misinterpretation of the earliest activation there. Nevertheless, the method may provide a visual impression of the local excitation direction and an approximation for the local conduction velocity.

Methods which directly access the catheter signals for analysis and conduction velocity determination [158, 160] rule out the user specific error in determination in activation time. On the other hand, the presented method works on existing LAT maps and can provide a picture of the complete atrial activation at once. A coupling of the automatic determination of the LAT from the catheter signals and the polynomial fitting procedure to determine the excitation direction and velocity could bring the advantages of both approaches together in the future. The present conduction velocity calculation method may also support methods to solve the inverse problem of ECG by providing boundary conditions for the solution space [287].

4.5 Patient & Volunteer Data

In the course of this work, the following datasets were acquired in different studies:

- one dataset with heart and thorax image data, BSPM data and intracardiac data,
- ten datasets comprising image data and BSPM data,
- three datasets containing image data from the atria and intracardiac data, and
- one dataset only of intracardiac data

Table 4.1 summarizes the acquired data. A number of datasets were not chosen for further processing due to low data quality (Sec. 4.5.1). For the remaining datasets, image data were segmented and anatomical models for the simulation of atrial electrophysiology and ECG computation were created (Sec. 7). ECG data were processed where available. Volunteer datasets (controls) were used to establish data acquisition and processing workflows and to create model databases (Sec. 7.3.1).

Table 4.1. Acquired patient and volunteer data sorted by completeness. Dataset with numbers in italics were not further processed.

Subject	Diagnosis	Sex	Age	Weight	Center	Date	models thorax / atria	LGE-MRI	LAT	ECG channels	ECG HR	ECG PWd
										thorax & atria datasets		
1[†]	control	male	26	79	Heidelberg	19.03.08	M / M&A	–	–	64 (FT)	69	107
2[†]	control	female	47	52	Heidelberg	15.08.07	M / M	–	–	64 (M)	81	95
3	control	male	27	100	London	16.07.09	M / A	–	–	64 (FT)	70	103
4	control	male	25	70	London	14.07.09	M / A	–	–	64 (OT)	53	97
5	control	male	38	90	London	25.02.10	M / A	–	–	83 (OT)	86	99
6[†]	LQT 2	female	17	66	Heidelberg	15.08.07	M / M	–	–	64 (M)	76	95
7[†]	LQT 1	female	50	79	Heidelberg	14.05.10	M / M	–	–	77 (M)	62	91
8	AF	male	66	100	London	14./15.07.09	M / A	pre, post	NavX	78 (OT)	62	176
										atria datasets with LAT maps		
9	AF	male	n/a	68	London	25.11.09	– / A/M	pre, post	NavX	3	n/a	107
9c						02.12.09	– / A	chronic	–	–	–	–
10	AF	n/a	n/a	105	London	27.07.11	– / A	chronic	Carto	3	n/a	118
11	AF	n/a	n/a	75	London	29.06.11	– / A	chronic	Carto	3	n/a	127
12	AF	n/a	n/a	n/a	Karlsruhe	02.11.11	– / –	–	NavX	–	–	–
										atria datasets		
13[‡]	pAF	female	50	n/a	Karlsruhe	n/a	– / A	–	–	–	–	–
14[‡]	pAF	female	73	n/a	Karlsruhe	n/a	– / A	–	–	–	–	–
15[‡]	pAF	male	58	n/a	Karlsruhe	n/a	– / A	–	–	–	–	–
16[‡]	pAF	male	75	n/a	Karlsruhe	n/a	– / A	–	–	–	–	–
17*	n/a	n/a	n/a	n/a	n/a	n/a	– / A	–	–	–	–	
18*	n/a	n/a	n/a	n/a	n/a	n/a	– / A	–	–	–	–	
19°	control	male	33	86	Heidelberg	n/a	– / M	–	–	–		
										Visible Human datasets		
20 (VM)	control	male	38	n/a	–	11.1994	M / M	–	–	–	–	–
21 (VF)	control	female	59	n/a	–	11.1995	M / M	–	–	–	–	–
										discarded datasets		
23	VT	male	n/a	70	London	27.04.09	– / –	–	–	–	–	–
24	control	male	32	75	London	28.04.09	– / –	–	–	64 (FT)	–	–
25	control	male	43	80	London	28.04.09	– / –	–	–	–	–	–
26	control	male	27	100	London	29.04.09	– / –	–	–	64 (FT)	–	–
27	AFlut	male	n/a	88	London	13.07.09	– / –	–	–	64 (FT)	–	–
28	AF	female	n/a	90	London	15.07.09	– / –	–	–	64 (OT)	–	–
29	AF	male	n/a	118	London	24.02.10	– / –	–	–	80 (OT)	–	–

[†]: Data which has been collected as part of the PhD thesis of Dr.-Ing. David Keller [267]. [‡]: Data which has been collected as part of the PhD thesis of Dr.-Ing. Frank Weber [268]. *: Data provided by Philips Research, Hamburg. °: Data provided by Thomas Fritz, IBT, KIT. ECG processing, cardiac segmentation and parts of the thorax segmentation were performed as part of this dissertation. M: manual segmentation, A: automatic segmentation, LQT: long QT syndrome, FT: Fastrak, OP: Optotrak, –: not available, x: available.

4.5.1 Data Selection

Seven datasets were excluded from post-processing (23–29). Dataset 23 was discarded because no thorax MRI could be acquired for this patient. Datasets 24 – 26 had low thorax MRI quality. In datasets 27 and 28, the BSPM recording failed. A very low signal to noise ratio made the data unusable. For the data acquisition of these datasets, the initial version of the BioSemi BSPM system was used (Sec. 4.2). Patient 29 was in constant AF during the MRI scans, BSPM acquisition and the EP procedure. The data was thus not selected for further use, as relevant mophological and ECG information could not be extracted from the data.

4.5.2 Discussion

The aim of the data acquisition was to acquire complete datasets comprising cardiac MRI/CT, thorax MRI, BSPM data and LAT maps. Seven datasets could not be used for further processing as some of the data were corrupted. Among these were mainly datasets from the time, in which the collaboration between King's College London and IBT started. During this phase, data acquisition protocols and workflows were tested.

5

Modeling of the Atria

5.1 Adaptations of the Courtemanche-Ramirez-Nattel Model

5.1.1 Adaptation to Variable Extracellular K^+ Concentrations

In contrast to other models of cardiac electrophysiology (e. g. [288]), the CRN model in its original formulation is not able to cope with variable extracellular potassium concentrations $[K^+]_o$. To allow for variable $[K^+]_o$ concentrations, the mathematical formulations of two potassium currents were changed to reflect such behavior by utilizing correction terms commonly used in models of ventricular electrophysiology. A square root term was added to the formulation of the time-independent K^+ current I_{K1} and to the formulation of the rapid delayed outward rectifier K^+ current I_{Kr} (highlighted in gray in equations 5.1 and 5.2). The correction term was formulated as in the Luo-Rudy model [195] and bases on experimental data [289].

$$I_{K1} = \frac{g_{K1} \cdot (V_m - E_K)}{1 + e^{(0.07(V_m + 80))}} \cdot \sqrt{\frac{[K^+]_o}{5.4}} \tag{5.1}$$

$$I_{Kr} = \frac{g_{Kr} \cdot X_r \cdot (V_m - E_K)}{1 + e^{(\frac{V_m + 15}{22.4})}} \cdot \sqrt{\frac{[K^+]_o}{5.4}} \tag{5.2}$$

g_{K1} and g_{Kr} are the maximum channel conductances for I_{K1} and I_{Kr}, respectively. V_m is the transmembrane voltage and E_K is the K^+ Nernst potential. The denominator in the new square root term (5.4) is the physiological value of $[K^+]_o$ in the CRN model, the unit of $[K^+]_o$ is mmol.

Table 5.1. Implementation of electrical remodeling due to persistent AF into the CRN model by adaptation of model parameters. Maximum ion channel conductances were changed [73, 246] based on measurements of [241, 244, 245]. Initialization values for Ca^{2+} and V_m needed to be adapted afterwards, as the CRN model was not in steady state due to the altered channel conductances. The effects are negligible for the simulation of the action potential, but significantly influence Ca^{2+} concentrations in the simulations.

Parameter	Healthy Model	Persistent AF
g_to	0.16520	0.05782
g_CaL	0.12375	0.04330
g_K1	0.090000	0.194625
Init_f_Ca	0.75607	0.81388
Init_Ca_i	0.000112836	0.000079981
Init_Ca_up	1.52919	0.96431
Init_Ca_rel	1.10817	0.55740
Init_Vm	-0.0808887	-0.0844802

5.1.2 Steady State Remodeling Conditions

As described in Section 3.2.2, ongoing AF alters the atrial electrophysiology. The CRN model has been adapted to reflect electrical remodeling in patients with persistent AF by changing three maximum ion channel conductances (Tab. 5.1) [73, 246]. With these changes, the CRN model is able to reproduce the action potential morphology previously described in the literature for AF patients. Action potential morphology changes only marginally during the course of multiple beats but calcium concentrations in the different model compartments change significantly (Fig. 5.1b). In combination with a tension development model, this can cause unsteady results (Fig. 5.1c). Adaptation of the calcium initialization parameters of the remodeling CRN model (Tab. 5.1) compensated for this transient behavior (Fig. 5.1).

5.2 Model of Heterogeneous Atrial Electrophysiology

In-silico models of the human atrial electrophysiology have been used for decades to investigate the role of atrial electrophysiology in the context of atrial arrhythmias and other cardiac pathologies [2, 5]. Besides pathological changes of the atrial electrophysiology, e.g. electrical remodeling (Sec. 3.2.2), the human atria show regional variations in action potential morphology and dynamics [75, 290–292]. In 3D atrial models, electrophysiological heterogeneities, mostly in the RA

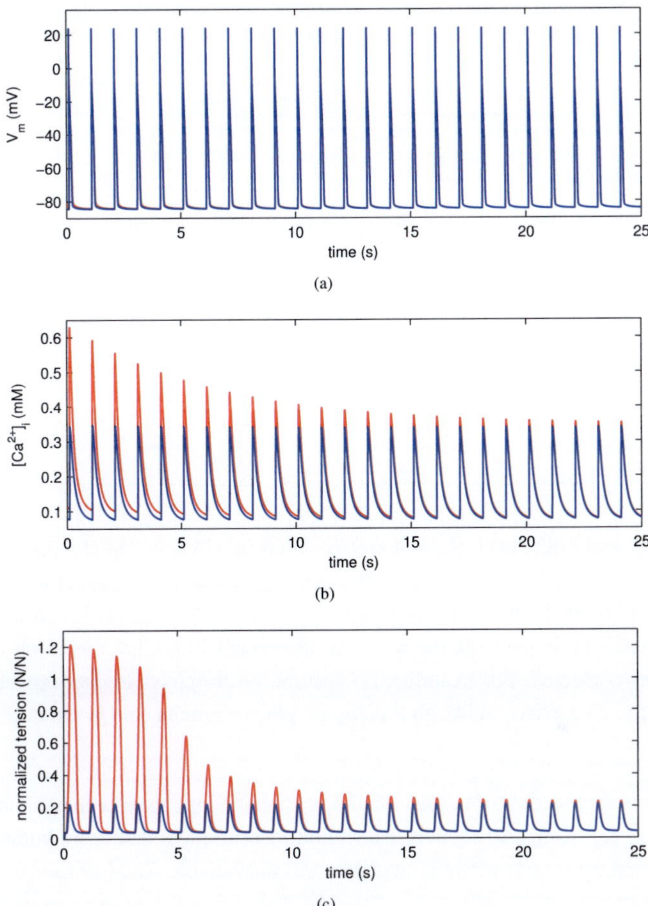

Fig. 5.1. Transient behavior into steady-state of the remodeling CRN model. Red curves show the transient behavior of the remodeling model presented in [73]. Blue curves display the transient behavior of of the model with altered initialization values. Normalized tension was computed using the Hybrid model adapted to the tension development of AF patients (Sec. 5.3).

(CT, PM, TVR), have been recognized (Tab. 3.1). In simple patch simulations with electrophysiological models of non-human mammals, also other regional differences in atrial electrophysiology have been investigated for their pro-arrhythmic character [68, 69, 239, 293]. More recently, Aslanidi et al. presented a 3D atria model, which also covered differences between right and left atrial electrophysiol-

Table 5.2. Overview of experimental studies of heterogeneous atrial electrophysiology. MC: multi-cell, SC: single-cell, n/a: information not available, RK: also measurements of recovery kinetics. Table adapted from [295].

Study	Species	Location	MC/SC	BCL (ms)	Channels
Burashnikov et al. 2004 [296]	canine (n/a)	RA	MC	700	I_{to}
Caballero et al. 2010 [247]	human (22)	RAA, LAA	n/a	n/a	I_{to}, I_{Ks}
Cha et al. 2005 [297]	canine (37)	PV, LA	MC	500–2000	$I_{to}, I_{CaL}, I_{K1}, I_{Kr}, I_{Ks}$
Ehrlich et al. 2003 [298]	canine (n/a)	PV, LA	MC	1000	$I_{to}, I_{CaL}, I_{K1}, I_{Kr}, I_{Ks}$
Feng et al. 1998 [234]	canine (19)	RA	MC	500, 1000, 10000	I_{CaL}, I_{to}, I_{Kr}
Gong et al. 2008 [299]	human (16)	Sep, RAA	MC	10000	$I_{Na}, I_{CaL}, I_{Kur}, I_{Na}$ (RK)
Melnyk et al. 2005 [300]	canine (n/a)	PV, LA	n/a	n/a	I_K
Li et al. 2001 [235]	canine (33)	LA, RA	SC, MC	167-1000	I_{Kr}
Qi et al. 1994 [301]	rabbit (n/a)	LA, RA	SC	1000	I_{to} (RK)
Voigt et al. 2010 [249]	human (43)	RAA, LAA	n/a	n/a	I_{K1}
Wand et al. 1999 [302]	human (26), rabbit (n/a)	RA	n/a	n/a	I_{to} (RK)

ogy [218] and Ridler et al. showed that gradients in APD may protect the atria from arrhythmia [221]. None of these studies investigated the tissue properties of the regional variations in atrial electrophysiology in detail or presented more extensive heterogeneities. Especially, the regional differences in the LA electrophysiology are often neglected. For example, no specific electrophysiological model of the area of the PVs exists, although this region plays a crucial role in the initiation of AF.

Up to now, five models of human atrial electrophysiology have been published [72, 227, 237, 238, 294]. All models use similar measurement data from human atrial cells, or mammal cells where human data was unavailable. The form and dynamics of the simulated action potentials varies quite significantly underneath the models (Sec. 3.2). This could be interpreted as additional evidence for regional differences in atrial electrophysiology. In the models as well as in the previously modeled heterogeneities in right atrial electrophysiology, mainly the repolarization phase of the action potential varies.

In this section a model of heterogeneous atrial electrophysiology in healthy and AF-remodeled state is presented. The electrophysiology in the different parts of the atria is examined for its pro-arrhythmic properties and an approach how to smoothly include electrophysiological heterogeneities into patient-specific atrial

Table 5.3. Relative values of ion channel conductivities of all implemented heterogeneities with respect to the original model of [72]. *: derived heterogeneity as a combination of LA conductivity and heterogeneous RA conductivity.

Heterogeneity	Source	g_{Na}	g_{to}	g_{CaL}	g_{Kr}	g_{Ks}	g_{Kur2}	g_{K1}
RA / PM	original model	1.0	1.0	1.0	1.0	1.0	1.0	1.0
CT upper endo	[234]	1.0	1.0	1.67	1.0	1.0	1.0	1.0
CT upper epi	[296]	1.0	0.50	1.67	1.0	1.0	1.0	1.0
CT lower endo	[296]	1.0	0.68	1.67	1.0	1.0	1.0	1.0
CT lower epi	[296]	1.0	0.34	1.67	1.0	1.0	1.0	1.0
BB (RA part)	[234, 296]	1.0	1.0	1.67	1.0	1.0	1.0	1.0
TVR	[234]	1.0	1.0	0.67	1.53	1.0	1.0	1.0
MVR*	[234, 235]	1.0	1.0	0.67	2.44	1.0	1.0	1.0
RAA	[234]	1.0	0.68	1.06	1.0	1.0	1.0	1.0
LAA*	[234, 235]	1.0	0.68	1.06	1.60	1.0	1.0	1.0
LA	[235]	1.0	1.0	1.0	1.60	1.0	1.0	1.0
PV	[297, 298, 300]	1.0	0.75	0.75	2.40	1.87	1.0	0.67
Sep	[299]	1.5	1.0	0.25	1.0	1.0	0.67	1.0

geometries is proposed. The simulations are validated against *in-vitro* cell measurements. Conclusions are drawn for a better understanding of the repolarization sequence in the ECG under persistent AF remodeling.

In models of cardiac electrophysiology, ionic currents across the cell membrane are described by the ion channel conductivity g_x and a product of gating variables γ_i (Sec. 3.2). The conductivity scales the amplitude of the ionic current and the gating variables describe the open probability of the channels and thus the kinetics of the ionic current. Most of the experimental data presented in the literature (Tab. 5.2) show that the ionic currents differ across the atria only in the amplitude of current density and not in current kinetics. In other cases it can also be necessary to change the mathematical formulations of the ion channels, e.g. to reproduce certain pathologies [93]. Although a number of models of the human atrial electrophysiology exist, the models of Courtemanche et al. and Nygren et al. are the most established ones and the majority of the newer models base upon their mathematical formulations [5].

Measurement data of human atrial electrophysiology are sparse, especially regarding action potential variations. Few studies investigated the ionic mechanisms underlying regional action potential variations in human due to ethical considerations. More studies investigated such effects in other mammals, mainly canine (Tab. 5.2). Canine cardiac electrophysiology is similar to human cardiac electrophysiology and therefore often used if human data is not available.

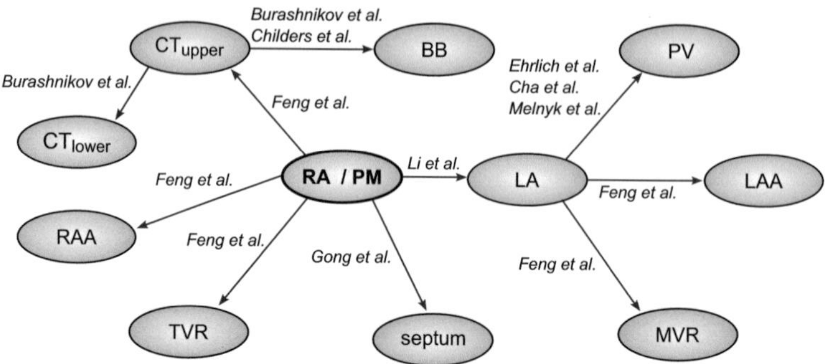

Fig. 5.2. Derivation of parameters sets from known electrophysiological parameters to model the electrophysiology of various atrial regions.

From the experimental data (Tab. 5.2), relative changes in maximum ion channel conductivities were extracted (Tab. 5.3). The current densities of the modeled currents are directly related to the value of the corresponding maximum channel conductance. Most experiments were conducted on mammal cells and all experiments were done *in-vitro* in different experimental settings. Relative changes in ion channel expressions between two preparations from different atrial cell types provided the most reliable data on how atrial electrophysiology might change regionally in the *in-vivo* human atria. Figure 5.2 provides an overview how the parameter sets for the various atrial regions were created.

Experimental data from the left mitral valve annulus musculature as well as from the left atrial appendage were not found. It was assumed that the relative difference between the left atrial myocardium and these structures is the same as the relative difference in ion channel conductances between respective structures in the right atrium. It was also assumed that the two basic models reproduce action potentials from the right atrial free wall (CAM and RA cells). The parameters for the upper part of the CT were also used for the right part of BB, as AP morphology and APD was shown to be similar in both structures [296, 303].

Three experimental studies were found describing the relative difference between LA and PV canine electrophysiology [297, 298, 300]. They found differences in four potassium currents (I_{to}, I_{K1}, I_{Kr}, I_{Ks}) and one calcium current (I_{CaL}). The tendency of change in all current densities was similar in all studies, but absolute values differed slightly. Melnyk et al. did not measure I_{to} and I_{CaL} current densities.

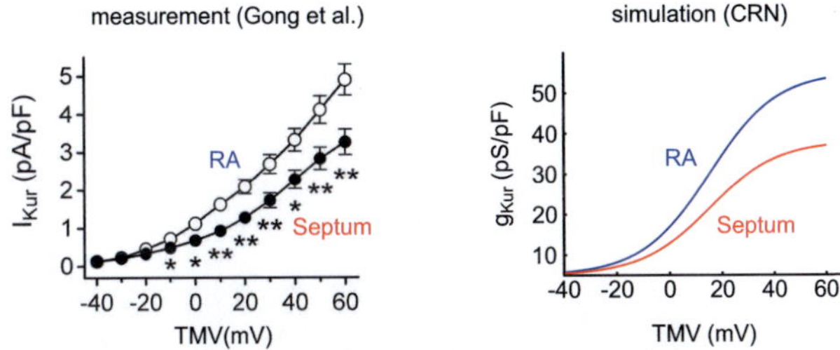

Fig. 5.3. Current density (channel conductance) – transmembrane voltage (TMV) relationship as observed between cells from the human right atrium (RA) and the septum by Gong et al. [299] and during simulation with the heterogeneous CRN model. In both cases curves split at voltages above -20 mV and the curve of the septal tissue runs flatter compared to the right atrial curve. Left figure adapted from [299].

To model the electrophysiology of the PVs, the measured relative differences to the LA electrophysiology were averaged.

The relative changes in ion current densities were introduced into the models of Courtemanche et al. and Nygren et al.. Measurements from human atrial septum cells revealed a decreased current density of the ultrarapid delayed rectifier K^+ current I_{Kur} for TMVs greater than -20 mV [299] (Fig. 5.3). In the CRN model, this was considered by adaptation of the parameter g_{Kur2} in the formulation of the maximum I_{Kur} conductance:

$$g_{Kur} = g_{Kur1} + \frac{g_{Kur2}}{1 + e^{\frac{V_m - 15mV}{-13mV}}} \tag{5.3}$$

This parameter determines the slope of the curve, if the channel conductance is plotted against the TMV (Fig. 5.3). The model of Nygren et al. does not include a similar formulation. It was therefore not possible to generate a parameter set for the septal region with this model.

Persistent AF alters atrial electrophysiology (Sec. 3.2.2). The parameter sets for the different atrial regions were coupled with a model of electrical remodeling due to persistent AF [73] to create a model of heterogeneous AF-remodeled atrial electrophysiology. The remodeling conditions were created by scaling two potassium and one calcium current ($I_{K1}\cdot2.1$, $I_{CaL}\cdot0.35$, $I_{to}\cdot0.35$) as described in [73]. These scaling factors were multiplied with the scaling factors of the various atrial regions (Tab. 5.3) for this purpose. Single-cell action potentials simulated with the CRN

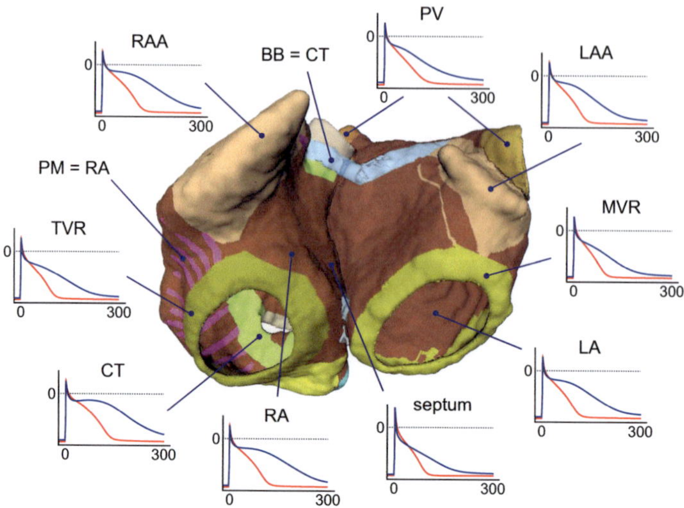

Fig. 5.4. Regional distribution of physiological (blue) and AF-remodeled (red) single-cell action potentials.

model and the different parameter setups in healthy and AF-remodeled state are shown in Fig. 5.4.

Simulation of action potentials with the model of Nygren et al. and the parameter sets for CT and PVs led to model instabilities. While a CT action potential could be simulated in a single-cell setup, simulation of coupled Nygren CT cells resulted in a drift of resting membrane potential. The drift was so strong, that after few stimulations no AP could be initiated. For Nygren PV cells, a similar behavior was also observed in single-cell and coupled cell simulations.

Atrial electrophysiology does not change abruptly, but shows transition gradients between adjacent regions [296, 303]. These gradients could ensure smooth transitions in ion channel expression or gradual mixture of cells with different channel expressions. No data were found to quantify these smooth changes. Smooth transition between neighboring regions with different channel expression was achieved in the geometrical atria models by applying Gaussian distance maps at region borders. The width were set depending on the region borders. The CT is known to have a rather isolated electrophysiological behavior. Therefore the width of the CT to other structures was set smaller than the gradient width between large structures, e.g. the valve rings and the atrial body. The widths of the transition gradients were

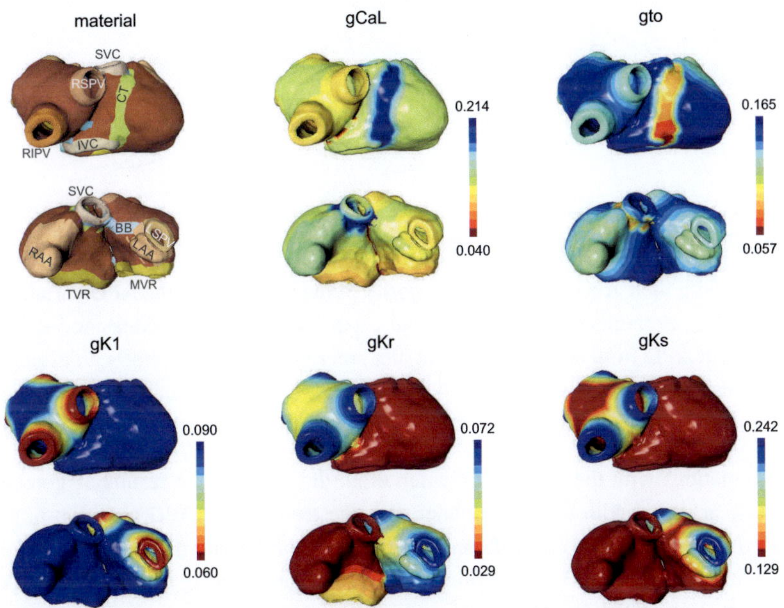

Fig. 5.5. Gradients of maximum ion channel conductances in model number 8 (Tab. 7.1). Figure adapted from [295].

set to LA-PV: 16.5 mm, RA-Septum: 3.3 mm, CT-RA: 6.6 mm, TVR-RA: 26.4 mm and MVR-LA: 26.4 mm. Figure 5.5 shows the gradients in a representative model of the model cohort.

5.2.1 Reproduction of *in-vitro* Experiments

In the majority of the experiments (Tab. 5.2), both a change in ion channel expressions and a change in APD were measured. The changes in ion channel expression were used to parameterize the model for different atrial regions. The relative APD difference between two regions was used to evaluate the simulation. For this, the experimental settings of the *in-vitro* studies (BCL, single-cell / multi-cell) were reproduced *in-silico* and the simulated APD change was compared to the experimental findings. Table 5.4 summarizes the results of the evaluation process. The tendency of APD variation could be reproduced in the simulations in nearly all

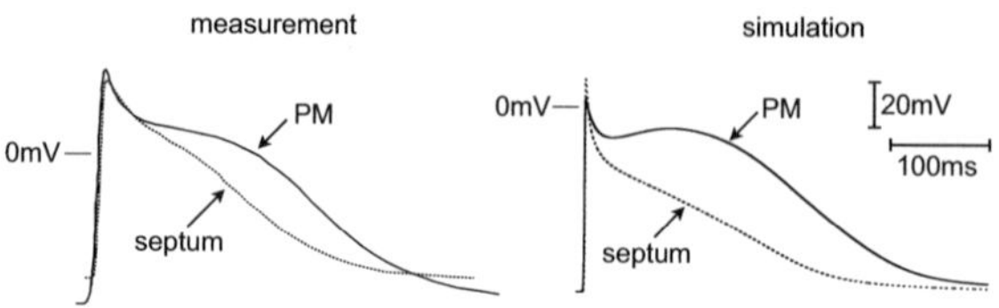

Fig. 5.6. Comparison of simulated and measured action potentials in the right atrium and the right atrial septum. Measurement data from [299]. Figure adapted from [295].

cases with the CRN model. In the majority of the simulations, the APD variations were less pronounced compared to the experimental findings. Measured differences in APD from human cells were met precisely. Figure 5.6 shows an example for such well correlating measured and simulated action potentials in the human right atrium and septum.

Simulations with the model of Nygren et al. did not produce results as good as those achieved with the CRN model. E.g. the tendency of APD variation in the RAA was not met and the difference between simulated and measured APD variation between left and right atrium was greater than by using the CRN model. A comparison of PV and septal cell could not be performed, as no parameter setup for the septum was available and simulations of PV action potential failed.

5.2.2 Detailed Analysis of Heterogeneous Electrophysiology

The electrophysiological behavior of atrial myocytes in different atrial regions were explored using a cable model. The cable consisted of 300 voxels (Δx=0.1 mm) and was stimulated at one side with different cycle lengths and stimulation protocols. Using this setup, APD, effective refractory period (ERP) and CV were determined depending on the preceding stimulation cycle length. The wave length was calculated as the product of the ERP and CV. The wavelength is a quantitative measure for the minimal length of a pathway in which a reentrant circuit can sustain. Simulations were carried out with the heterogeneous models of Courtemanche et al. and Nygren et al.. Furthermore, simulations were performed for the AF-remodeled heterogeneous CRN model.

The comparison of the models of Courtemanche et al. and Nygren et al. revealed that i) the action potential shape is triangular in the Nygren model and the CRN model has a more pronounced plateau phase. ii) The morphology of the APD and

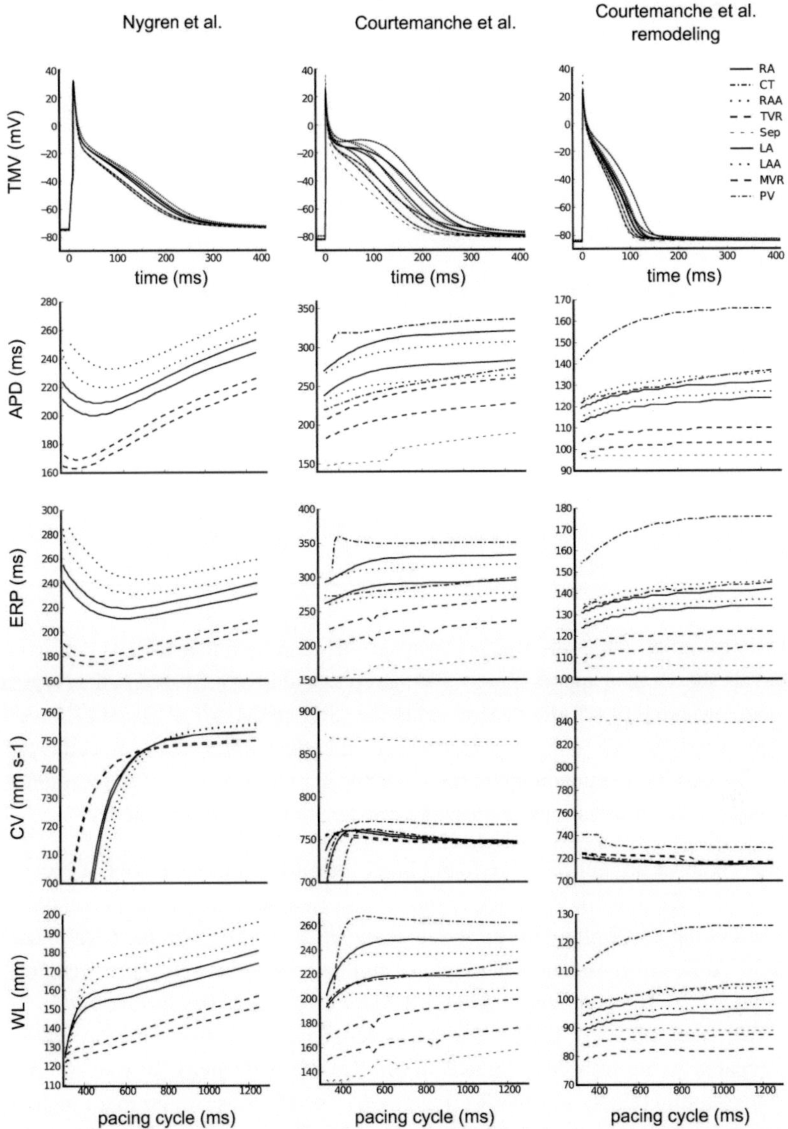

Fig. 5.7. Simulated electrophysiological properties in various regions of the atria. Action potentials were simulated in single-cell simulations and other properties were determined using a cable setup. The y-axes are differently scaled in all three columns to highlight the regional differences in electrophysiological properties. Figure adapted from [295].

Table 5.4. Relative changes of APD between two atrial regions in simulations compared to measurements. Compared were APD_{95} for measurements of Feng et al. [234] and APD_{90} for all other measurements, as measurements were quantified by the authors this way. Table adapted from [295].

Heterogeneity	ΔAPD (%)			BCL (ms)	Setting	Source
	Measurement	CRN	Nyg.			
CT (w.r.t. PM)	44	17	14	500	SC	[234]
	42	23	4	1 000	SC	
	34	3	22	10 000	SC	
RAA (w.r.t. PM)	2	2	2	500	SC	
	-5	0	4	1 000	SC	
	-14	-3	7	10 000	SC	
TVR (w.r.t. PM)	-8	-20	1	500	SC	
	-16	-22	-7	1 000	SC	
	-26	-8	-14	10 000	SC	
LA (w.r.t. PM)	-14	-9	-13	250	SC	[235]
	-14	-14	-3	330	SC	
	-12	-13	-3	500	SC	
	-11	-13	-3	1 000	SC	
	-19	-11	-5	400	MC	
	-19	-12	-3	1 000	MC	
PV (w.r.t. LA)	-16	-11	n/a	500	MC	[297]
	-11	-6	n/a	1 000	MC	
	-5	2	n/a	2 000	MC	
Sep (w.r.t. PM)	-29	-30	n/a	10 000	MC	[299]

ERP restitution curves differed between the CRN and Nygren model. At pacing intervals shorter than approximately 500 ms, APD and ERP increased in the Nygren model, whereas they further decreased in the CRN model. iii) APD and ERP were significantly shorter in the Nygren model than in the CRN model in all cell types. iv) CV was in the same range in both models, but the steep part of the restitution curve covered a wider range of pacing cycle length in the Nygren model.

Inter-regional differences were found in all measures. i) In the model of Nygren et al., the triangular morphology of the action potential is present in all cell types, whereas the morphology of the action potentials generated with the CRN model shows different morphologies ranging from spike-and-dome (CT) to triangular (septum). ii) The variation in APD, ERP and wavelength was greater within the CRN model. iii) APD, ERP, CV and wavelength were smaller in the LA regions compared to the respective regions in the RA. iv) APD and ERP were shortest in the septum (CRN) and MVR (Nygren). v) The CV was similar in all regions, except for the septum setup. This was the only setup in which the maximum ion channel conductance of a depolarizing sodium current was altered. vi) The slope of the wavelength restitution curves produced with the Nygren model are steeper

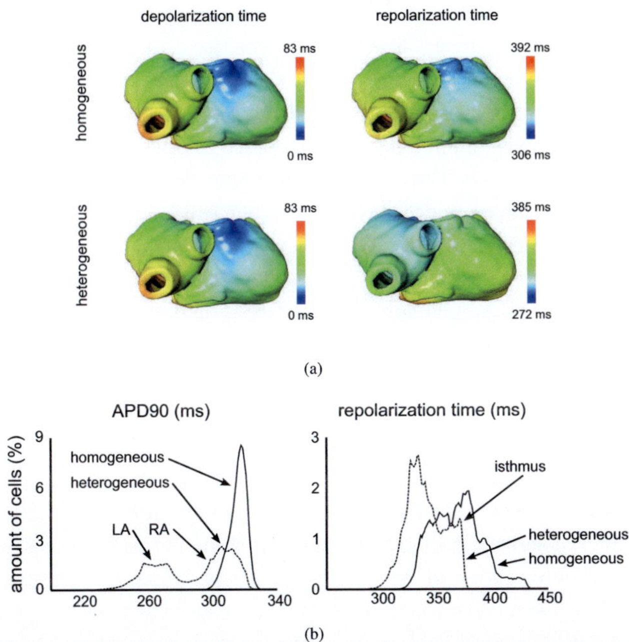

Fig. 5.8. a) Time of start of depolarization, APD and time of start of repolarization. Representative example (Model 5). BCL of 800 ms. b) Representative histogram plots of the distribution of APD and repolarisation time (model 5). In the homogeneous simulations (solid), a normal distribution of APDs around the general APD of the CRN model could be observed. Introduction of heterogeneities (dashed) led to a bi-maximum distribution of APDs centered around the APD of the left and right atrial model. Overall, the distribution widened as expected. Distribution of the repolarization time shifted towards shorter times. In most models a small plateau at the right side was present (arrow). This was due to isthmus-caused late-activated cells in the RA with a relative long APD due to their location near the CT.

than the curves produced with the CRN model. vii) In general, APD was decreased with increased distance to the sinus node, with the exception of the LAA setup.

Combination of the heterogeneous CRN model and a model of electrical remodeling significantly reduced APD, ERP and wavelength. The differences between atrial cell types were preserved. CV was not significantly changed. The morphology of the action potential was triangular in all cell types. The relative difference between APD, ERP and wavelength of the CT and the other cell types was larger under remodeling conditions compared to the regular model.

In eight atrial models (Tab. 7.1, 1-8), sinus node activity (BCL=800 ms) was simulated to evaluate the impact of the heterogeneities on the atrial depolarization and repolarization sequence. In the 3D atria models, depolarization time was determined in each node as the time at which the TMV was greater than -6 mV for the first time. Repolarization time was determined as time at which the TMV was to 90% repolarized again. The depolarization sequence showed no significant differences between the homogeneous and heterogeneous simulations (Fig. 5.8(a)). Repolarization followed the depolarization sequence in the homogeneous simulations, thus beginning near the sinus node and ending in the posterior left atrium. In the heterogeneous simulations, first repolarization occurred in the anterior left atrium while the depolarization was still ongoing. The last region to repolarize was in the inferior right atrium, near the area of the inferior isthmus. This region was slowly activated due to the reduced CV in the right inferior isthmus and had a long APD due to the proximity to the CT. The histogram of the APD and repolarization time distribution (Fig. 5.8(b)) revealed that the heterogeneities flatten the APD distribution. APD values accumulated around the APDs of the left and right atrium. The repolarization time histogram was shifted to earlier times, as the repolarization started earlier in the heterogeneous simulations. A plateau at the late end of the histogram curve was present. This was caused by the late repolarization of the right atrial inferior isthmus area. The repolarization sequence was overall more synchronous in the heterogeneous simulations.

5.2.3 Discussion

In this section, a model of heterogeneous healthy and AF-remodeled atrial electrophysiology was presented. The CRN model was thereby more suitable for the reproduction of the measured changes in APD between different atrial regions than the model of Nygren et al.. The latter model also showed instabilities for the CT and PV parameter setups. The shorter APD and ERP as well as the steeper wavelength restitution produced by the Nygren model might explain why the Nygren model is used more frequently to simulate atrial fibrillation in 3D atrial models [2].

The observed variety in action potential morphologies in the atria might also serve as an explanation that the plurality of models of atrial electrophysiology base on similar measurement data, but reproduce different action potential shapes. It could be that each model describes the action potential from a different atrial region, or that measurement data from different regions were mixed in the development process of the models. Both investigated models were identified to reproduce RA action potentials in their original formulations. It may be arguable that the Nygren model rather reproduces the LA action potential, as its action potential morphology is rather triangular. It is yet to be investigated if a different parameterization for various atrial regions beginning from the LA would result in a better outcome for the model of Nygren et al.. As the changes in ion current densities for the RA regions will be exaggerated in this case, the observed problems in simulating CT action potentials using the Nygren model could then increase as well.

Only very sparse experimental data was available on regional heterogeneities of human atrial electrophysiology. Nevertheless, experimental results not applicable to humans were neglected in this study. E. g. measured differences in rabbit I_{to} recovery kinetics [301] were not implemented, because no changes in APD were expected due to much faster I_{to} recovery in human cells [302]. All other human and mammal data were used, although single-cell preparations may alter I_{Kr} expression [304]. Thus, experimental data from multi-cell preparations might be more reliable. The measurement of EP properties of atrial cells is also affected by the method of cell isolation and cell temperature [305]. Therefore only relative changes of ionic current densities of cells in the same experimental setup were used. There was an uncertainty within some of the experimental studies, as not all studies provided precise information on the region of cell extraction. E. g. sometimes the RAA was defined as the rough RA wall, and other times it was defined as the triangular extension of the RA next to the aortic arch. If it was not clear from the experimental data where cells were exactly extracted, parameter setups for both anchor regions were created and simulated. The configuration with better match to the simulation data was chosen afterwards. APs simulated with altered channel conductances reproduced the relative changes in APD of the atrial region in the experimental data.

For the right atrial septum, Gong et al. were able to show that besides the alteration in ion channel expression, also the channel kinetics were different [299]. The CRN model was adapted to reflect these changes using the Powell-algorithm. The optimized parameters ($\tau_{act,original} = 0.135$ ms, $\tau_{act,septum} = 0.265$ ms, recovery time constant $+40\%$) had only little influence on the simulated action potential and were thus neglected for the heterogeneous 3D model.

Atrial regions in the left atrium had a shorter wavelength compared to right atrial parts. This hints that the LA may provide a better substrate for the development and maintenance of atrial arrhythmia than the RA [1]. The regional variations in ion channel conductivities resulted in a gradient of decreasing APD with increasing distance to the sinus node in the models. This finding supports the hypothesis of the presence of such gradients from Ridler et al. [221]. It is also in agreement with experimental studies [291, 306] and *in-vivo* intracardiac measurements from the RA [307], which showed decreasing APD along the CT and other RA parts. The repolarization sequence in heterogeneous bi-atrial sinus rhythm simulations was more synchronous than in the respective homogeneous simulations.

Although the repolarization sequence in the heterogeneous models was more synchronous, repolarization did not start in all atrial regions at the same time, e.g. the LA started to repolarize before the RA in most models. Changes in interatrial CV, e.g. via Bachmann's bundle, could compensate for this difference and thus produce a smoother repolarization sequence. An additional gradient in APD in the LA from between the anterior and posterior region, similar to the gradient along the CT in the RA, could further synchronize the LA repolarization. No experimental data were found to approve or disprove this hypothesis.

The effect of the electrical remodeling on APD, ERP and wavelength was much larger than the effects of regional variations in electrophysiological properties, although these were also present under electrical remodeling. This finding is in agreement with canine experimental results from Cha et al. [297]. Cha et al. found an APD difference between LA and PV cells, which was decreased under remodeling conditions, although the remodeling acted in the same way on both tissue regions.

As described in Section 3.2.2, electrical remodeling due to persistent AF has been implemented in electrophysiological models in various ways. The described model used the remodeling implementation proposed by Seemann et al. [73], which altered I_{CaL}, I_{to} and I_{K1}. Other implementations might vary the model behavior. Nevertheless, the major impact of electrical remodeling (short APD, ERP and wavelength) is similar in all remodeling model variants. The tendency of the present simulation outcome is therefore not likely to change with a different remodeling implementation. The reduction in CV due to a reduced gap junction density under AF remodeling [242, 243] was neglected in the present simulations, as the aim was to investigate the influence of changes in membrane electrophysiology only.

5.3 Model of Atrial Tension Development

The electrical excitation of cardiomyocytes causes the development of an active force in the cells (Sec. 2.1.2). The electro-mechanical coupling depends on the intracellular calcium concentration. For ventricular cells, various models of the tension development have been used to study ventricular tension and subsequently deformation [308]. These models neglected the atrial compartments of the heart so far, as neither a tension model was available to compute the local tension in the atria, nor sufficiently detailed geometrical models of the atria were available for the simulation of the atrial deformation. Some groups proposed to use a prescribed low blood pressure as boundary condition for the AV junction or to model atrial contraction by using a 0D model (a time-dependent ODE) [309]. Although such approach might be sufficient for ventricular deformation simulations, the influence of pathological alterations of atrial electrophysiology or anatomy on the cardiac function cannot be investigated. In this section, a methodology to model the atrial tension development under healthy and persistent AF remodeled conditions is proposed. The model is named atrial hybrid tension development (aHTD) model following the naming of the hybrid tension development (HTD) model which was used for a basis.

Schotten et al. measured the tension development in atrial muscle preparations [74, 310] and showed that the tension development is faster in the atria than in the ventricles. They also revealed a decrease in tension in chronic AF patients [74]. To model the atrial tension development, the HTD model, which had been proposed to simulate ventricular tension [308, 311], was adapted to the atrial experimental data. The HTD model consists of three coupled Markov chain models with 14 state variables (Fig. 5.9). These describe the three components involved in the tension development process: troponin, tropomyosin, and actin-myosin interaction. The model computes a tension normalized to the maximum tension during resting stretch. The HTD model was coupled to a model of the atrial electrophysiology via the intracellular calcium concentration $[Ca^{2+}]_i$.

In a first step, the experimentally determined maximum amplitude of atrial tension $F_{max,healthy} = 11.27\,mN/mm^2$ was normalized to the maximum of the tension $F_{max,HTD,CRN} = 0.86\,N/N$ produced by the HTD model coupled to the CRN model. This yielded the following relationship:

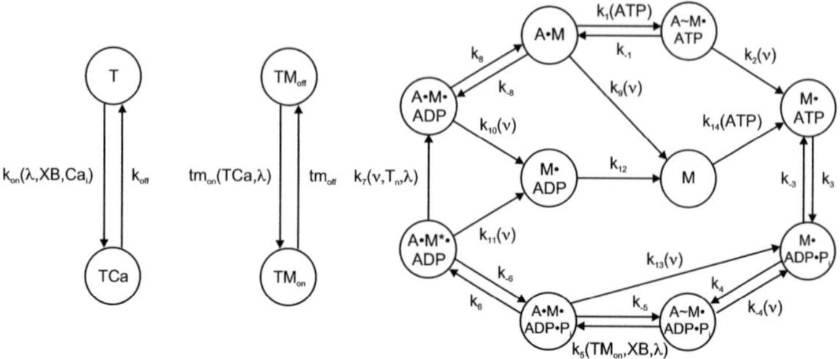

Fig. 5.9. Schematic structure of the hybrid tension development (HTD) model. Figure from [308].

$$F_{meas} = \frac{F_{max,healthy}}{F_{max,HTD}} \cdot F_{HTD} = 13.10 \frac{\frac{mN}{mm^2}}{\frac{N}{N}} \cdot F_{HTD} \tag{5.4}$$

The AF remodeled curve was then adjusted by the same factor

$$F_{max,AF} = 2.74 \frac{mN}{mm^2} \mathrel{\hat{=}} 0.21 \frac{N}{N}. \tag{5.5}$$

The adjustment preserved the same relative difference between amplitudes of the healthy and the AF curve

$$\frac{F_{max,AF}}{F_{max,healthy}} = \frac{2.74 \frac{mN}{mm^2}}{11.27 \frac{mN}{mm^2}} = \frac{0.21 \frac{N}{N}}{0.86 \frac{N}{N}} = 0.24. \tag{5.6}$$

The HTD model was adapted to the experimental data iteratively using the Powell algorithm [312]. As the duration of the optimization process largely depends on the number of free parameters, it is desirable to only include those parameters having a reasonable influence on the model output. To determine these parameters, tension curves were simulated with variation of all model parameters independently by a factor of two initially. Those parameters which showed a significant influence on the tension curve morphology were selected for the adaptation process. Parameters contributing to the cross-bridge cycle were included into the optimization process independently from the sensitivity analysis. The parameter for the maximum tension during resting stretch was left constant. Altogether, 16 parameters were considered for the optimization process (Tab. 5.5).

Table 5.5. Parameters of the HTD model which were chosen for variation to adapt the model to measurement data from the atria. Original parameter values as well as adapted values to fit the tension development in the healthy and AF remodeled atria are listed.

Parameter	HTD	aHTD	aHTD AF
k_{on}	40	26.6	46.4
k_{off}	40	55.1	74.2
tm_{on}	12	28.9	26.2
tm_{off}	35	56.9	45.8
$TMon_{coop}$	2	5.31	9.63
$TMon_{pow}$	4	1.66	1.61
k_5	25	9.4	13.6
k_{m5}	8	$2.6 \cdot 10^4$	292
$k5_{stretch}$	1	3.37	1.33
$k5_{xb}$	1.5	4.65	5.35
k_6	50	$9.8 \cdot 10^4$	$2.1 \cdot 10^3$
k_{m6}	20	585	$3.2 \cdot 10^3$
k_7	30	13.2	56.3
$k7_{base}$	2.25	2.18	2.88
$k7_{stretch}$	1	1.41	5.91
$k7_{force}$	1	$6.0 \cdot 10^{-7}$	22.3

For the optimization process, $[Ca^{2+}]_i$ was computed in an isolated atrial myocyte using the CRN model. Subsequently, beginning with the standard parameters of the HTD model (Tab. 5.5), tension curves were computed. A fitness value was calculated as the mean squared error between the measured and simulated tension curves and the aHTD parameters were altered following the Powell algorithm to iteratively find the minimum in the multi-dimensional parameter space.

The same procedure was afterwards applied to fit the HTD model to the AF remodeled tension curves reported in [74]. The initial implementation of the persistent-AF remodeling [73] did not produce stable $[Ca^{2+}]_i$. This did not influence the action potential significantly, but lead to false results in the computation of the tension based on the calcium concentration. Therefore, the steady-state persistent-AF remodeling CRN implementation introduced in Section 5.1.2 was used to calculate $[Ca^{2+}]_i$ for the adaptation of the HTD model to the measured AF remodeled tension curve.

The aHTD model parameters for healthy and AF remodeled atrial tension are given in Table 5.5. The final root mean squared error (RMSE) was $1.68 \cdot 10^{-2}$ for the healthy case and $2.80 \cdot 10^{-3}$ for the AF case. The RMSE is unitless, as the tension curves were normalized for the HTD model. Figure 5.10(a) shows the measured

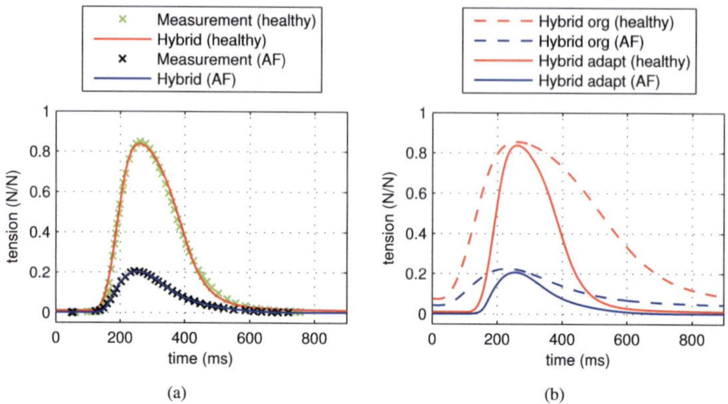

Fig. 5.10. Tension development in the atria. a) Simulated normalized tension with the Hybrid model [311] which was iteratevily adapted to measurement data from [74]. b) Comparison of normalized tension simulated with the original and the adapted Hybrid model.

atrial tension development curves and the curves simulated with the two aHTD model parameter sets. Figure 5.10(b) shows an overlay of the tension curves simulated with the original HTD model coupled to the healthy and AF-remodeled CRN model. The tension development was notably shorter compared to the ventricular setting. The amplitude of the AF-remodeled tension curve was approximately 25% of the amplitude of the healthy curve.

In a further step, the aHTD model was coupled to the heterogeneous model of atrial electrophysiology introduced in Section 5.2 to evaluate eventual spatial gradients in atrial tension development. Figure 5.11(b) shows the normalized courses of the sinus rhythm action potential, $[Ca^{2+}]_i$ and tension in the different atrial regions. Only minor variations in tension between the RA, LA and appendages were observed. However, the amplitude of the tension curve is increased by 25% in the CT and BB. Maximum tension in the region of the openings of the atria to the ventricles and connecting veins (PVs) is reduced by -75% compared to the regular myocardium.

At last, sinus rhythm excitation was simulated in a 3D bi-atrial model (model 19, Tab. 7.1) using the heterogeneous CRN model (Sec. 5.2) coupled with the aHTD model. Gradients between adjacent atrial regions were created to ensure a smooth transition between region with different electrophysiological properties. Tension curves were extracted from simulation results at 50 site equally distributed over the

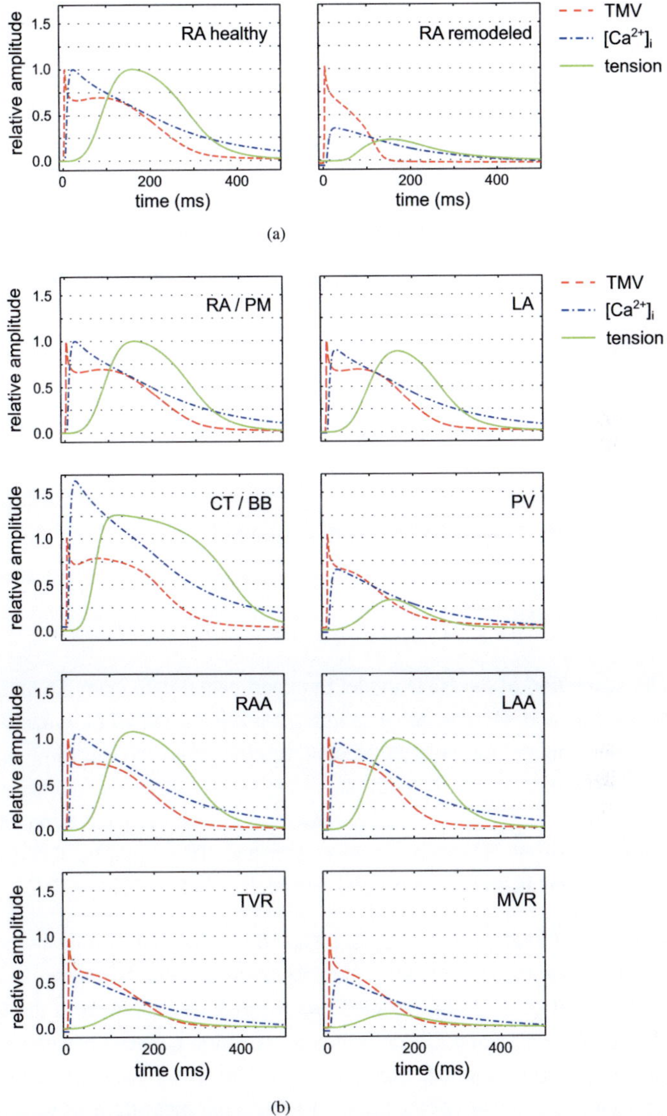

Fig. 5.11. a) Curves for healthy and persistent AF remodeled right atrial tissue (Sec. 5.1.2). b) Simulated (single-cell) transmembrane voltage (dashed red), $[Ca^{2+}]_i$ (dash-dotted blue) and tension (solid green) in various atrial cell types (Sec. 5.2). Curves are normalized to curves in healthy right atrial myocytes. The regional curves for electrical remodeling are not shown, as these did not show a relevant difference to the RA remodeling data.

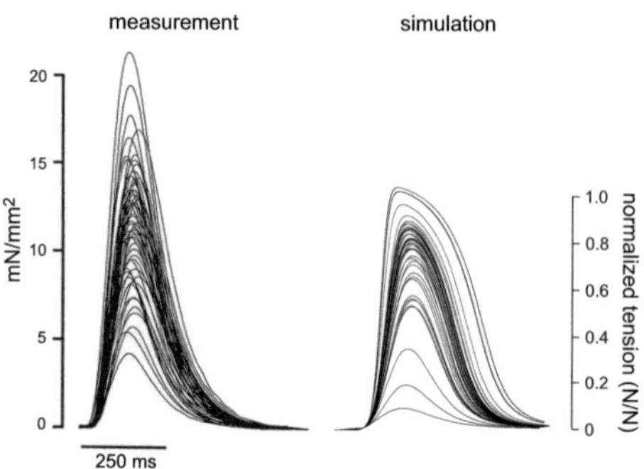

Fig. 5.12. Superimposed tension curves from measurements and a 3D gradual heterogeneous simulation (model 19, Tab. 7.1). Tension curves were extracted in 50 equally distributed locations in the right atrial part of the 3D model. Time scale of the simulated curves is the same as for the measured curves. The relationship between the normalized tension and the measured tension is given in Equation 5.4. Left figure adapted from [75] based on data from [310].

right atrium. The curves were aligned at their onset (F = 0.01). A broad spectrum of tension curves was observed (Fig. 5.12, right).

5.3.1 Discussion

The adaptation of the HTD model to experimental data from the human atria under healthy and AF remodeled conditions using a standard optimization algorithm provided a model for the atrial tension development. The workflow can be applied to other tension model / electrophysiological model combinations, once desired, as the relevant subset of model parameter was identified and a framework for the adaptation of the HTD model was established. This could be also used to couple other implementations of electrical remodeling to the HTD model.

The adapted model parameters k_{m5}, k_6 and k_{m6} showed significant variation from the initial parameter value after the adaptation process. Parameter k_{m5} adds to the inhibition of cross-bridge enabled cross-bridge development in the HTD model. The parameters k_6 and k_{m6} describe the transition between a state contributing to the final tension magnitude and a non-contributing state within the cross-bridge cycle. k_6 and k_{m6} are out of the range of ventricular measurement values [313–315]. It is not clear whether the cross-bridge cycle in the atria differs from the ventricular one and thus the Markov chain formulation needs to be changed or if measurements in atrial myocytes would reveal a different data range. The other adapted parameters are within the measurement ranges [308] or in reasonable proximity to the original values if no measurement data were available.

The Powell algorithm may lead to a local minimum instead of the global minimum. Other optimization approaches, e.g. the particle swarm algorithm [316], might provide more reliability in this respect. In the present case, the assumption was made that the atrial tension development should be closely related to the ventricular tension development. Therefore, the minimum was expected to be in close proximity to the ventricular HTD model parameters and thus the use of the Powell algorithm should be valid.

The aHTD model was coupled to the heterogeneous CRN model and additionally also reflected AF remodeling. The effects of AF-remodeling outruled the effects of regional variations in atrial electrophysiology when both influences were coupled. This was to be expected, as electrical remodeling also outruled the effects of heterogeneous ion channel expression (Sec. 5.2). Eventual regional variations in the electro-mechanical coupling were not included into the heterogeneous model, as respective experimental data were not available.

The use of the aHTD model parameters for healthy atrial tension coupled with the AF-remodeled CRN model and thus reduced $[Ca^{2+}]_i$ did not produce senseful tension results (not shown). This indicates that the remodeling of the tension development, as observed in the measurements, is not solely caused from electrical remodeling processes. AF tension curves rather result at least from the combination of remodeling of the electrophysiology and remodeling of the electro-mechanical coupling.

In agreement with the combined cause of the AF tension curves, the duration of the tension development curve was not reduced in the measurements (and thus also not in the aHTD model), in contrast to a significantly reduced APD and refractory period. Under AF-remodeling conditions, the duration of the tension development was almost twice as long as the ERP. This implies that myocytes can depolarize again while still developing tension during AF. A more detailed analysis of the electro-mechanical coupling under such extreme conditions is needed to understand effects resulting or influencing such behavior.

Coupling of adapted HTD model with the heterogeneous model of atrial electrophysiology (Sec. 5.2) led to differences in maximum tension output, but did not significantly alter the tension development onset or duration. Schotten et al. provided an averaged tension curve for human right atrial tissue in [74] which was used to fit the HTD model parameters. The initial measurements on the other hand showed a broader variance in tension amplitude [310]. The simulated regional variations in tension amplitude suggest that the variance of measured curves were caused by using tissue samples from different right atrial locations in addition to using samples from different individuals. Figure 5.12 shows the measured data and the simulated curves side-by-side.

Maximum tension in regions close to the orifices of the PVs and the valve annuli was strongly decreased. Although regional measurement data of the human atrial tension development are missing, these results make sense, as the atrial myocardium is not contracting significantly in these regions, as can be seen e.g. in cine-MRI data. The heart skeleton (AV plane) and the attachment of the PV to the left atrium do not allow for a great deformation in these regions. A regular contraction of the surrounding atrial myocytes would work against the static properties of these structures. It is therefore reasonable to assume, that over the human evolution, tension development in myocytes near valve opening is markedly reduced.

Notably, the morphology of the tension curve in the CT differs from the other regional curve morphologies and could not be recognized in the experimental data. Further region specific variations in the electro-mechanical coupling process could be responsible for additional changes in the tension curve.

Up to date, elastomechanical simulations of the human atria were restricted to the investigation of passive atrial deformation and stress without a link to the atrial electrophysiology [317–319]. The atrial hybrid tension development model (aHTD) now allows for the simulation of the electro-mechanical coupling in the human atria. The aHTD model enables *in-silico* investigations of active atrial elastomechanics and subsequently atrial blood flow. Among others, the risk of blood clot formation during AF could be estimated or the effects of atrial fibrosis on the ejection fraction could be examined model-based. In a future step, the mechano-electrical coupling in the atria [320–322] could be included into the modeling workflow to better understand atrial arrhythmia caused by hypertension or atrial dilation. First attempts have been undertaken to use atrial models to investigate the role of stretch-sensitive channels in the initiation and maintenance of AF [323–325]. These investigations were limited to the simulation of the atrial electrical activation under pre-defined conditions for the stretch-activated effects.

The aHTD model together with volumetric anisotropic atrial models (Sec. 5.4) now provide all prerequisites for the simulation of both the passive and active atrial deformation resulting from electrical excitation of the atria and ventricles. A first application of such an 3D atrial elastomechanical model with electro-mechanical coupling is presented in Chapter 14.

The adaptation process which has been established to fit the HTD model to atrial measurement data by altering a subset of model parameters can in the future be used to adapt the HTD model to other situations. The use of other atrial or ventricular cell models, different implementation of electrical remodeling or new experimental data of the cardiac tension development may require such parameter adaptation and may now be done in a reasonable time as the essential parameters and a suitable parameter fitting framework are available.

5.4 Modeling Fiber Architecture in Patient-Specific Models

Atrial fiber architecture (Sec. 2.1.1.3) determines the atrial activation pattern [10, 26, 89] and plays a crucial role in the development and maintenance of AF [32]. It therefore also influences ablation strategies for different atrial arrhythmias [38, 66].

Various *in-silico* studies have investigated the impact of local fiber architecture on the excitation pattern in patch preparations (e. g. [67, 68]). In these studies, fiber orientation was transferred from image data directly into the model. Detailed 3D data about fiber architecture in the whole human atria is sparse (Sec. 2.1.1.3) and volumetric data, e.g. diffusion tensor MRI (DTI), is not available for the atria in contrast to the ventricles [166, 326–330]. This is mainly due to the DTI resolution, which is in the order of magnitude of the atrial wall thickness (Sec. 2.1.1.4) and thus partial volume effects outrule the real image information. Additionally, the thin wall makes it difficult to fixate the atria with a physiological filling. Fiber orientation in patient-specific ventricle models, can be included rule-based (rules from Streeter et al. [167, 331]), based on the adaptation of atlases [330, 332] and recently also from *in-vivo* interpolated DTI [328]. In human atrial models, fiber architecture was commonly included based on anatomical studies (Sec. 2.1.1.3). Thereby, fiber orientation was placed manually in each computational node [92, 93, 163–165]. In some cases simple interpolation approaches were

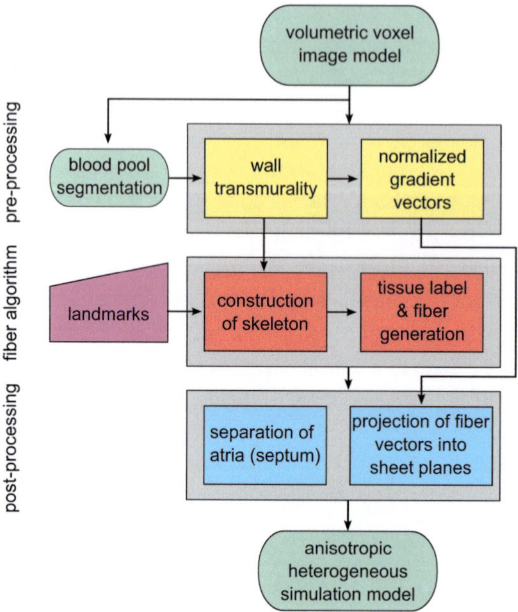

Fig. 5.13. Workflow for the generation of rule-based fiber architecture in atrial geometries. Green: input / output, pink: eventual user interaction needed, yellow: pre-processing, red: fiber-placement, blue: post-processing.

also utilized [94, 168]. As the majority of these models represent only a monolayer 3D surface [93, 163–165], the complex multi-layer atrial fiber architecture could not be reproduced realistically. Volumetric models [92, 94] also assumed a constant transmural fiber orientation in the atrial wall and covered only parts of the atria with fiber orientation (e.g. only LA or CT/PM/BB, Figs. 5.14(a), 5.14(b)).

Fiber orientation in the heart of the Visible Human datasets was previously determined using techniques similar to those which were used to extract ventricular and skeletal muscle fibers [161, 162] (Figs. 5.14(c) & 5.14(d)). The quality of the extracted fiber orientation is questionable, as the approach was created for ventricular fiber orientation and the atrial results were not validated.

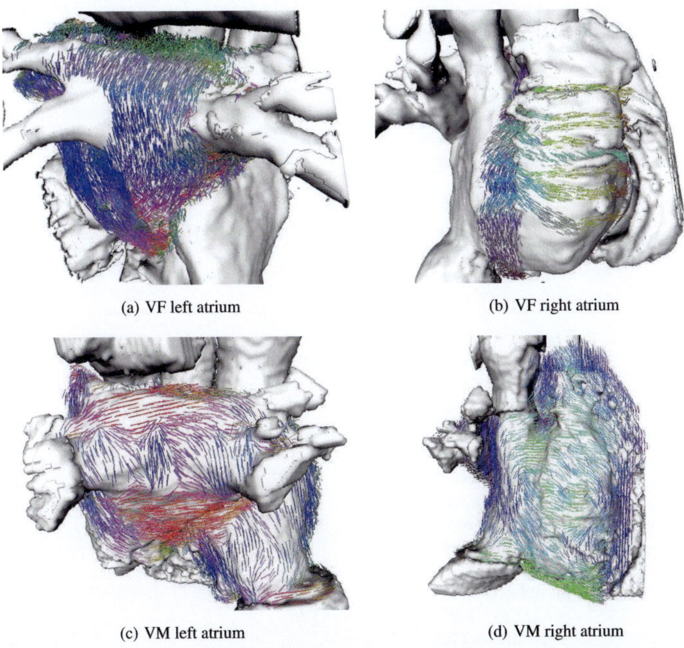

(a) VF left atrium   (b) VF right atrium

(c) VM left atrium   (d) VM right atrium

Fig. 5.14. Fiber in models of the Visible Human datasets. Local fiber orientation is shown as colored cylinder. The color corresponds to the orientation of the fiber. a) Left atrial fiber orientation as modeled by Plank et al. [94]. b) Right atrial fiber orientation as modeled by Seemann et al. [92]. c,d) Left and right atrial fiber architecture as extracted from image data [161, 162].

5.4.1 Semi-Automatic Rule-Based Placement of Fibers

It is necessary to incorporate human fiber architecture into patient-specific atrial models with minimal user interaction. As no atlas data was available, a rule-based approach was chosen to model human atrial fiber architecture in patient geometries. Rules (Fig. 2.3) were extracted from anatomical literature (Sec. 2.1.1.3), photographic images of atrial specimen (Fig. 2.4) and previously described fiber orientation (Fig. 2.3).

The transfer of the fiber rules into geometric models was realized by identifying 22 landmarks in the atrial anatomy. These were used as seed points for an augmented fast-marching level set shortest path method for 3D lattices based on [259]. With

Table 5.6. Description of the locations of the 22 atrial landmarks required for the creation of rule-based fiber orientation in the atria.

	Right Atrium
R1	orifice of SVC, posterior wall, border of LA and RA
R2	orifice of SVC, anterior wall, border of LA and RA
R3	orifice of SVC, right lateral posterior, border of SVC and RAA
R4	orifice of IVC, right lateral
R5	orifice of IVC, left lateral at anterior wall
R6	tip of RAA, if multiple, tip with greatest distance to R3
R7	orifice of TV, inferior of RAA
R8	orifice of TV, right lateral
R9	orifice of TV, minimal distance to LA

	Left Atrium
L1	orifice of MV, left lateral at anterior wall, inferior of LAA
L2	orifice of MV, posterior wall, centered between L1 and L3
L3	orifice of MV, right lateral at anterior wall, minimal distance to RA
L4	onset of RSPV, left lateral superior on anterior wall
L5	onset of LSPV, anterior wall between LAA and LSPV
L6	right lateral between RPVs
L7	left lateral between LPVs
L8	right lateral inferior of RIPV
L9	left lateral inferior of LIPV
L10	left lateral on anterior wall at onset of LAA
L11	tip of LAA
L12	left of LPVs, inferior of L7, onset of LAA
L13	right inferior of RIPV near septum

this, a skeleton was constructed and fiber orientation was interpolated along and between the borders of the skeleton. Fig. 5.13 summarizes the workflow.

The method required a pre-processing of the models (Fig. 5.13, yellow). The left atrial wall was separated into two transmural regions by computing a static electrical field. By solving Poinsson's equation, a virtual steady state potential distribution between the endocardial surface (blood pool) and the epicardial surface in LA was calculated. The wall was then separated at the mid potential into a subendocardial layer and a subepicardial layer. Similarly, a $0.33 - 0.66$ mm thick epicardial layer was marked in the RA. The 22 landmarks (Tab. 5.6, Fig. 5.15 blue circles) were marked in each model manually.

From the seed points, an initial skeleton was constructed by connecting the points with the shortest geometric path between them (Fig. 5.15). The fast-marching shortest path approach was tuned in one aspect. If multiple shortest paths were found, the path with the least distance to the direct line between the two connected

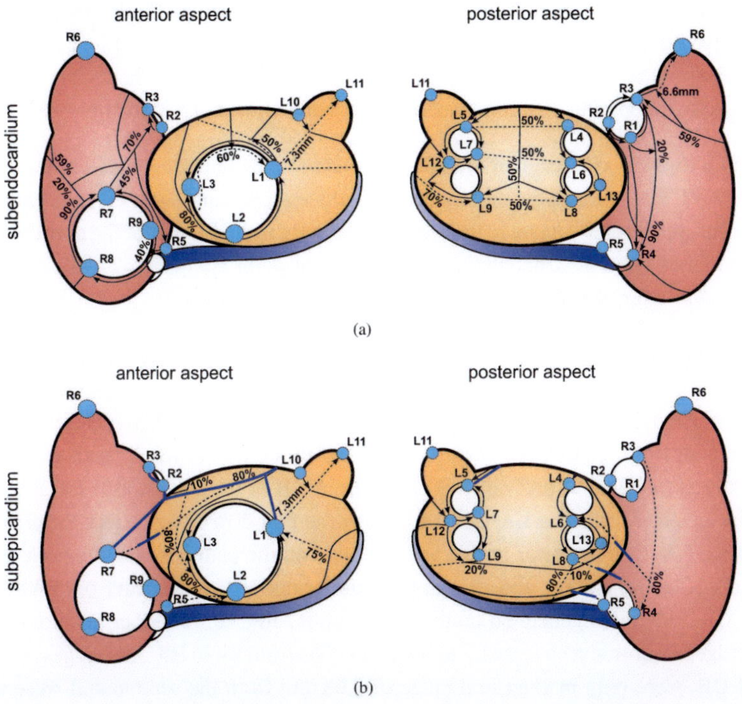

Fig. 5.15. Location of seed points (blue circles) and schematic view of the skeleton created by fast-marching shortest path between those. Dashed lines were used for auxiliary purposes only and did not contribute to the fiber interpolation later on. Blue lines indicate interatrial connections. Figure modified from [216].

points was used. Bundles aligned along such paths were dilated at the centerline of the path to receive a finite bundle thickness within the atrial wall. Fiber orientation was set along the extent of the centerline. A spatial low pass was applied by averaging the orientations within a radius of 1.65 mm. The orientations in the centerline were then interpolated into the dilated region.

The shortest paths algorithm was partly restricted to certain tissue classes (e.g. right / left atrium or subendocardial / subepicardial LA layer). As a restriction for the shortest path CT centerline (between R3 and R4, Tab. 5.6, Fig. 5.15), a 5.28 mm wide corridor needed to be created around a 3D plane defined by the points R3, R4 and the point 40% down the path between R4 and R5. The width of the CT diminishes linearly from R3 to R4 (3.96 mm at R3, 2.64 mm at R4).

The CT extended further towards R5 with a constant width. 15 pectinate muscles started equidistantly from the CT centerline between 20% and 90% of its extent. These connected equidistantly to the path between R7 and R8. The upper pectinate muscle, the septum spurium, was set to be 3 mm wide and 5 mm long.

Circular fiber orientation around the valve annuli, the pulmonary vein orifices and the SVC was realized by the connection of three seed points each. Fiber orientation in the appendages was set to be circulating around the main axis of the appendages. The sinus node was modeled as an ellipsoid at point R3 ($4\times2\times2$ mm). Its main axis was aligned at the local CT fiber orientation at this site.

Five interatrial connections were realized. Bachmann's bundle started from R3, passed R2 and then extended towards L5. The main component of the BB was set to be 4.29 mm broad. A thinner right inferior branch of BB was modeled as the path from R7 to the breakthrough of BB to the LA. A left inferior branch of BB was modeled as path from 80% extent of BB in the LA towards L1. BB was disconnected from the LA myocardium in the first 20% of its extent towards L5 in the LA. One posterior bridge was modeled on the path between the point 80% from R3 to R4 and L6, and another posterior bridge lay on the path from R4 to R8. A CS bridge was generated on the path from R7 to L10. A further lower anterior bridge was introduced from L2 to the point 40% from R4 to R5. All bridges, except of BB, were only marked in a range of 1.65 mm from the septum and were also marked 1.65 mm thick.

Fiber orientation in voxels which had not been marked as bundles before, was interpolated from the fiber orientation surrounding the isotropic region. These regions were marked by using a symmetric region growing [333]. This was e.g. done for the septo-atrial bundle in the subendocardial layer of the LA as the region enclosed by L4, L6, L8, L9, L7, L5, L1 and L3. A region superior of the IVC orifice in the RA, was excluded from the fiber placement algorithm, as fiber orientation in this area is not clearly described in the literature. The electrical properties were set to be isotropic in this region.

Different regions in the atrial models were also labeled during the fiber generation process. In the models, the following tissue labels were distinguished besides the original tissue labels RA and LA: SN, CT, PM, BB, RAA, LAA, TVR, MVR and IS.

The anisotropic models underwent a post-processing afterwards (Fig. 5.13, blue). At the border of the models, fiber artifacts were observed in the form of fibers pointing outside the atrial tissue. As such fiber alignment is not realistic, myocardial sheets in the atria were defined, similar to those found in the ventricles [334]. The normal to the sheet plane was set to be in the direction of the local gradient from the endocardial surface to epicardial surface. The local fiber orientation was then projected into the sheet plane to generate an orthonormal system of the fiber, sheet and sheet normal vector. In a further post-processing step, the right and left atrium were separated by a 0.66 mm thick layer of isolating tissue. Interatrial bridges were thereby not interrupted.

The method has been applied to 20 patient-specific models (Tab. 4.1). In the models, well aligned fiber orientation along the main bundles (CT, BB, PM) could be observed. The CT was crossed epicardially by the intercaval bundle (Fig. 5.16a). The fibers of this merged with the fibers of the PMs, which then ended perpendicular in the TVR (Fig. 5.16b). Fibers encircled the orifices of the valve annuli and the PV ostia as well as the SVC. The appendages showed circular fiber orientation, with some discretization artifacts at the thin parts. Septo-pulmonary and septo-atrial bundles overlayed another in the box between the PVs in the LA and on the anterior side of the LA (Fig. 5.16c,d). The method was constructed to generate fiber orientations in left atrial geometries with four PV ostia, but it was able to handle variation in the number of PVs. In such cases, two or more adjacent PVs were circulated by the fibers, instead of each individual PVs being circulated.

Fig. 5.17 provides a comparison of rule-based fiber architecture in three representative models to fiber orientations reported in anatomical literature. The camera view of the models was adjusted to reflect the camera view of the anatomical photographs. The modeled fiber architecture was in good agreement with the fiber architecture of the pictured specimen.

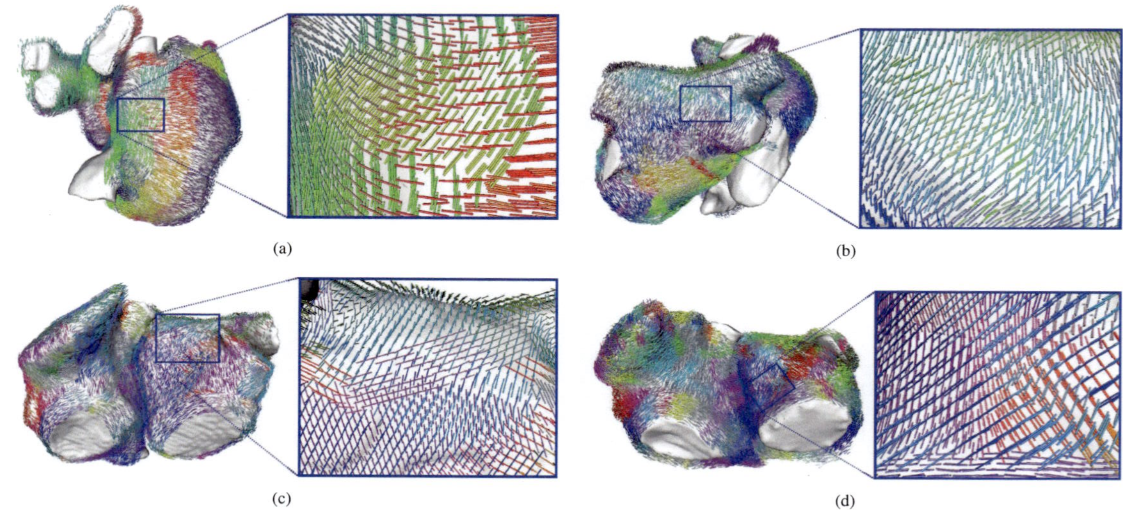

Fig. 5.16. Close ups of fiber architecture. a) Model 9, intercaval bundle crosses the CT epicardially and joins with the PMs. b) Model 3, overlap of septo-pulmonary and septo-atrial bundle in the box between the four PVs. c) Model 4, BB. d) Model 5, overlap of septo-pulmonary and septo-atrial bundle near the anterior side of the mitral valve annulus musculature. Figure adapted from [216].

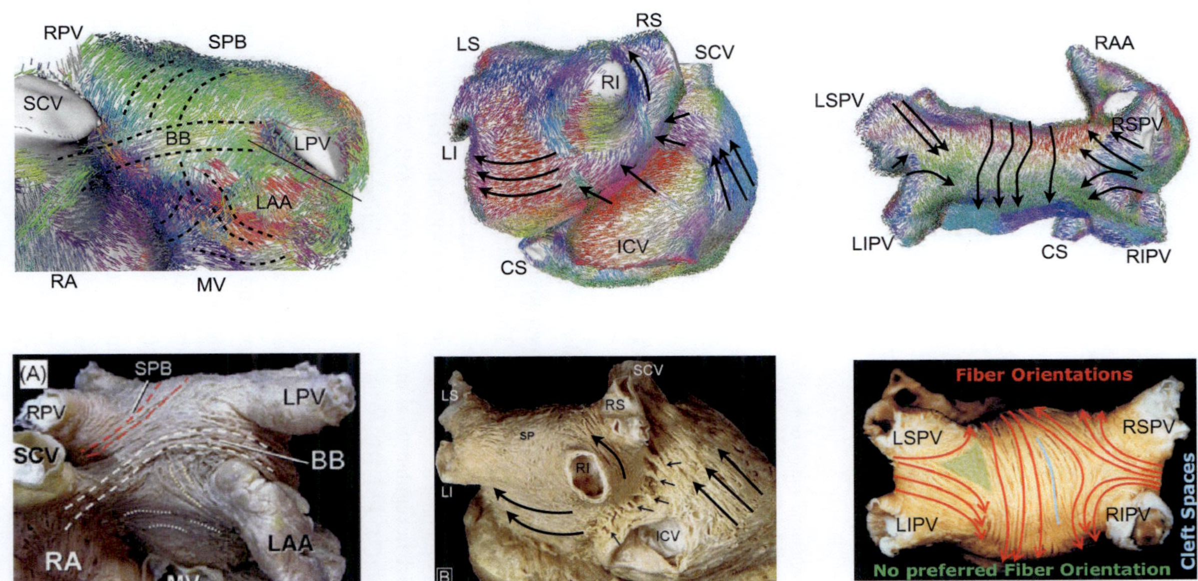

Fig. 5.17. Comparison of semi-automatically generated fibers in three exemplary patient models (5, 8, 9) to annotated anatomical photographs from the literature [10, 66, 94]. Camera views of the models were adjusted to reflect the camera views on the photographs. Abbreviations were used as in the literature figures. S/ICV: superior/inferior caval vein, SP(B): septo-pulmonary bundle, LS/I: left superior/inferior pulmonary vein, RS/I: right superior/inferior pulmonary vein.

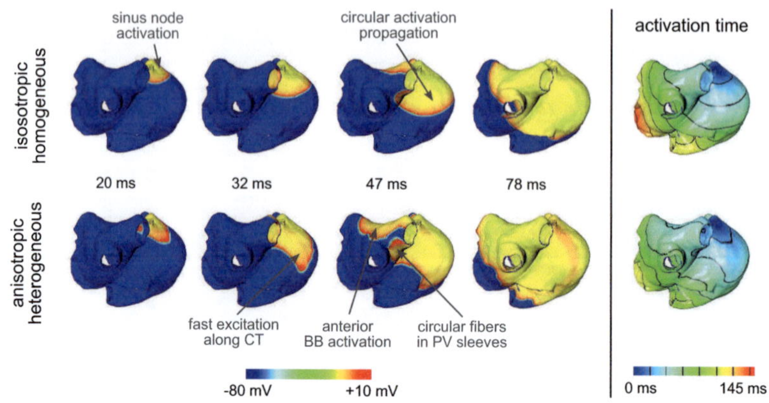

Fig. 5.18. Simulated transmembrane voltage and activation times in an isotropic, homogeneous and an anisotropic, heterogeneous model 3. Figure adapted from [216].

5.4.2 Influence on the Atrial Excitation Sequence

Models 3 and 8 were chosen randomly for simulation of sinus rhythm excitation using the CRN model for heterogeneous human atrial electrophysiology [92] and the monodomain equation for cell coupling. The monodomain tissue conductivities were determined to reproduce heterogeneous conduction velocities of a well established atria model [91]. The ratio of longitudinal to transversal tissue conductivity was set to 3:1 in regular atrial myocardium and 1:7 to 1:20 in fast conducting atrial bundles (CT, PM, BB). The right atrial inferior isthmus was set to be isotropic.

For model 3, two simulations were performed. One isotropic simulation, neglecting tissue heterogeneities and fiber orientation, and one regular anisotropic heterogeneous simulation. In both simulations, the isolating layer in the septum was neglected. In the isotropic simulation, the excitation wave spread circular over both atria centered around the sinus node (Fig. 5.18, top row). In the anisotropic simulation, the atrial excitation sequence was mainly determined by a fast conduction along BB and CT. The LA was activated by two wavefronts, one coming from the BB the other one coming from the posterior bridge. In the heterogeneous model,

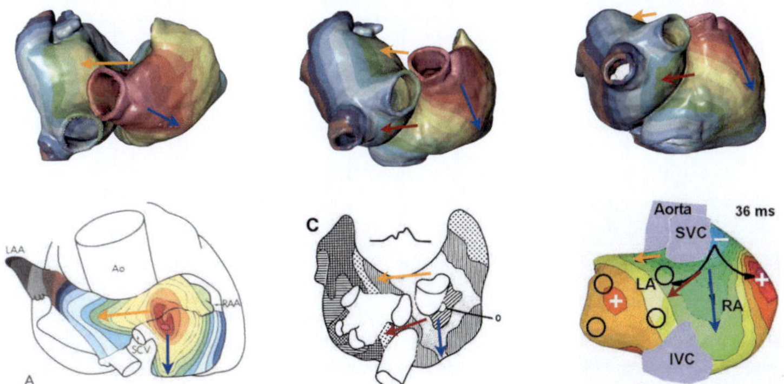

Fig. 5.19. Comparison of the simulated activation sequences in model 8 and atrial activation sequences reported in literature [85–87].

the prominent plateau phase of the CT was clearly visible in the RA (Fig. 5.18, bottom row).

For model 8, only an anisotropic, heterogeneous simulation was performed. The computed atrial excitation sequence was then compared against measurements of the atrial excitation pattern (Fig. 5.19). In both models, the atrial excitation pattern was determined mainly by the fast conduction along the CT and BB. This behavior was also observed experimentally before (Fig. 5.19, bottom row). The curvature of the wavefront was stronger in the regions of CT and BB. This highlights the importance to include such structures in patient-specific models and also provides evidence for the correct location of the modeled fiber bundles.

5.4.3 Comparison of Rule-Based Fibers to Existing Models

Validation of the outcome of the rule-based fiber modeling procedure is an essential task, but difficult to realize as no ground-truth 3D data was available. The rule-based fibers were therefore evaluated against i) photographic images, ii) against existing fiber models in the Visible Female (VF) atria model and iii) against interpolated fiber orientation obtained from image data in the Visible Man (VM) atria model. Furthermore, sinus rhythm simulations were performed and the activation patterns were compared to measured activation patterns.

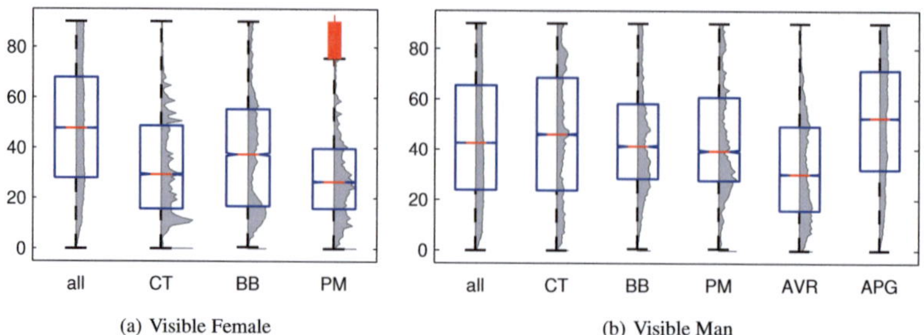

Fig. 5.20. Combined boxplot (blue, red, dashed black) and histogram (grey) of the angular errors between rule-based fiber orientation and pre-existing fiber orientation in the atrial models of the Visible Man and Visible Female dataset. Errors are shown for all material and certain atrial structures.

For the atrial models of the Visible Human datasets, a quantitative error between the rule-based and pre-existing fiber orientations was calculated. The angle E_i between two fiber orientation vectors $F_{i,1}$ and $F_{i,2}$ in voxel i was calculated as

$$E_i = \arccos\left(\frac{\mathbf{F_{i,1}} \cdot \mathbf{F_{i,2}}}{\|\mathbf{F_{i,1}}\|\|\mathbf{F_{i,2}}\|}\right). \tag{5.7}$$

If E_i was greater than $90°$, it was corrected by

$$E_i = 180° - E_i \quad \forall\, E_i > 90°. \tag{5.8}$$

An error $E_i = 90°$ then represents orthogonal fiber orientation, whereas errors $E_i \leq 45°$ represent fiber orientations having similar major axis.

In the VF atria model [94, 335], we observed similarities in the fiber direction in the major atrial bundles, CT, BB and PM (accumulation of fiber errors below $45°$ in the histogram, Fig. 5.20(a)). The left atrial fiber architecture in the VF dataset used in [94] had only little agreement with the rule-based fiber orientation leading to a very broad distribution of fiber errors overall (Fig. 5.20(a)). The latter was caused, among others, by the fact, that the fibers from Plank et al. did not represent the two layers of fibers in the LA and that the LAA had a different fiber orientation. Fiber orientation extracted from image data in the VM atria had a low resolution (Δx=1.0 mm). The local change in fiber orientation was therefore coarser compared to the changes achieved by fiber placement in higher resolution models (Δx0.33 mm). This led to a wide distribution of fiber angles between the image-based and rule-based fiber orientations overall. Better agreement of the fiber

Table 5.7. Difference between microscopy and rule-based fiber orientation in the sheep atria model.

Region	Original Fibers				Filtered Original Fibers			
	Q_{45}	$Q_{22.5}$	RMSE	MAE±STD	Q_{45}	$Q_{22.5}$	RMSE	MAE±STD
All	47.6	18.9	53.4	47.6±24.1	53.5	23.7	50.5	44.3±24.3
BB	73.2	36.9	40.7	34.1±22.2	78.2	42.7	38.5	31.6±22.0
CS	51.3	21.8	51.4	45.4±24.0	66.0	38.7	42.5	35.9±22.8
CT	55.2	25.2	50.1	43.5±24.9	60.0	33.2	46.7	39.3±25.2
PM	61.4	30.0	47.9	41.0±24.9	72.2	34.2	39.6	33.9±20.6
MVR	57.5	30.6	49.2	41.9±25.7	69.3	36.8	43.2	36.1±23.7
TVR	60.4	28.7	47.5	40.8±24.4	75.5	41.7	38.3	31.8±21.4
LAA	46.8	16.9	53.5	48.1±23.4	54.2	26.2	49.3	43.0±24.0
RAA	31.5	8.5	61.0	56.8±22.3	30.6	7.3	61.5	57.4±22.0

orientations were found in the region of the AVR, PM and BB. This led to an accumulation of fiber errors below 45° in the histograms (Fig. 5.20(b)).

5.4.4 Anisotropic Model of Sheep Atria

Dr. Jichao Zhao and Prof. Dr. Bruce Smaill from the Bioengineering Institute of the University of Auckland, New Zealand, provided a high resolution model of sheep atria [96]. Fiber orientation was extracted from a stack of microscopic images (Δx=50 μm), which were aligned and post-processed previously. Fiber orientation was extracted from the 3D data using a gradient-based structure tensor approach [336]. The structure tensor matrix J was constructed as the tensor product of the gradient vector components I

$$J = \begin{pmatrix} I_x \cdot I_x & I_x \cdot I_y & I_x \cdot I_z \\ I_y \cdot I_x & I_y \cdot I_y & I_y \cdot I_z \\ I_z \cdot I_x & I_z \cdot I_y & I_z \cdot I_z \end{pmatrix} \tag{5.9}$$

Local fiber orientation was extracted as the Eigenvector of the tensor with the largest Eigenvalue, hence the direction which had the greatest signal variation. A more detailed description of model generation procedure can be found in [96, 337]. The model was provided with a voxel resolution of Δx=300 μm and contained tissue labels (RA, LA, CT, PM, BB, PVs) as well as fiber orientation in each voxel.

In a first step, the sheep atrial model was compared to the well established VF dataset [92] with rule-based fibers [216] (see previous section for more details). The anatomical features and gross fiber architecture showed qualitatively a good

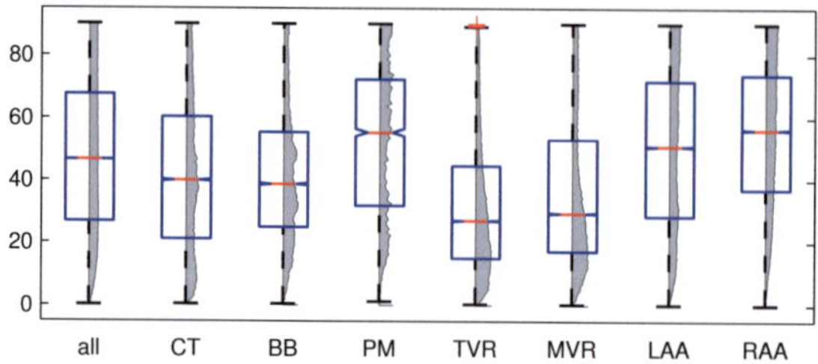

Fig. 5.21. Quantitative comparison between rule-based and microscopy fibers in the geometry of the Auckland sheep atria.

correspondence (Fig. 5.22, left & center) [337]. In the sheep model, fiber orientation was more chaotic in the appendages and less organized in the rest of the atria as expected. This might partly also be caused by the direction extraction method from the image data [96]. Also, simulation of sinus rhythm in both models showed a similar atrial excitation sequence (Fig. 5.22, right) [337].

In the fiber data provided for the sheep model, artifacts and distortions in fiber orientation at the model borders were observed. Therefore a spatial low-pass filter was designed to smooth the fiber vector field. Fiber orientation defined by the angles ϕ and θ in each node covered by the $3{\times}3{\times}3$ filter kernel was transformed into a symmetric structure tensor T_s.

$$T_s = \left(R \cdot F \cdot R^T \right)$$

$$R = \begin{pmatrix} \cos\phi & \sin\phi & 0 \\ -\sin\phi & \cos\phi & 0 \\ 0 & 0 & 1 \end{pmatrix} \begin{pmatrix} \cos\theta & 0 & \sin\theta \\ 0 & 1 & 0 \\ -\sin\theta & 0 & \cos\theta \end{pmatrix} \qquad F = \begin{pmatrix} 1 & 0 & 0 \\ 0 & 0 & 0 \\ 0 & 0 & 0 \end{pmatrix} \qquad (5.10)$$

These tensors were summed up and the Eigenvector which corresponds to the largest Eigenvalue of the tensor was then used as fiber vector for the voxel at the filter kernel center.

In a second step, the sheep model was augmented with a second fiber vector field, by generating rule-based fiber orientation as described in Sec. 5.4.1. Notably, the sheep atrial geometry differs significantly from the human atrial geometry. The inferior caval vein did not connect from inferior to the right atrium but rather from posterior. The two right pulmonary veins had a common trunk and only small ostia. The left pulmonary veins showed a different connection pattern. The appendages

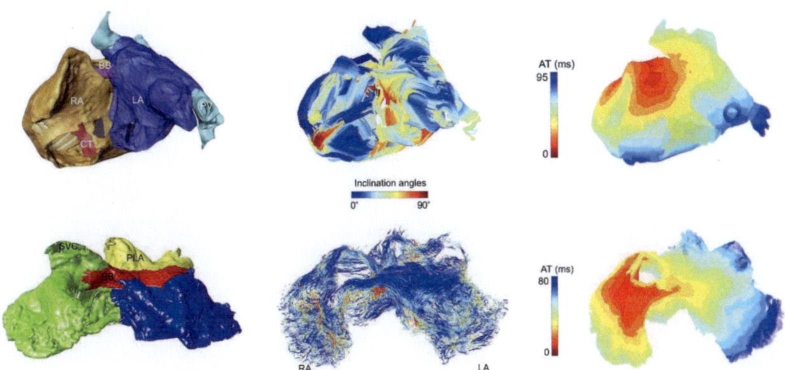

Fig. 5.22. Comparison between Visible Female atria with rule based fibers (top row) and Auckland sheep atria with measured fibers (bottom row). Left: material labels; Middle: fiber orientation; Right: sinus rhythm activation times. Figures adapted from [337].

were the dominating structures in both atria. Nevertheless, the necessary landmark locations could be identified on the sheep geometry.

The difference between both fiber vector field could be expressed by the angle between local fiber orientation vectors E_i (Eqns. 5.7 and 5.8). The mean absolute angle error (MAE) or the root mean squared angle error (RMSE) were not the best suitable error measures, as the error distribution was not expected to be Gaussian. Therefore, two quality measures Q were defined. They represent the relative amount of error angles below two threshold (45° and 22.5°).

$$Q_{45} = \frac{|\{x \mid \forall E_i < 45°\}|}{|\{x \mid \forall E_i\}|} \cdot 100$$

$$Q_{22.5} = \frac{|\{x \mid \forall E_i < 22.5°\}|}{|\{x \mid \forall E_i\}|} \cdot 100 \tag{5.11}$$

Vectors with an angle of less than 45° between them point approximately into the same direction and vectors with an angle error below 22.5° may be considered to be nearly similar, as local jitter in the measured fiber orientation was present even after application of the spatial low pass filter.

Table 5.7 sums up the RMSE, MAE and quality measures Q_{45} and $Q_{22.5}$ between the measured and the rule-based fiber orientation in the sheep atrial model for various regions of the model. Histograms and box plots of the error distribution are

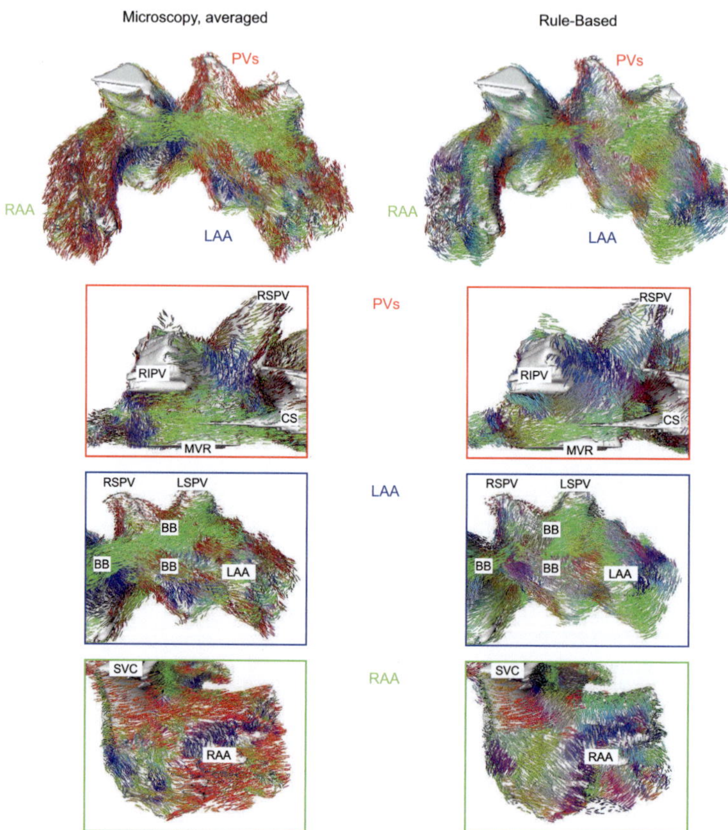

Fig. 5.23. Visual comparison of microscopy fiber orientation (averaged) and rule-based fiber orientation in the Auckland sheep atria. Fiber orientation is color encoded. Circular fiber orientation around the PVs and linear fibers orientation between PVs as well as fiber orientation along the rim of the MVR can be observed in both models (PVs inlay). BB is less wide in the rule-based model compared to the microscopy fiber model, but extent and overall orientation of the bundle is similar (LAA inlay). Fibers encircle the SVC in both cases. Fibers from microscopic images in the RAA are mainly oriented towards the tip of the RAA, whereas fibers in the rule-based model show a rather circular pattern around the RAA (RAA inlay).

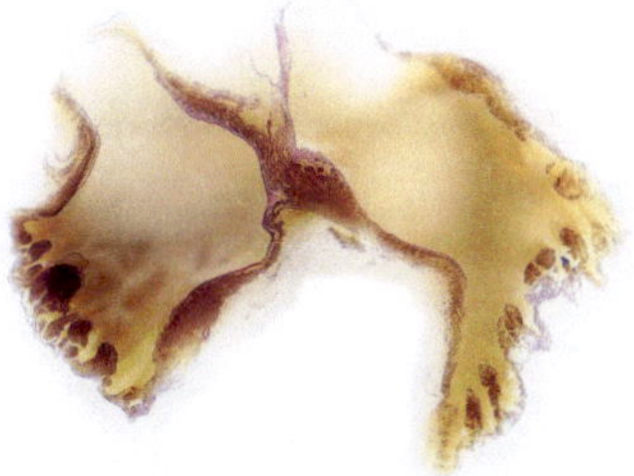

Fig. 5.24. Photographic image of a slice of the Auckland sheep atria [96]. Figure provided by Jichao Zhao, Auckland Bioengineering Institute.

shown in Fig. 5.21. The regions automatically created by the algorithm described in Section 5.4.1 were used for the purpose of regional fiber comparison. Good correspondences of fiber orientation could be observed in BB and the valve rings as well as in the PMs. The CT was a rather prominent muscular bundle in the rule-based model, whereas in the original model, PMs dominated the fiber orientation in this area of the RA. Nearly no agreement of fiber orientation was present in the region of the RAA. Measured fiber orientation in the RAA pointed towards the tip of the RAA, whereas rule-based fibers encircled the RAA leading to angle errors accumulation between $45°$ and $90°$ in the histogram. Fiber orientation in the LAA was in better agreement. In other human anatomical studies, organized circular fiber orientation and near chaotic arragement of myofibers were observed in the atrial appendages depending on the individual patient [10]. Figure 5.23 provides a qualitative, visual comparison of the measured and rule-based fiber orientations in different regions of the model.

Overall, the rule-based fiber modeling approach was able to reproduce the global fiber orientation in the sheep atria, although the sheep anatomy was significantly different compared to the human atria.

In a third step, sinus rhythm activity was simulated in the sheep atrial model in different setups (Tab. 5.8). Thereby, always pairs of simulations were run. In each pair the excitation in the microscopy and rule-based fiber orientation were compared. The CRN model of human atrial electrophysiology and the monodomain equation for cell coupling were solved to simulate the atrial excitation. The ratio of longitudinal and transversal tissue conductivity was set to 10:1 (in one case to 3:1), which results in a conduction velocity anisotropy ratio of approximately 3:1

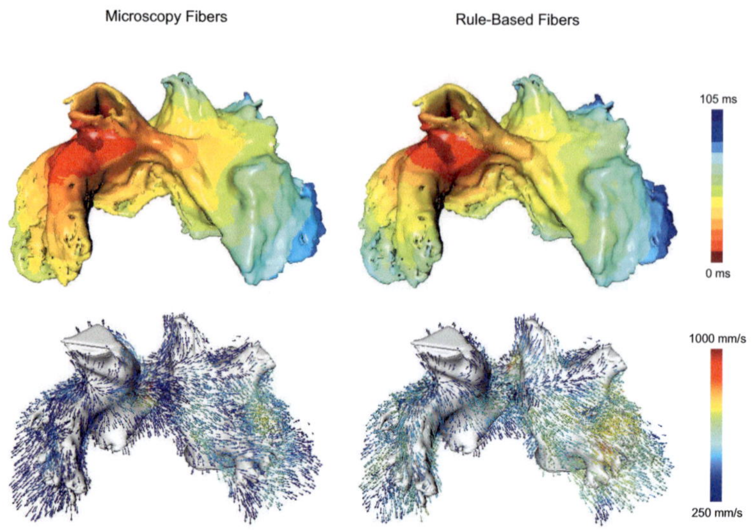

Fig. 5.25. Comparison of local activation times and conduction velocity vectors in simulation No. 1 (Tab. 5.8).

(1.75:1). The sinus stimulus was applied at the sinus node location of the original Auckland sheep model.

In two pairs of simulations (Tab. 5.8, No. 1 & 2), conduction velocity variations and electrophysiological heterogeneities were neglected to determine the influence of the differences in fiber orientation on the sinus rhythm excitation sequence. The anisotropy ratio was set to 10:1 in the first simulation pair and to 3:1 in the second pair of simulations. Table 5.8 provides the quantitative differences between both models. Figure 5.25 shows the isochronal LAT maps and the conduction velocity vector fields extracted from the LAT data (see Sec. 4.4 for details). The simulations with both fiber orientation setups were in good agreement, although the anisotropy ratio was rather pronounced in the first setup. The errors in LATs and in excitation propagation direction were smaller in the low anisotropy simulation pair.

The introduction of regional conduction velocity heterogeneity into both models (Tab. 5.8, No. 3 & 4) led to a better correspondence of the sinus activation sequence between both fiber orientation models. This showed that the regional variations in conduction velocity outruled the effect of local fiber orientation variations during

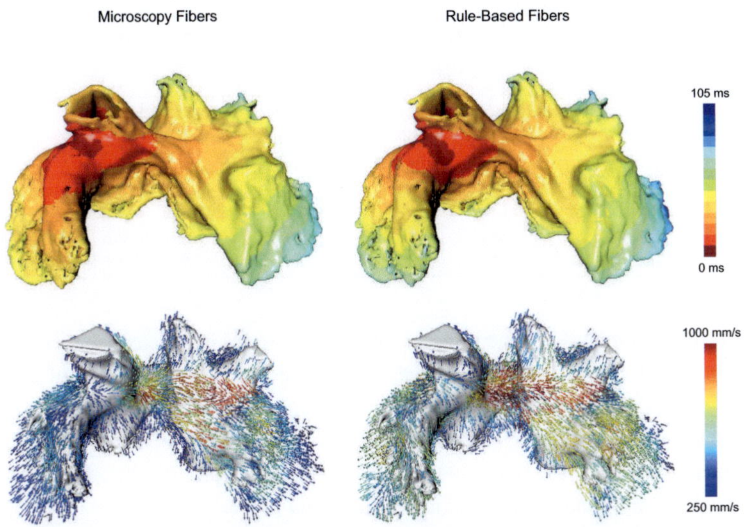

Fig. 5.26. Comparison of local activation times and conduction velocity vectors in simulation No. 5 (Tab. 5.8).

sinus rhythm activity. The error between simulations with models having different conduction velocity regions and different fibers (Tab. 5.8, No. 5) was also smaller compared to the homogeneous, anisotropic simulations (No. 1).

5.4.5 Discussion

In this section, a semi-automatic method to introduce local multi-layer fiber orientation into patient-specific atrial models was introduced. The presented method was partly based on user-provided data. Although the locations of the anatomical landmarks were described in detail, it cannot be outruled that an unexperienced user may provide false landmark locations. For twelve models, at least two experts selected landmark locations independently. Although locations varied, the resulting fiber orientation fields did not differ significantly. Additional uncertainties may be introduced if more than four PVs are present in the patient LA geometry. In this case, two or more PVs need to be considered as one PV and landmark locations need to be set accordingly.

Table 5.8. Quantitative comparison between simulations with the original sheep fiber orientation and the rule-based fiber orientation. M: taken from microscopy model, R: taken from rule-based model, H: homogeneous.

	Microscopy Fibers		Rule-Based Fibers		Activation Time	CV vectors		
	Anisotropy	Tissue	Anisotropy	Tissue	RMSE (ms)	MAE ($^\circ$)	Q_{45} (%)	$Q_{22.5}$ (%)
1	10:1	H	10:1	H	7.0±13.8	33.2±21.4	79.3	33.8
2	3:1	H	3:1	H	4.3±11.4	21.5±15.6	93.6	62.7
3	10:1	M	10:1	M	5.9±8.1	28.9±18.2	85.8	41.8
4	10:1	R	10:1	R	5.1±6.8	28.5±19.3	86.5	44.6
5	10:1	M	10:1	R	8.1±12.7	32.3±21.5	79.0	36.9

A further automation of the process could be achieved by implementation of the landmark locations into mean models for the automatic segmentation of cardiac images [338]. This would also eliminate user-introduced errors and ensure reproducible results. Alternatively, the generation of fiber architecture could be performed directly in volumetric mean models [60] and could then be adapted to the patient without running the fiber algorithm again [339].

Despite the questionable reliability of the VM and VF LA data, the rule-based fiber architecture was able to reproduce fiber orientation in the major atrial bundles, as described by Seemann et al. [92] and as observable in the VM RA. Additionally, fibers were arranged circularly around the valve annuli in the VM dataset. An arrangement which was recreated by the fiber rules as well.

Other groups have presented anisotropic singular anatomy models (Tab. 3.1). Fiber orientation was thereby placed manually into 3D surface models based on similar data as used for the presented semi-automatic method. These models usually contain less artifacts but cannot reflect the complex transmurally differing fiber orientation in the human atria.

Although it is not clear how sheep fiber architecture corresponds to human fiber architecture, application of the rule-based fiber algorithm led to a good approximation of the sheep fiber orientation and also allowed realistic simulation of sinus rhythm activity in the sheep model. The electrically isolating layer, as introduced by the rule-based fiber algorithm, was neglected in all simulations, because in the original sheep image data (Fig. 5.24) no such layer could be identified. It is not clear whether sheep atria differ in this respect to human atria, or if a different staining technique would make such layer better visible in the image data.

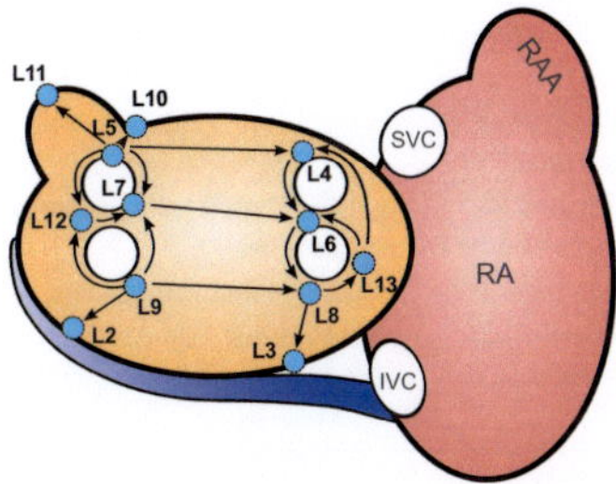

Fig. 5.27. Scheme for the creation of generic ablation lesion pattern in patient specific atrial geometries using shortest paths. Figure adapted from [340].

The applied tissue conductivity anisotropy ratio of 10:1 was rather strong for atrial simulation. Usually, such strong anisotropy is set in fast-conducting bundles in established atrial models (Sec. 2.1.3) or to model fibrotic tissue [94]. The effect of the conduction velocity heterogeneities outruled the effects of the differences in fiber orientation during sinus rhythm simulation. In contrast, fiber architecture may play a more important role in the simulation of atrial arrhythmia [68, 94, 96], although electrophysiological variations will also contribute to the development and maintenance of these arrhythmia.

5.5 Modeling Generic Lesion Patterns in Patient-Specific Models

In the literature, various approaches for successful RF ablation of AF are described. Commonly, a three step approach serves as a basis to cure patients from AF (Sec. 2.3). Initially, the pulmonary veins are isolated, then linear lesions are placed to prevent reentry circuits and at last, additional punctual lesions may be placed in regions showing abnormal electrical activity [1]. Modeling studies from various research groups have investigated the influence on different ablation lesions on the impact of the left atrial excitation sequence (Sec. 3.1.3). In this section a method is presented to incorporate different ablation lesion patterns into patient-specific atrial models semi-automatically.

Ablation lesion patterns were reconstructed based on the left atrial anatomical landmarks also used for the semi-automatic placement of atrial fiber orientation (Sec. 5.4). Landmarks were connected to create center lines for ablation lesions,

Table 5.9. Overview of implemented ablation lesion patterns.

pattern	PV isolation	linear lesions	reference
A	–	set of linear lesions	[341]
B	individual circumferential PVI	–	[342]
C	pairwise circumferential PVI	–	[343]
D	pairwise circumferential PVI	roof line, mitral isthmus line	[344]
E	individual circumferential PVI	posterior line	[345]
F	individual circumferential PVI	roof line, miral isthmus line, low posterior line	[346]
G	single circumferential lesion	mitral isthmus line	[347]
H	single circumferential lesion	mitral isthmus line, second line to MV	[348]
I		Cox Maze III	[349]

by using the fast-marching level set shortest path method described in the previous section. Figure 5.27 shows schematically the skeleton created from the landmark and the shortest paths. Table 5.6 provides distinct information about the location of the landmarks.

To model different ablation lesion patterns, different parts of the skeleton were used. At each point along these paths, a 5 mm [136] (Sec. 2.3.3) wide ablation lesion was applied to the atrial tissue. Ablation lesion patterns were taken from literature descriptions [1, 187, 350] and are listed in Table 5.9.

Tissue conductivity was set to zero in the ablation scars (Sec. 2.3.1, 3.1.3). Additionally, tissue anisotropy can be removed, as observed in experiments. To model the acute effects of RF ablation, a border zone of slow conduction or diffusive fibrosis may be set around the ablation lesions (Sec. 2.3.1). The method was applied to 18 atrial models (1-11 & 13-19, Tab. 7.1). Figure 5.28 shows an example of lesion patterns in a left atrial model.

5.5.1 Discussion

In this section a method to create a broad range of atrial ablation patterns in patient-specific atrial geometries was presented. The method was derived from the fiber placement algorithm (Sec. 5.4). It uses the same user-defined landmarks and was thus also prone to the same user-introduced uncertainties discussed previously (Sec. 5.4.5).

Tissue conductivity in the ablation lesions was reduced to zero. This approach is commonly undertaken in modeling RF lesions in the atria (Sec. 3.1.3). Neverthe-

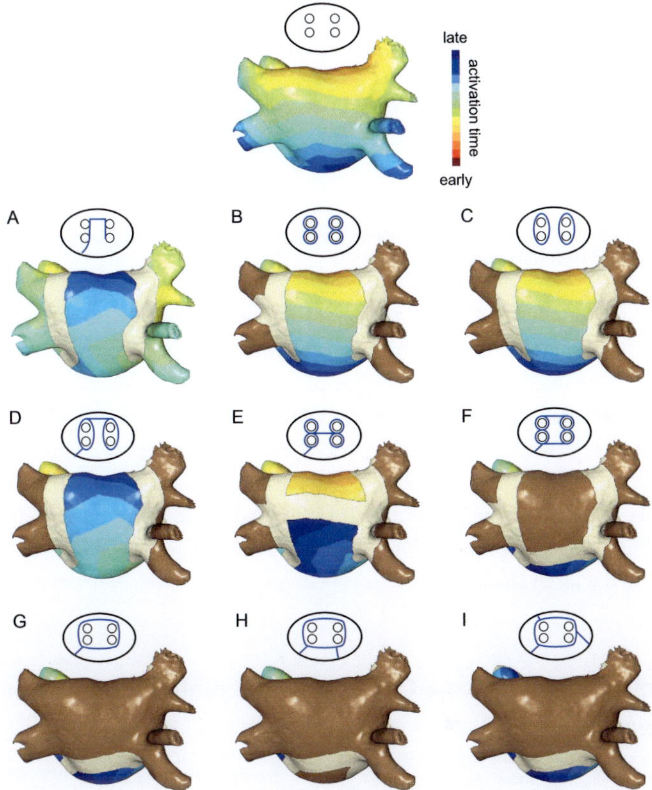

Fig. 5.28. Generic ablation lesion patterns in a patient specific left atrial model (subject 13, Tab. 7.1). Scar tissue is shown in pale white. Tissue which was isolated from sinus rhythm activation is shown in brown. Atrial activation was calculated using the fast-marching level set approach (Sec. 5.7). Figure adapted from [340].

less, the method also allows for a consideration of more details with respect to the microstructure of the ablation scars and the border zone (Sec. 2.3.2). Conduction velocity and electrical anisotropy could be reduced in the modeled border zone. During bidomain simulations, different conductivity tensors for the intra and extra-cellular space could be considered, e.g. setting only the intra-cellular conductivity to zero in scar tissue. These factors are independent of the geometrical modeling of the ablation lesion patterns.

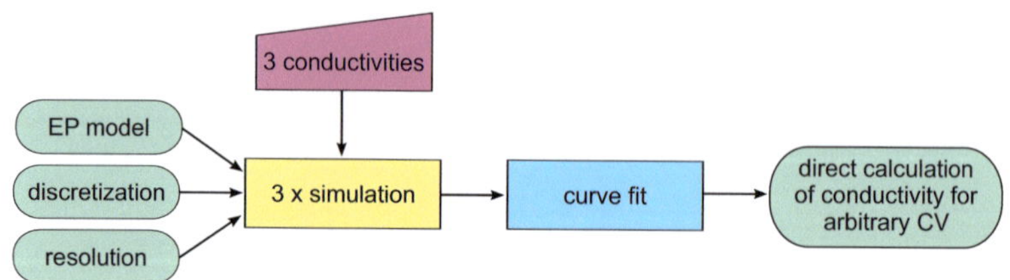

Fig. 5.29. Workflow for the determination of the monodomain tissue conductivity from only three simulations in a desired simulation setup.

Punctual ablation lesions placed in regions of abnormal electrical activity or with the aim to change the local dominant activation frequency (Sec. 2.3) were not included in the method. The placement of such ablation points is based on electrophysiological measurements and depends on the patients electrophysiology and may also change in the course of the intervention. The number of additional lesions and their location is therefore not predictable. The evaluation of the influence of randomly placed ablation lesion in left atrial models on the atrial excitation may pose an alternative approach to investigate the general influence of such ablation lesions on the atrial function.

5.6 Calculation of Tissue Conductivity to match Conduction Velocity

The monodomain tissue conductivity depends on various simulation parameters (e.g. electrophysiological model, model resolution, discretization method). The correct tissue conductivity for a desired conduction velocity in a certain simulation settings is usually determined iteratively employing optimization algorithms which vary the conductivity until the desired conduction velocity is matched. In this section a method to fit the tissue conductivity to a given conduction velocity with only three simulations is presented (Fig. 5.29).

The monodomain tissue conductivity and the conduction velocity can be approximated by a quadratic relationship [351, 352]:

$$\sigma_{discretization,resolution} = a \cdot CV^2 + b \cdot CV + c. \tag{5.12}$$

Table 5.10. Resulting coefficients of equation 5.12. FD: finite difference, FE: finite elements.

discretization	resolution	a	b	c
	mm	$s^2 m^{-3} S$	$sm^{-2} S$	$m^{-1} S$
FD	0.10	0.121	0.008	0.00
FD	0.20	0.119	0.018	$1.63 \cdot 10^{-4}$
FD	0.33	0.120	0.027	$1.52 \cdot 10^{-3}$
FD	0.40	0.119	0.034	$2.39 \cdot 10^{-3}$
FE	0.10	0.174	0.015	$-2.63 \cdot 10^{-4}$
FE	0.20	0.182	0.021	$8.80 \cdot 10^{-4}$
FE	0.33	0.179	0.036	$3.32 \cdot 10^{-3}$
FE	0.40	0.172	0.051	$3.76 \cdot 10^{-3}$

Figure 5.30 provides an example of this relationship. The unknown coefficients a, b and c can be determined if at least three conductivity – velocity value pairs are known. These can be created by choosing three tissue conductivities, running one simulation each and measuring the resulting conduction velocity (Fig. 5.29). The pairs can then be used to solve the linear system of equations for the three coefficients a, b and c, which are specific to the simulation setup. If more than three simulations were performed the over-estimated system can be solved with a regular least-squares approach. From there on, the conductivity for any desired conduction velocity for the same simulation setup can be calculated analytically.

To test the algorithm, 186 isotropic cable simulations using the CRN model were performed with different spatial resolution, discretization method and tissue conductivity. The procedure was done for the complete dataset and only for a subset of three value pairs of each dataset. Figure 5.30 provides an overview of the simulation results and the curve fits and Table 5.10 lists the resulting coefficients. The RMSE of the reduced fit was slightly worse than the RMSE of the original data fit (Tab. 5.11), but still provides a precise estimation of the actual conductivity.

To determine anisotropic tissue conductivities the following relationship can be set up

$$\frac{\sigma_{long}}{\sigma_{trans}} = \frac{a \cdot CV_{long}^2 + b \cdot CV_{long} + c}{a \cdot CV_{trans}^2 + b \cdot CV_{trans} + c}. \tag{5.13}$$

The coefficients a, b and c are the same for both fiber orientations, as fiber orientation is reflected in the excitation propagation models as anisotropic conductivity. The conductivity–CV relationship is thereby not altered. The quadratic term of the equation has a larger weighting than the linear part (Tab. 5.10). The relation can therefore be approximated to

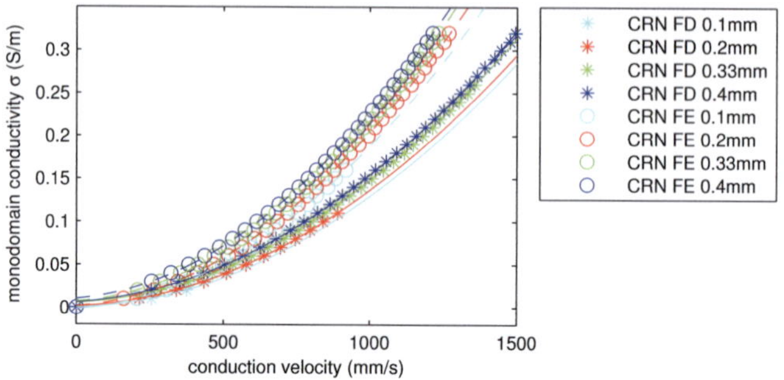

Fig. 5.30. Results of quadratic fits to simulation outcome.

$$\frac{\sigma_{long}}{\sigma_{trans}} \approx \left(\frac{CV_{long}}{CV_{trans}}\right)^2. \tag{5.14}$$

Using this relationship, conductivities for any desired anisotropic conduction can be calculated directly from the initially determined (longitudinal or transversal) tissue conductivity.

5.6.1 Discussion

In this section a simple method to determine the monodomain conductivity for desired conduction velocities was presented. Previously, if the conductivity for a specific conduction velocity needed to be estimated, an iterative simulation approach was employed to find a conductivity which resulted in the desired conduction velocity. This approach often required a large number of simulations and only provided an estimate of the conductivity. With the approach presented in this section, only three simulations and a curve fit are needed for a precise determination of the monodomain conductivity. The three initial conductivities should be chosen to cover the greatest possible conduction velocity range without running into discretization or numerical problems. A good choice was σ=0.05, 0.15, 0.25. For the future, the approach enables a regional adaptation of the conduction velocity in whole atria monodomain simulations.

Table 5.11. Error of quadratic fit to various sets of simulations. FD: finite difference, FE: finite elements.

EP model	resolution (mm)	discretization	number of simulations	RMSE (S/m)	RMSE 3 (S/m)
CRN	0.1	FD	3	0.000000	0.000000
CRN	0.2	FD	12	0.000300	0.000608
CRN	0.33	FD	26	0.000717	0.001225
CRN	0.4	FD	32	0.000835	0.001369
CRN	0.1	FE	17	0.000576	0.000845
CRN	0.2	FE	33	0.000799	0.001215
CRN	0.33	FE	32	0.001029	0.001481
CRN	0.4	FE	31	0.001274	0.002000

5.7 Fast-Marching Method to Compute Activation Times

Level set methods calculate arrival times at discrete points in 3D geometrical objects (Sec 3.3.3). These arrival times can be interpreted as activation times in a cardiac model [197, 264]. Level set methods thereby reflect a monotonously expanding front and are usually easier to handle than e. g. full Eikonal models [262, 263]. On the downside, level set methods neglect diffusion processes and cell electrophysiology.

Based on an implementation of fast-marching level set method for structured grids [259], an anisotropic, heterogeneous fast-marching method was developed [225]. The method calculated activation equivalent arrival times and also supports multiple starting points. The method was by order of magnitude faster than the monodomain approach and easier to handle than e. g. the adaptive cellular automaton [162]. It produced similar results for sinus rhythm excitation as the monodomain approach both in the isotropic and anisotropic case. Body surface ECGs also showed a good similarity to the ECGs computed from the monodomain solution (Sec. 9.1.2).

The distance between two points in 3D space can be expressed as the L2 norm in Cartesian coordinates:

$$d_i = \sqrt{(\Delta x)^2 + (\Delta y)^2 + (\Delta z)^2}. \tag{5.15}$$

This distance corresponds to the arrival time in the level set approach. Anisotropy could be introduced by weighting the isotropic distance d_i depending on the fiber angle in the starting coordinate and the angle between the two points:

$$d_a = \left\| \left((R \cdot W \cdot R^T)^{-1} \right) \cdot \begin{pmatrix} \Delta x \\ \Delta y \\ \Delta z \end{pmatrix} \right\| . \tag{5.16}$$

W is a tensor containing the conduction velocity (CV) along and across the fiber orientation and R is a rotation matrix (Eq. 5.20).

$$W = \begin{pmatrix} CV_{long} & 0 & 0 \\ 0 & CV_{trans} & 0 \\ 0 & 0 & CV_{trans} \end{pmatrix} \tag{5.17}$$

The longitudinal CV (CV_{long}) can be expressed as a product of the anisotropy ratio for the tissue k and the transversal CV (CV_{trans}). For regular sinus rhythm simulations k was set to $1{:}\sqrt{3}$, to reflect the commonly used monodomain tissue conductivity ratio of 1:3. For fast conducting bundles, k was set to 1:3 (corresponding to 1:9 in monodomain tissue conductivity).

$$CV_{long} = k \cdot CV_{trans} \tag{5.18}$$

$$W = \begin{pmatrix} k \cdot CV_{trans} & 0 & 0 \\ 0 & CV_{trans} & 0 \\ 0 & 0 & CV_{trans} \end{pmatrix} \tag{5.19}$$

Fiber orientation in the starting point can be expressed in spherical coordinates by two angles ϕ and θ, which vary between 0 and π radians. This led to the rotation matrix R

$$R = \begin{pmatrix} \cos\phi & \sin\phi & 0 \\ -\sin\phi & \cos\phi & 0 \\ 0 & 0 & 1 \end{pmatrix} \begin{pmatrix} \cos\theta & 0 & \sin\theta \\ 0 & 1 & 0 \\ -\sin\theta & 0 & \cos\theta \end{pmatrix} \tag{5.20}$$

Heterogeneous CV between different tissue types was included in the distance calculation by weighting the anisotropic distance norm between two coordinates d_a with the reciprocal of the ratio between the CV in a specific tissue and the CV in the regular myocardium.

$$d_{a,t} = w_t \cdot d_a \tag{5.21}$$

Inter-tissue CV ratios w_t were set based on the monodomain tissue conductivities used in [225]:

$$w_{CAM} = 1.000$$
$$w_{CT} = 0.756$$
$$w_{PM} = 0.740$$
$$w_{BB} = 0.599$$
$$w_{IS} = 1.940$$

Body surface ECG signals were forward computed in a three step process. 1) For each time step (e.g. each millisecond), the activation time results were converted to binary data. The threshold was chosen as the respective time step. Each computational node with activation time shorter or equal to the threshold was set to +5mV, all other nodes were set to have a value of -82 mV. From this binary model, a regular two step forward calculation was performed (Sec. 3.4).

5.7.1 Evaluation

To evaluate the correspondence of the activation times calculated from the fast-marching method to the monodomain solution, both types of simulations were performed on models 1-8 (Tab. 7.1). The monodomain simulation setup is described in more detail in [225]. The mean absolute error was 11.9 ± 13.2 ms. Synchronization of the latest activation times resulted in a mean RMSE of 4.25 ms for the isochrones in the models. The error between the conduction velocity vectors (Sec. 4.4) had a mean $Q_{22.5}$ of 62.6% and a Q_{45} of 92.4%. Figure 5.31 shows two representative examples of the simulations.

5.7.2 Discussion

In this section, a method to calculate LATs based on the fast-marching level set method was presented. The algorithm was able to reproduce anisotropic and heterogeneous conduction velocity and may thus be used to simulate the complex atrial excitation sequence.

The method was tested on eight patient-specific atrial models and compared to the monodomain solutions. Activation times calculated with both method showed a good qualitative and quantitative correspondence, although the time for complete atrial activation varied between the solutions. The mean RMSE error was

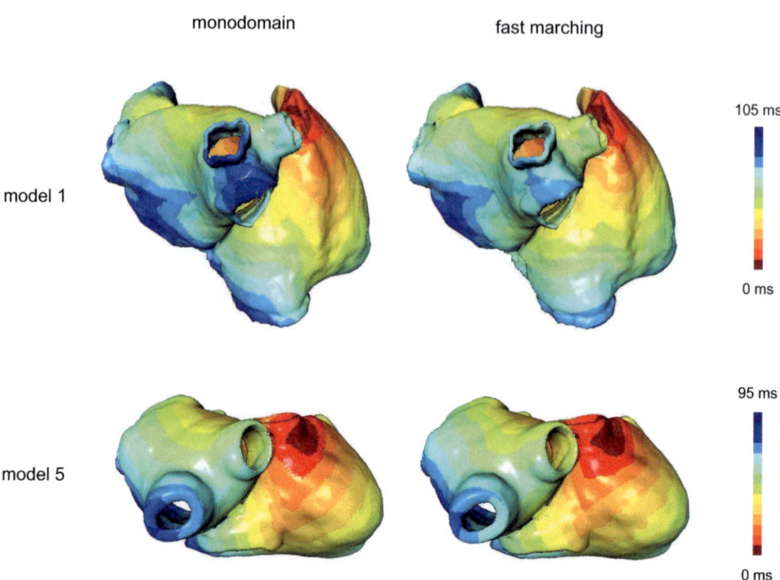

Fig. 5.31. Two representative examples of monodomain and fast-marching simulation of sinus rhythm in two patient models (Tab. 7.1). The excitation sequence and the isochrones correspond well. Local isochrone curvature varies to some extent.

mainly caused by differences in activation times in very thin tissue areas, where the excitation propagation was slowed down in the monodomain solution. Additionally, in the monodomain simulations a broader range of anisotropy ratios were used compared to the fast-marching approach. Differences in anisotropy ratios and local conduction velocity variations also account for variations in wavefront curvature and thus conduction velocity vector directions. The fast-marching calculations were by magnitudes faster than the monodomain computations (minutes vs. hours) and the method was easier to use than the adaptive cellular automaton. The latter is an important factor to use the simulation environment for high-throughput simulations.

The fast-marching approach was not able to account for diffusion processes. It may therefore not be able to reproduce wavefront curvature dependent conduction velocity alterations. The method was able to reproduce multiple wavefronts, but could not reproduce reentries, as it does not include repolarization. Sermesant et al.

have presented a similar method including cell repolarization [264]. The method introduced in this section could in the future be extended accordingly.

Other groups have also shown approaches to include diffusion processes into Eikonal models [260, 261] which may be used to model arrhythmic behavior [262, 263]. Such approaches might be useful for arrhythmia simulations, but will slow down the computation compared to the pure fast-marching approach, which was not desirable for this study. The advantage in computational speed was used for a high throughput simulation study with multiple models (Sec. 9.1.2). A graph based method for a yet faster computations of the ventricular activation sequence was presented recently [265]. The adaptation of this method to the atrial sinus rhythm excitation could further speed up the computations. On the other hand, the forward-calculation of the ECG currently consumes the vast portion of computational time.

Part III

Model Personalization

The best material model for a cat is another, or preferably the same cat.

A. Rosenblueth & N. Wiener

Role of Models in Science

6

Personalization Framework

The use of atrial models in a clinical environment requires a personalization of the existing general models. Three different types of models need to be distinguished in the context of modeling and simulating the cardiac electrical activity: i) Anatomical models, ii) cellular electrophysiological models and iii) models of the excitation propagation. The personalization of the anatomy is mostly a task of image acquisition, segmentation and post-processing. The personalization of the cellular electrophysiological model requires the acquisition, processing and interpretation of electrical cardiac signals. The personalization of the excitation propagation requires anatomical and electrophysiological data acquisition and a personalization of both other models (Fig. 6.1).

The main task of anatomical model personalization is the segmentation of the 3D atrial geometry from clinical image data (MRI, CT), but it also covers the extrac-

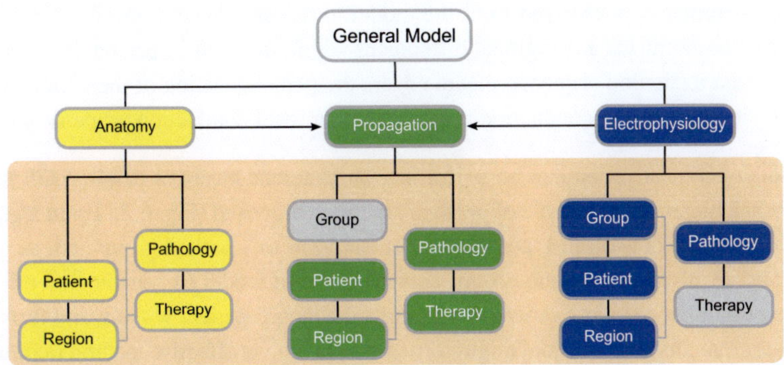

Fig. 6.1. Hierarchical structure of the model personalization with respect to atrial anatomy, electrophysiology and excitation propagation.

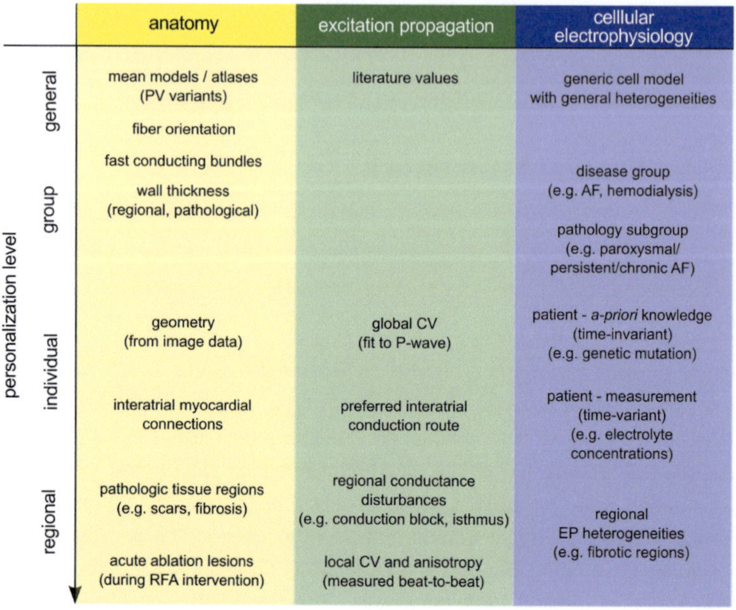

	anatomy	excitation propagation	celllular electrophysiology
general	mean models / atlases (PV variants) fiber orientation	literature values	generic cell model with general heterogeneities
group	fast conducting bundles wall thickness (regional, pathological)		disease group (e.g. AF, hemodialysis) pathology subgroup (e.g. paroxysmal/ persistent/chronic AF)
individual	geometry (from image data)	global CV (fit to P-wave)	patient - *a-priori* knowledge (time-invariant) (e.g. genetic mutation)
	interatrial myocardial connections	preferred interatrial conduction route	patient - measurement (time-variant) (e.g. electrolyte concentrations)
regional	pathologic tissue regions (e.g. scars, fibrosis)	regional conductance disturbances (e.g. conduction block, isthmus)	regional EP heterogeneities (e.g. fibrotic regions)
	acute ablation lesions (during RFA intervention)	local CV and anisotropy (measured beat-to-beat)	

(row labels read top-to-bottom along the left axis: personalization level)

Fig. 6.2. Personalization of atrial models can be divided into three different categories each covering levels of detail from generic models to regional variations. Figure adapted from [340].

tion of further patient-specific information from such images, as e.g. regions of pathological tissue. Atrial cellular electrophysiology models have various parameters which can be adapted to a patient or patient group. Usually a combination of such parameters is changed to fit the model to diseases, electrophysiological *in-vitro* measurements and regional differences within the atria. A personalization of the atrial excitation sequence model covers the personalization of the conduction velocity, specialized conduction routes and other local conduction properties.

Model personalization can be performed on different levels of detail. With this work, a hierarchical scheme of personalization is proposed (Fig. 6.2). From a general level, models of all types can be personalized to a patient group and, as appropriate, also a patient sub-group. This first level of model personalization offers the possibility to adjust general models to a diagnosed patient without any further measurements and therefore might form a simple but yet effective method in future clinical application. Such model personalization was formerly often used to investigate general effects of diseases *in-silico* without the direct link to the real patient (e.g. [2]). On a next level, the models can be adapted to the individual patient by taking global measurements into account (e.g. [225]). The most detailed level of model personalization covers the adaption of the models in specific regions of the atria. This step usually requires invasive measurement data (e.g. [223]). Figure 6.2 provides examples for personalization of each of the three model types at the different levels of detail.

The types and levels of model personalization need to be serialized into a continuous workflow, to allow for a use of the personalized models in a clinical environment. In this thesis a sequential and flexible framework of model personalization is proposed to create, evaluate and eventually validate personalized atrial models (Fig. 6.3). This framework couples *a-priori* knowledge about the atrial anatomy, electrophysiology and excitation sequence, with measurement data from various entities and with different level of detail to create patient-specific models for the model-based evaluation and planning of RFA in patients with AF. The workflow was used on seven volunteer datasets and four AF patient datasets so far (1–11, Tab. 7.1). Not all datasets underwent the complete workflow, as some data were not acquired for some AF patients. The steps of the personalization workflow are described in more detail in the next three chapters.

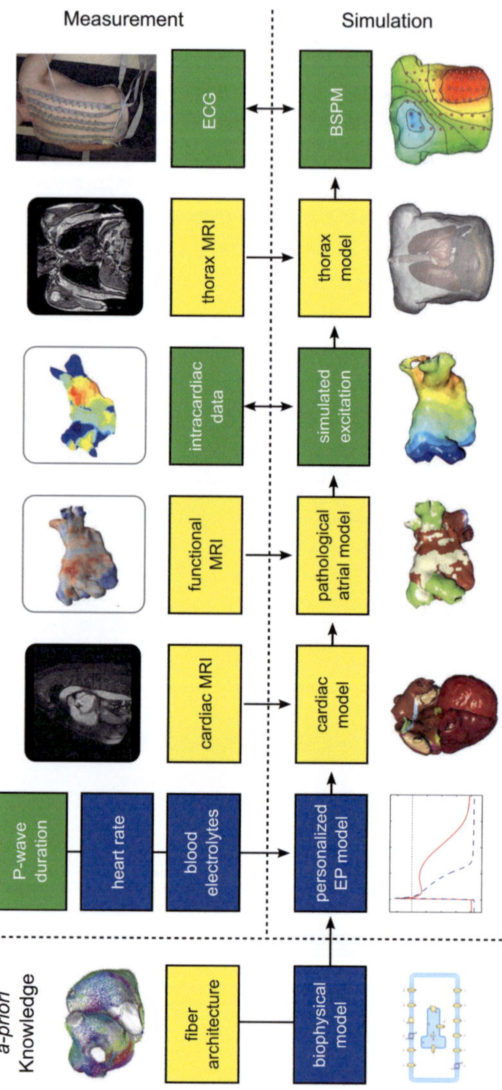

Fig. 6.3. Generalized framework for the creation and validation of personalized atrial models. Background colors of the individual steps refer to the color scheme used in Fig. 6.1. Figure adapted from [225].

Personalization of Anatomy

Over the past decades, modeling studies have used standard anatomical models of the atria and thorax. The atrial geometry was thereby either approximated by simple geometric shapes, e.g. a "peanut", or was derived from clinical image data. Models of the segmented Visible Man [353, 354] and Visible Female [355, 356] datasets were also frequently used, especially for whole body ECG computation studies. For the computation of atrial ECG signals, various groups also used simple homogeneous thorax models, sometimes including lung compartments [217–219, 229–231]. Table 3.1 provides an overview of atrial modeling studies using generalized anatomical models or the Visible Human atrial models. The creation process of these models often lasted months, and the models were refined over several years and several studies.

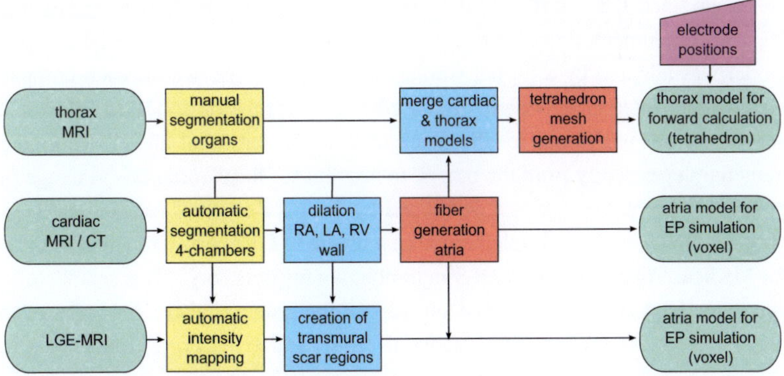

Fig. 7.1. Workflow of the creation of patient-specific atria and thorax models. The automatic segmentation of the MRI data was realized by software developed by Philips Research. The LGE-intensity mapping was performed at King's College London, UK.

Today, cardiac modeling is moving towards clinical applications. Methods to create patient-specific models are needed to create new anatomical models in a reasonable time with minimal user interaction. Atrial simulation studies put special requirements on the anatomical model (Sec. 2.1.1.5). Specifically, pulmonary vein drainage pattern, atrial wall thickness, fiber orientation, isolating septal layer and region classifications need to be included. In this chapter, a workflow to create patient-specific atrial and thorax models from clinical image data is proposed and the results of the modeling process are shown. The workflow is based on methods to segment the atrial endocardium. These were developed at Philips Research Europe, Aachen and Hamburg, in collaboration with the IBT. These segmentation methods were not part of this PhD thesis, but received collaborative input from the work described in this thesis. The methods are described here for the matter of completeness. Figure 7.1 schematically summarizes the workflow. All model generation steps after the segmentations were developed as part of the present thesis.

7.1 Creation of Simulation-Ready Atrial Models from MRI

The workflow to create simulation-ready anatomical atrial models from image data was comprised of three steps. In a first step, the endocardial surfaces of the whole heart (4 chambers, great vessel stubs) were extracted from the image data using an automatic approach [357–359]. Thereby, a statistical shape model was adapted to the image stack. The location of the heart in the image was initially determined using a 3D implementation of the generalized Hough transform. Then, the complete model was adapted by a global similarity transform. Afterwards, each compartment was adjusted independently by an affine transformation and last a deformable adaptation was performed locally. Each node in the statistical model was thereby trained independently from the others, to account for local variations in image parameters. The framework was developed by Philips Research, Aachen and Hamburg, Germany, and was implemented as a plugin to GIMIAS (Graphical Interface for Medical Image Analysis and Simulation, www.gimias.org). This segmentation approach defined the endocardial surfaces of the four chambers, the stubs of the major vessels and the left ventricular wall from cardiac MRI or CT data.

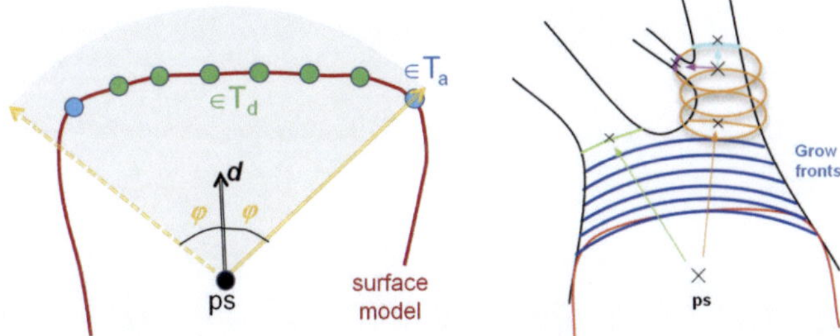

Fig. 7.2. Hybrid segmentation approach as presented in [48]. Initially, the left atrial body is segmented without PV orifices (left). Then a guided region growing is conducted in the area where PV ostia are commonly present (right). The region growing can further split the PV segmentation into multiple branches. Figure from [48].

Segmentation of Atrial Chambers

The segmentation algorithm has been previously adapted to better handle regional variations in atrial anatomy. A hybrid method to detect and segment variable pulmonary vein drainage patterns in the left atrium from CT data is used [48, 360]. The method bases on the approach of Lorenz and von Berg [357]. In this approach, the left atrial endocardium is segmented automatically, but the PV ostia are left out. A directed region growing in the region in which the ostia were expected are used to segment the PVs and early branches. Models 13–16 (Tab. 7.1) were segmented this way.

In a second effort, a method to segmented both atria from CT data was developed at Philips Research. Wall thickness as well as myocardial structures relevant for electrophysiological simulations were added to the statistical shape model [60, 361] in collaboration with this thesis. The method also based on the model proposed by Lorenz and von Berg [357]. Although it neglected variations in PV drainage pattern, it solved the problems arising with the segmentation of volumetric models (e.g. overlap of epicardial and endocardial surfaces). The atrial wall thickness could not be extracted from the CT data reliably. Therefore, regional atrial wall thickness was extracted from a set of high resolution MRI images and a statistical regional wall thickness was determined and augmented the statistical shape model. The output meshes also included an isolating layer at the septum. The electrophysiological structures in the myocardium were marked in the statistical shape model using the fiber placement algorithm presented in Section 5.4.

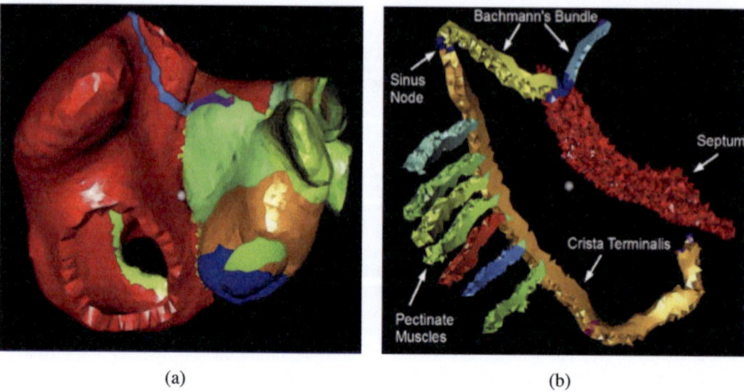

(a)   (b)

Fig. 7.3. Volumetric statistical shape model with electrophysiological structures based on [216, 362] as presented in [60]. Figure from [60].

The segmentation of cardiac MRI data is in some aspects more challenging than the segmentation of CT data. Especially vessels in close proximity to the cardiac chambers are often hard to separate from each other and from the myocardium. Therefore, the hybrid method presented by Hanna et al. [48] to segment CT data failed to segment MRI images. On the other hand, statistical shape models can segment the endocardium reliably, but cannot account for anatomical variations in PVs. Therefore, a segmentation workflow was developed at Philips Research in collaboration with this thesis, which used a family of statistical shape models to segment the left atrial endocardium from cardiac MRI data. Three mean models were created and trained, which covered the vast majority of observed pulmonary vein drainage patterns [363, 364]. Each image dataset was segmented three times, once with each mean model. A support vector machine was then used to determine the most suitable segmentation from the results. This approach was used to segment the image data for models 10 and 11 (Tab. 7.1).

Processing of Segmentations for the Simulation Use

As the second step in the workflow to create simulation-ready atrial models, the endocardial surface models were transferred into a high-resolution structured grid (Δx=0.33 mm). Based upon literature values for the atrial wall thickness (Sec. 2.1.1.4), the right ventricle and the endocardia of both atria were dilated. This resulted in a volumetric four chamber model. If the original segmentation of the endocardial wall did not include tissue labels for the cardiac vessels, the os-

Fig. 7.4. Three different statistical shape models of the left atrium as presented in [363, 364]. Left: regular model with four PVs, middle: the left PVs have a common trunk, right: model with extra right middle PV.

tia and stubs of the PVs and caval veins were labeled manually in the volumetric model. Afterwards, atrial fiber orientation was semi-automatically introduced into the atrial models, as described in Section 5.4. The semi-automatic segmentation also introduced tissue labels for right atrium, left atrium, SN, CT, PM, BB (R/L), II, R/L blood pool, Septum, TVR, MVR, middle posterior interatrial bridge, low posterior interatrial bridge, CS bridge and low anterior interatrial bridge into the atrial models. The resulting volumetric, heterogeneous, anisotropic atrial model could be used for the simulation of the electrical activity in the patient-specific atria.

If desired, the simulation-ready patient atrial models were triangulated into tetra-hedron meshes using CGAL (see next section for more details). These models could then be used with other simulation environments [365, 366], e.g. Chaste (Cancer, Heart and Soft Tissue Environment, www.cs.ox.ac.uk/chaste) [367], or for elastomechanical simulations (Chap. 14). An example for such a patient-specific, heterogeneous, anisotropic atrial tetrahedron model can be publicly retrieved from the Anatomical Model Data Base (AMDB, http://amdb.isd.kcl.ac.uk) [368].

7.1.1 Late Gadolinium Enhancement MRI Models

The cardiac segmentation process was augmented with structural tissue data retrieved from LGE-MRI (Sec. 4.1). First, the endocardial surface in the LGE-MRI needed to be registered with the endocardium segmented from the cardiac morphology scan. Second, LGE-intensity was calculated as maximum intensity projection of the image intensity within 3 mm surrounding the endocardial surface [270]. A direct transfer of the LGE-MRI intensity was not possible, as i) the resolution of the LGE-MRI was too coarse (commonly approximately 4 mm) and ii) the LGE image contains quite a few artifacts and noise. As the atrial myocardial wall cannot be extracted from the MRI data, the maximum atrial wall thickness of 3 mm was used as radius for the interpolation. The procedure is described in more detail in [270].

The LGE-intensity on the endocardial surface was transferred into the volumetric atrial models in three steps. First, the intensity values were binarized. Values above the threshold were projected along a gradient vector from the endocardium to the epicardium into the atrial wall. Last, a morphological closing operator with a structure element size of $3\times3\times3$ voxel was applied to the model to close small gaps. Those arose from the difference in endocardial surface to epicardial surface ratio. The technical realization of the projection along the transmural gradient was as follows. First, a circumsphere around each selected triangle of the endocardial surface mesh was created. Then, each voxel in the volumetric atrial model which lay on the border between the left atrial blood pool and the myocardium and which was enclosed by the sphere was marked according to the binarized intensity value [369].

This method can be used to recreate regions of ablation scars (post-ablation LGE-MRI), fibrosis (pre-ablation LGE-MRI) and/or edema (post-ablation T2-weighted MRI) in volumetric atrial models (Sec. 4.1). Scars created from RF ablations were transferred for models 8, 9, 9c, 10 and 11 (Tab. 4.1 and Tab. 7.1). More details are provided in Chapter 13.

7.1.2 Discussion

In this section, a workflow to create patient-specific simulation-ready atrial models with minimal user interaction was presented. The workflow was optimized for

MRI data. Within the euHeart project, MRI was the imaging modality of choice, as it does not rely on ionizing radiation. Nevertheless, also a number of models were segmented from CT data in collaboration with Dr.-Ing. Frank M. Weber [268]. The initial segmentation was thereby performed using software tools developed by Philips Research Aachen and Hamburg. Other groups also provide solutions to segment the cardiac chambers from CT [370], 4D CT [371], cone-beam CT [372], and TEE [373]. The modular setup of the developed framework allows to also use these tools instead of the Philips software.

The atrial anatomy requires segmentation approaches to cope with variability in PV drainage pattern. In this section two approaches to account for this were presented and used to generate patient-specific atrial models. Another approach, proposed by Zheng et al. [374], utilizes a left atrial model to segment ungated C-arm CT data during EP interventions. Such approach could enable a model generation without previous imaging sessions, but may not provide functional data about pathological tissue in the atria.

The introduced atrial fiber orientation and atrial region labels were not specific to the individual patient, but based on rules. Although some of the prominent bundles may be visible for an expert eye in high resolution CT data [12, 36], they are usually occluded in MRI and regular CT data and thus cannot be segmented for each patient individually. The rule-based structures reflect the atrial excitation sequence better than isotropic models, which was shown in Section 5.4.2. Coupling of the anatomical model with electrical measurements from the atria, e.g. LAT maps, could allow a personalization of the locations of the fast conducting bundles in the future, as proposed in Section 9.1.1.

Similarly, the atrial wall cannot be imaged with MRI data yet. First results from manual determination of atrial wall thickness from sparse high resolution MRI slices were presented in [60]. Until reliable imaging techniques for the atrial wall become available, statistical atlas data provides the best approximation to build volumetric atrial models.

The multi-model approach introduced by Kutra et al. [363] also includes an automatic assessment of the quality of the segmentation result. In the future, the analysis of the local segmentation error could provide the user with a reliability score for the segmentation and could show where manual corrections are needed.

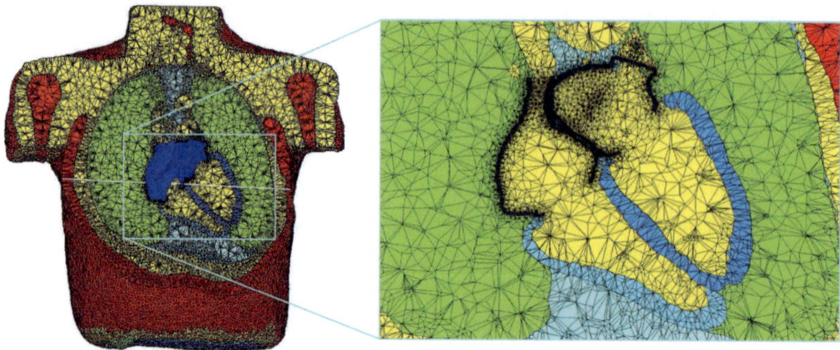

Fig. 7.5. Example of a tetrahedron thorax model (model 4). The thorax was cut open along the coronal plane and a right lateral axial plane. The atria were not cut for better impression of their location in the thorax. Tissue classes are colored to highlight the organs. The node distances at organ borders is smaller than within the organs (see inlay). Mean node distance in the atria is 0.83 mm.

The transfer of regional structural information of the myocardium from LGE-MRI still bears some uncertainties. As the atrial wall cannot be imaged with MRI techniques yet, LGE intensity in the surrounding of the maximum atrial wall thickness was analyzed. This might also include high LGE-MRI intensity arising from other structures, e.g. the esophagus next to the left atrial posterior wall. A comparison of LGE-MRI segmentation techniques revealed a high inter-expert variability in manual LGE-MRI segmentation as well as a broad range of segmentation results using a variety of segmentation techniques [274]. One of the major problems for these issues lay in the imaging quality for the atrial wall and the LGE-MRI. Once these problems are better resolved, the transfer of structural information from LGE-MRI and T2w-MRI into the models will become more reliable without needing to adjust the transfer method.

The method to transfer structural information from LGE-MRI introduces error at the borders between healthy and pathological tissue due to the use of circumspheres around triangles in the endocardial surface mesh. The magnitude of the error depended on the resolution of the endocardial surface mesh and was negligible small in the present datasets (≤ 0.66 mm). The resolution of the surface mesh can easily be increased by refining the mesh and thus, the error could be suppressed further.

7.2 Manual Segmentation of Thorax MRI Data

The forward-calculation of ECG signal from simulations of atrial excitation requires a thorax model of the same patient. In this section a manual approach to create thorax models from MRI data is presented. Whole thorax MRI data were segmented manually using deformable triangle meshes [375]. Thereby, a sphere was placed into the center of the organ. The radius of the sphere was chosen such that it resembled the mean radius of the organ. The mesh surface was then dragged locally to overlay the border of the specific organ. This method is independent of the imaging modality and image quality, but the user has to identify the correct border of the organ by eye. Organs with a non-spherical shape were segmented using multiple meshes, which were merged afterwards. E.g. the liver could be segmented using two meshes to cover the organ inside and parts of the bowel, and a third mesh covered only parts of the bowel next to the liver. The two volumes enclosed by the inside meshes were added and the volume of the third mesh was subtracted afterwards, such that only the volume of the liver remained.

For some organs, e.g. the lungs, region growing [333] or level set methods were also used for the segmentation process. The latter were realized in a software tool provided by Hans Barschdorf, Philips Research Hamburg, Germany, for the euHeart project. Using this software, organs were segmented by user-selected points inside, outside and on the border of an organ. Common level set methods [258] implemented in the Visualization Toolkit (VTK) and Insight Segmentation and Registration Toolkit (ITK) were then used to create a triangle surface mesh around the organ from these points. In all thorax models lung, kidneys, liver and spleen were segmented. Depending on the case, the remaining thorax volume was additionally classified into fat, bone, muscle, blood in the aorta, bowel or bowel content.

The surface meshes were then used to label different organs and tissue types in a voxel volume of the patient's thorax. The voxel representation of the patient's thorax was merged with the segmented cardiac voxel data. From the lattice data, a tetrahedron model of the thorax and heart of the patient was created using the Computational Geometry Algorithms Library (CGAL, www.cgal.org)) [376]. CGAL employs the 3D Delaunay triangulation [377] to create a tetrahedron model from structured grid data. Borders between organs were triangulated as smooth surfaces and thus contained more nodes compared to the inner volume of an organ or tissue region. The atria were meshed finer than the rest of the organs. At the end, the atria included approximately as many nodes as the rest of the thorax (Tab 7.1). Fiber orientation in the atria was transferred from the voxel model into the tetrahedron model afterwards. All other organs were set to have isotropic conduction properties. Figure 7.5 provides an example for a tetrahedron thorax model constructed in this manner. Table 7.1 provides more detail on the created models.

7.2.1 Discussion

In this section a workflow to create patient-specific thorax models from respective MRI data was presented. The manual segmentation using deformable surface meshes was a time consuming process. The VTK/IKT tool provided by Philips sped up the process significantly (5 - 15 minutes per organ). Nevertheless, the process might be prone to inter-user variability in the segmentation outcome. An automatic segmentation approach using statistical shape models, similar to the approaches presented for the cardiac segmentation could overcome this limitation. Although a number of segmentation approaches for organs other than the heart are described in literature [378], no approach to segment the whole thorax is present. For most soft-tissue structures desired for the forward calculation (lungs, abdominal organs), no model-based segmentation pipeline from MRI data is available [378]. The combination of multiple MRI sequences allows for a fast and automatic generation of muscle-fat-lung models [379, 380] on the cost of longer MRI acquisition sessions. This approach could in the future be used to create patient-specific thorax models for ECG computations with least possible user interaction.

The utilized software suites are just two examples of segmentation software. Other free tools, e.g. CardioViz3D [381, 382], Seg3D [383] or 3DSlicer [384], are also based on VTK/ITK. Although some of these provide more functionality than the tools used here, the present software could be better included into the technical workflow, because the software was either developed in-house or in close collaboration with the IBT.

Not all thorax organs are needed for a reliable simulation of the atrial ECG [71]. In the segmented thorax models, organs which significantly influence the ECG computation were included. Additionally, the kidneys and spleen were segmented, as these organs were clearly delineated in the image data and segmentation could be done in a short amount of time.

Except for the atria, all organs in the thorax models were set to have isotropic electrical properties. Fiber orientation in these organs cannot be retrieved from image data in a reasonable time in the clinical setting. Rule-based approaches to include fiber orientation in the skeletal muscles were shown to produce ECGs worse than those computed with isotropic models [385]. Anisotropy in the thorax was therefore neglected for the forward calculation.

For the future, further automation in methods to segment thorax MRI data are desired to allow for a smooth model generation workflow, which could also be introduced into clinical procedures.

7.3 Anatomical Models from Patient Data

Using the workflow and techniques presented above, 22 patient-specific atrial and 8 thorax models were created. Table 7.1 summarizes the properties of the anatomical models. Figure 7.7 to 7.14 provide visualizations of the thorax images as well as the thorax, heart and atria models of subjects 1–8. Figures 7.15 and 7.16 show visualizations of the atrial models and the modeled atrial fiber orientation in all cases.

Thorax MRI data of subjects 1, 2, 6 and 7 were segmented by Dr.-Ing. David Keller as part of his PhD work [267]. For model 1 there is also a high resolution dynamic ventricle model available [386, 387]. The detailed models of the atria and the whole thorax models of these patients were created as part of the present study. The geometrical models of the Visible Human data (20, 21) were created previously as well [92, 353–356]. The geometry of the sheep atria were created at the Auckland Bioengineering Institute, New Zealand, and provided by Dr. Jichao Zhao and Prof. Dr. Bruce Smaill [96]. For all models, rule-based fiber orientation was created as described in Section 5.4 as part of this thesis.

From the thorax models, the organ volumes of the lung, liver, spleen and kidneys were calculated (Fig. 7.6(a)). The mean volumes were $3385\,cm^3$ (lung), $1749\,cm^3$ (liver), $278\,cm^3$ (spleen) and $360\,cm^3$ (kidneys). The lung volume showed considerable inter-model variation of two to more than five liters. This was probably caused by the strong differences in patient/volunteer height and weight (Tab. 4.1). All values were within physiological range.

Table 7.1. Model data. Model number corresponds to dataset number in Table 4.1. Mean node distances (Dist) given in millimeters.

| | Cardiac Model | | | | | | Thorax Model | | | | | |
| | Atria | | | Ventricles | | | Total | | Organs | | Atria | |
Model	add. Tissues	Voxel	add. Bridges	Kind	Fibers	additional Organs	Elements	Nodes	Nodes	Dist	Nodes	Dist
1	PVs, CS, SVC, IVC	$1.3\cdot10^6$	–	simulation	y	intestine content	$1.9\cdot10^6$	$3.3\cdot10^5$	$2.4\cdot10^5$	4.47	$2.1\cdot10^5$	0.83
2	PVs, CS, SVC, IVC	$1.2\cdot10^6$	–	tissue	n	–	$1.6\cdot10^6$	$2.8\cdot10^5$	$2.1\cdot10^5$	4.91	$1.7\cdot10^5$	0.83
3	PVs, CS, SVC, IVC	$2.2\cdot10^6$	–	tissue	n	spine, intestine content	$2.9\cdot10^6$	$4.8\cdot10^5$	$3.0\cdot10^5$	4.76	$3.3\cdot10^5$	0.83
4	PVs, CS, SVC, IVC	$1.1\cdot10^6$	–	tissue	n	spine, composite tissue, stomach	$2.1\cdot10^6$	$3.6\cdot10^5$	$2.9\cdot10^5$	4.69	$2.0\cdot10^5$	0.84
5	PVs, CS, SVC, IVC	$1.8\cdot10^6$	–	tissue	n	spine, intestine content	$2.4\cdot10^6$	$4.2\cdot10^5$	$2.5\cdot10^5$	5.48	$2.7\cdot10^5$	0.83
6	PVs, SVC, IVC	$5.4\cdot10^5$	–	tissue	n	–	$1.1\cdot10^6$	$2.0\cdot10^5$	$1.7\cdot10^5$	6.03	$8.1\cdot10^4$	0.83
7	PVs, CS, SVC, IVC	$2.6\cdot10^6$	–	tissue	n	spine	$2.3\cdot10^6$	$4.0\cdot10^5$	$2.2\cdot10^5$	5.15	$2.6\cdot10^5$	0.83
8	PVs, SVC, IVC	$2.1\cdot10^6$	long BB	tissue	n	spine, intestine content	$2.9\cdot10^6$	$5.0\cdot10^5$	$3.3\cdot10^5$	4.90	$3.3\cdot10^5$	0.83
9	PVs, SVC, IVC	$1.7\cdot10^6$	FO	–	–	–	–	–	–	–	–	–
9c	PVs, SVC, IVC	$1.5\cdot10^6$	–	–	–	–	–	–	–	–	–	–
10	PVs	$2.3\cdot10^6$	short BB	–	–	–	–	–	–	–	–	–
11	PVs	$1.4\cdot10^6$	FO	–	–	–	–	–	–	–	–	–
13	PVs	$1.8\cdot10^6$	–	–	–	–	–	–	–	–	–	–
14	PVs	$2.9\cdot10^6$	–	–	–	–	–	–	–	–	–	–
15	PVs	$2.7\cdot10^6$	–	–	–	–	–	–	–	–	–	–
16	PVs	$1.7\cdot10^6$	–	–	–	–	–	–	–	–	–	–
17	PVs, CS, SVC, IVC	$1.9\cdot10^6$	–	–	–	–	–	–	–	–	–	–
18	PVs, CS, SVC, IVC	$1.4\cdot10^6$	–	–	–	–	–	–	–	–	–	–
19	PVs, CS, SVC, IVC	$3.5\cdot10^6$	–	simulation	y	–	–	–	–	–	–	–
20 (VM)	–	$2.4\cdot10^6$	–	–	–	–	–	–	–	–	–	–
21 (VF)	PVs	$1.9\cdot10^6$	FO	–	–	–	–	–	–	–	–	–
22 (sheep)	CS, PVarea	$9.3\cdot10^5$	–	–	–	–	–	–	–	–	–	–
mean		$1.9\cdot10^6$					$2.2\cdot10^6$	$3.7\cdot10^5$	$2.5\cdot10^5$	5.05	$2.3\cdot10^5$	0.83

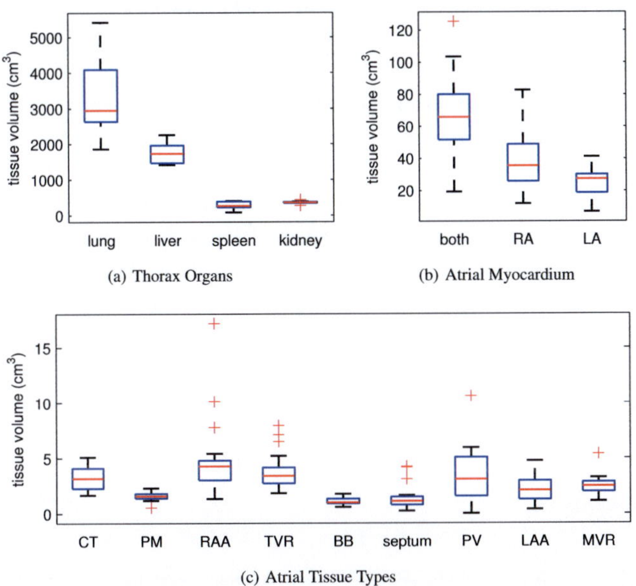

(a) Thorax Organs

(b) Atrial Myocardium

(c) Atrial Tissue Types

Fig. 7.6. Volumes of organs and atrial myocardium in the models.

In the models, the right atrial myocardium had a greater volume and thus mass compared to the left atrial myocardium (Fig 7.6(b)). In the right atrium, the most dominant structures in terms of relative myocardial volume were the RAA and the TVR (Fig 7.6(c)). In the left atrium, the PV volume showed a great intermodel variation. This was caused by the different segmentation approaches for the PV region. The CT was the largest fast specialized conducting bundle in both atria.

7.3.1 Use of Models and Model Cohorts

The models presented in the previous section were used for different simulation studies and collaborations. This section briefly describes which models were used for the different studies and also introduces three subgroups of models which were used in some of the investigations.

The Visible Human models (20, 21) and the Auckland sheep model (22) [96] were used to evaluate the outcome of the semi-automatic placement of atrial fiber orientation (Sec. 5.4.3, 5.4.4). Both Visible Human models were used to investigate the impact of fast-conducting bundles on the sinus rhythm excitation in man [388]. The Visible Female model with rule-based fibers was also used in the studies presented in [366, 389] and for a comparison of the anisotropic sinus rhythm excitation sequence in sheep atria compared to human atria [337]. For the hemodialysis study (Chap. 15), model 4 was used. This model was also used to refine methods to reconstruct the atrial excitation sequence from BSPM data [390, 391] and to simulate atrial fibrillation (Chap. 11). Model 8 was used to develop conduction velocity constraints for a solution of the inverse problem of ECG [287]. Models 17 and 18 were used to augment statistical shape models with fiber orientation in the atria [339]. Model 19 was used to create a deformation model of the atria (Sec. 14).

Model Cohorts

Three cohorts of models were created, which were used for different studies. Using a cohort of models instead of just using one model, which is common practice in atrial modeling (see Tab. 3.1), had the advantage that potential geometry-specific errors were averaged in the final results. The variation in the results of individual models in the studies presented in Chapters 9 and 5.2 foster this approach. Furthermore, a cohort of models allows the development and evaluation of model-processing techniques, which can be later on applied to new, patient-specific models.

1. Whole Thorax Model

This model cohort comprised the atrial and thorax models of subjects 1–8 (Tab. 7.1). These models are the most complete ones, as realistic simulation of the atrial excitation can be performed and the results can be forward calculated to body surface ECGs. This multi-scale simulation approach allows the investigation of e.g. pathological changes in cell electrophysiology from cell to ECG. The model cohort was used to evaluate the adaptation of the global conduction velocity in models to the P-wave duration (Sec. 9.3) and to investigate the effects of variations in the inter-atrial conduction on the ECG signals (Sec. 9.1.2). Furthermore, the models were used to study the atrial repolarization sequence in the ECG (Chap. 10).

2. AF Atria Models

The second cohort was made up of all models from AF patients (subjects 8–11). For these individuals, also left atrial LAT maps (Tab. 4.1) were available. The cohort was used to determine the individual region of first left atrial activation and to personalize the atrial simulations accordingly (Sec. 9.1.1).

3. Anisotropic Atria Models

The largest cohort of models was comprised of models 1–11 and 13–19 and represented all patient-specific human atrial geometries. The cohort was used to evaluate the fiber orientation algorithm (Sec. 5.4) and the algorithm to create generic ablation lesion patterns (Sec. 5.5).

7.3.2 Discussion

Using the segmentation and post-processing techniques presented in this chapter, more than twenty models of human atria and eight models of patient and volunteer thoraxes were created. The models were grouped into three model cohorts for the further utilization in different studies. A number of the models were also supplied to collaboration partners and used for different applications.

Atlases and statistical models play an increasing role in cardiac modeling [392]. Atlases are generated from image data of various sources, and are used to determine representations of the mean organ anatomy [393]. They can also be used to investigate the variability of anatomical structures, e.g. local fiber orientation [329]. Data for cardiac atlases can be acquired in the clinic, or directly be accessed from web-based databases, e.g the Cardiac Atlas Project (CAP) [394]. The anatomical databases usually provide quick access to a large number of annotated data. The model cohorts presented here were used to rule out geometry-specific effects in the outcome of multi-model simulation studies.

Publicly accessible datasets can enable the verification of modeling and simulation workflows using reference datasets [395]. Similarly, reference models can be used to cross-validate numerical methods [396]. For such purposes, an Anatomical Model Database (AMDB) [368] was set up as part of the euHeart and Virtual

Physiological Human projects. The anisotropic atrial model 18 was made publicly available on the AMDB.

Statistical shape models are currently widely used for image segmentation. In the future, they might also support clinical therapy planning and medical training. The present results helped to augment the statistical shape models with anatomical features relevant for electrophysiological simulations. The augmented models may in the future provide more confidence in the direct clinical use and medical training if combined with methods for the fast simulation of the atrial excitation.

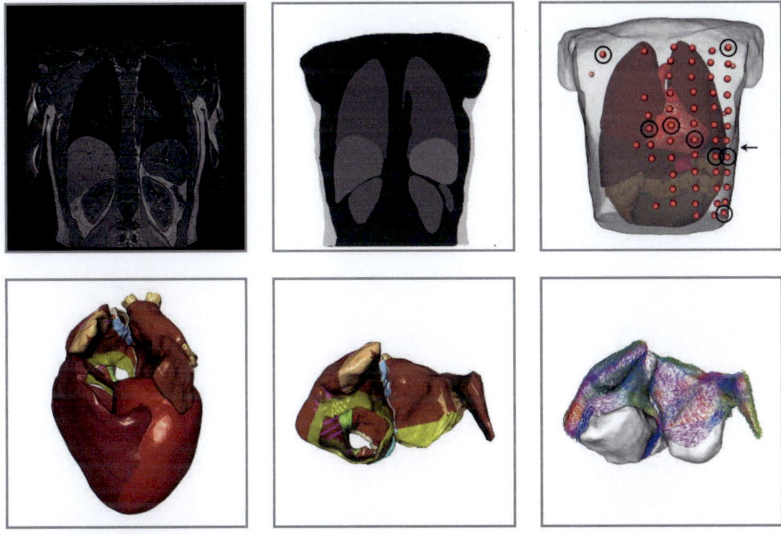

Fig. 7.7. MRI, segmentation and 3D models of subject 1 (Tab. 4.1). Circles mark electrodes used for standard ECG. Arrows point at standard ECC electrodes on the left side and the back.

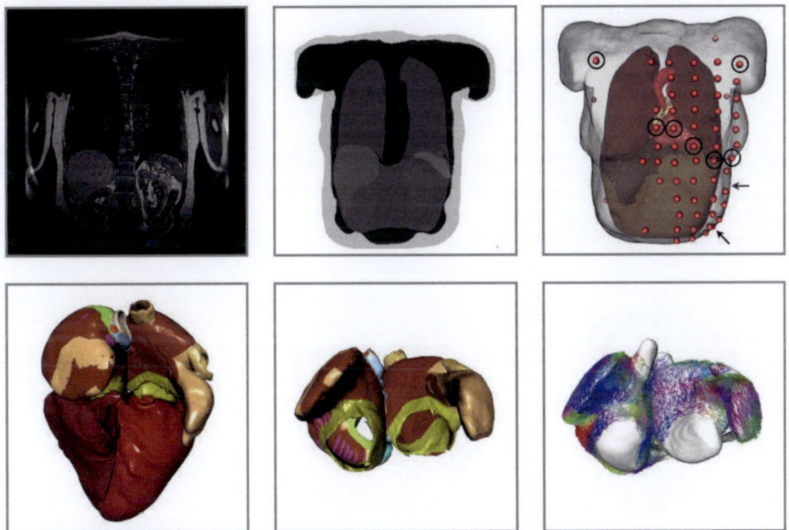

Fig. 7.8. MRI, segmentation and 3D models of subject 2 (Tab. 4.1). Circles mark electrodes used for standard ECG. Arrows point at standard ECC electrodes on the left side and the back.

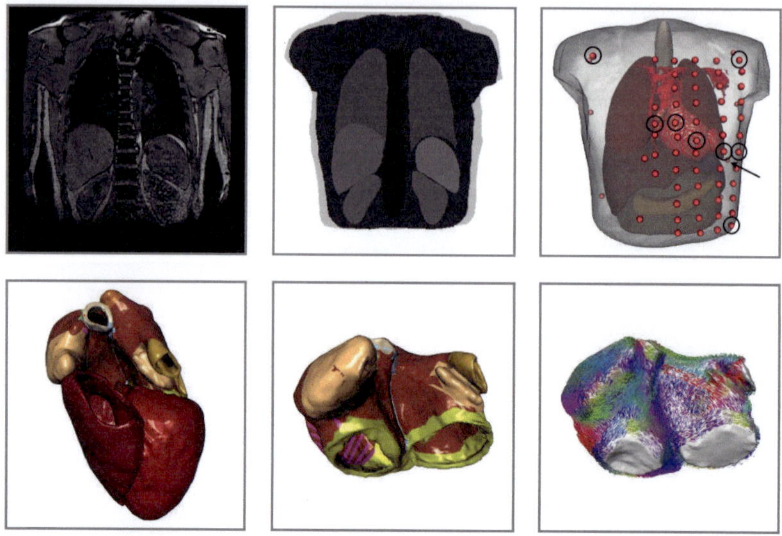

Fig. 7.9. MRI, segmentation and 3D models of subject 3 (Tab. 4.1). Circles mark electrodes used for standard ECG. Arrows point at standard ECC electrodes on the left side and the back.

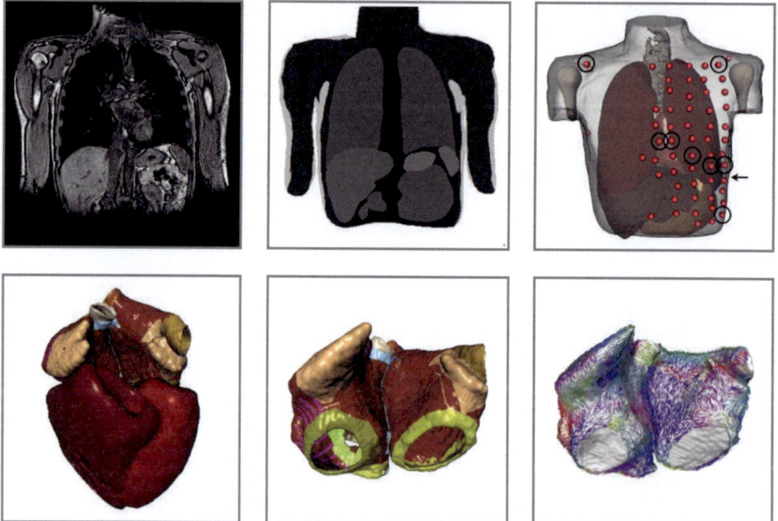

Fig. 7.10. MRI, segmentation and 3D models of subject 4 (Tab. 4.1). Circles mark electrodes used for standard ECG. Arrows point at standard ECC electrodes on the left side and the back.

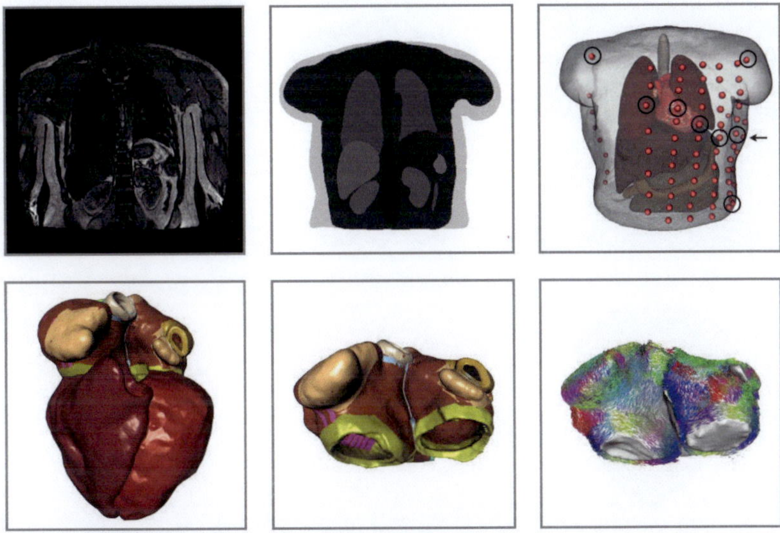

Fig. 7.11. MRI, segmentation and 3D models of subject 5 (Tab. 4.1). Circles mark electrodes used for standard ECG. Arrows point at standard ECC electrodes on the left side and the back.

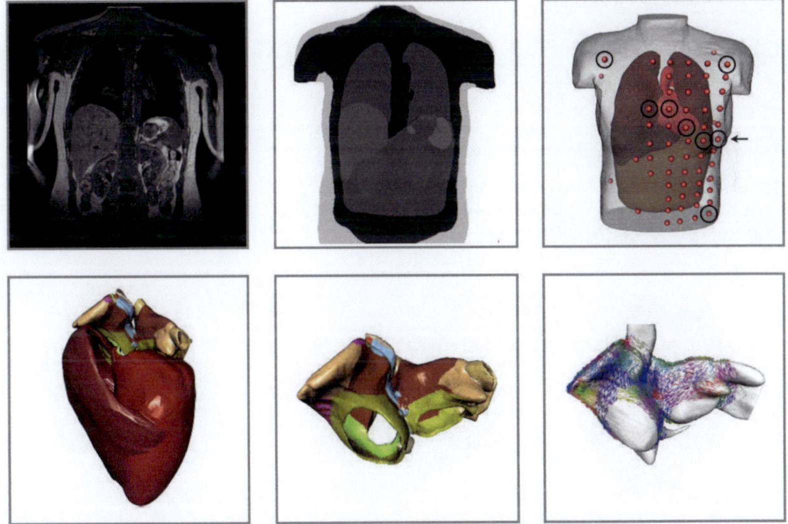

Fig. 7.12. MRI, segmentation and 3D models of subject 6 (Tab. 4.1). Circles mark electrodes used for standard ECG. Arrows point at standard ECC electrodes on the left side and the back.

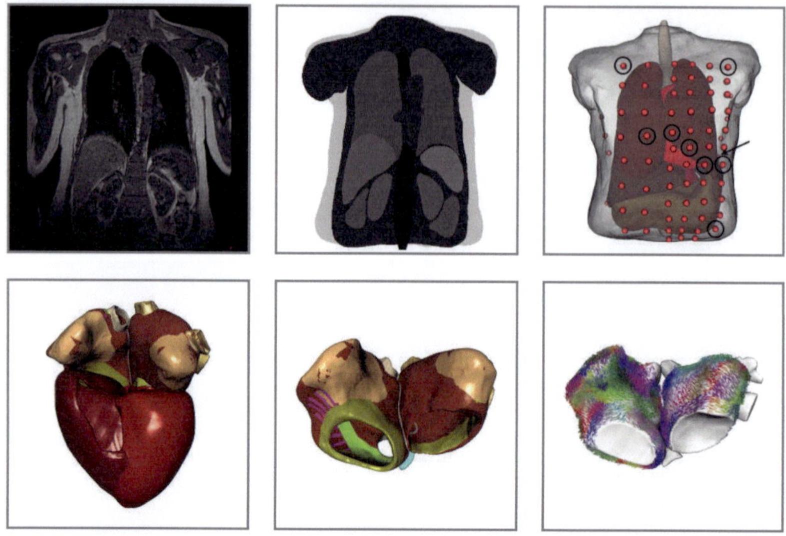

Fig. 7.13. MRI, segmentation and 3D models of subject 7 (Tab. 4.1). Circles mark electrodes used for standard ECG. Arrows point at standard ECC electrodes on the left side and the back.

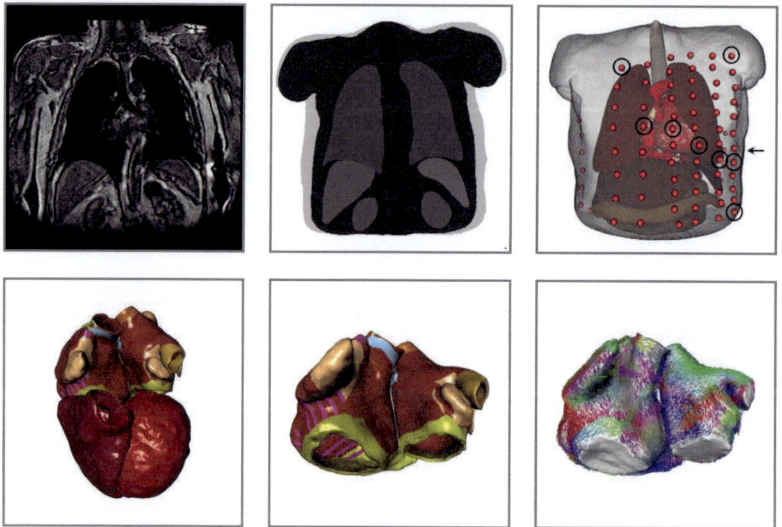

Fig. 7.14. MRI, segmentation and 3D models of subject 8 (Tab. 4.1). Circles mark electrodes used for standard ECG. Arrows point at standard ECC electrodes on the left side and the back.

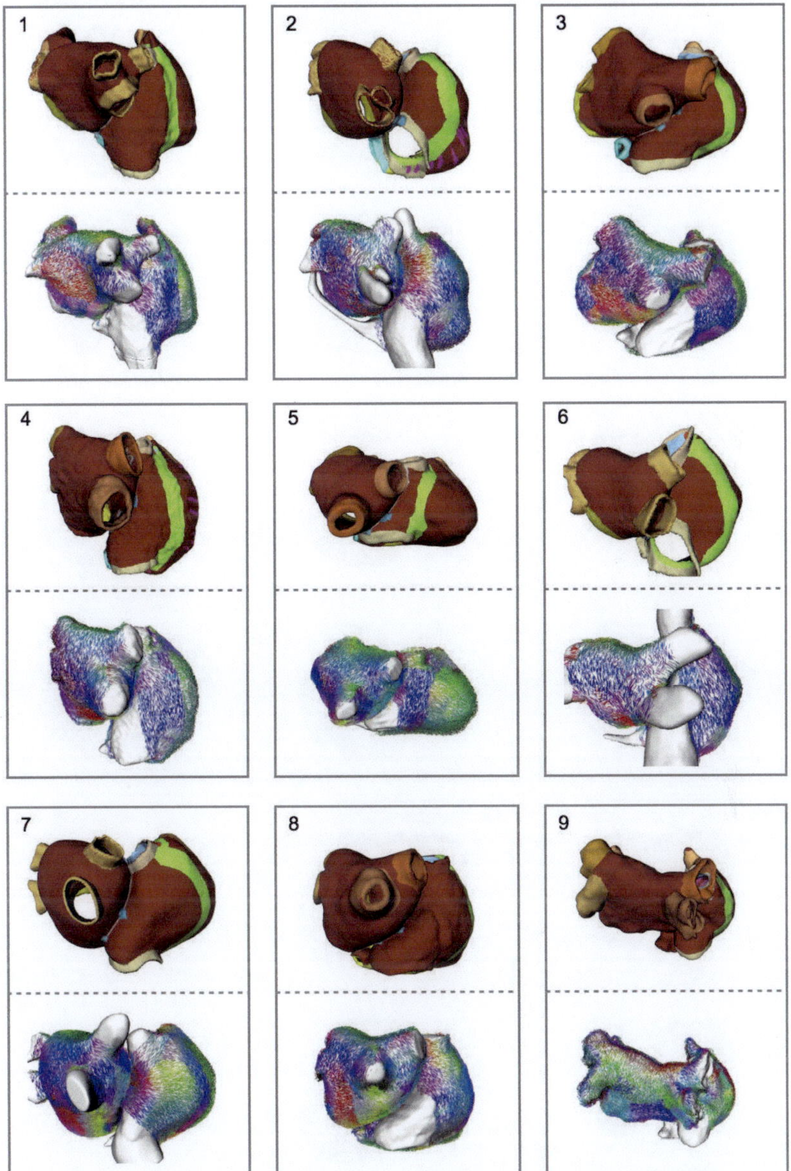

Fig. 7.15. Posterior view of atria models 1–9 augmented with fiber architecture (Tab. 4.1).

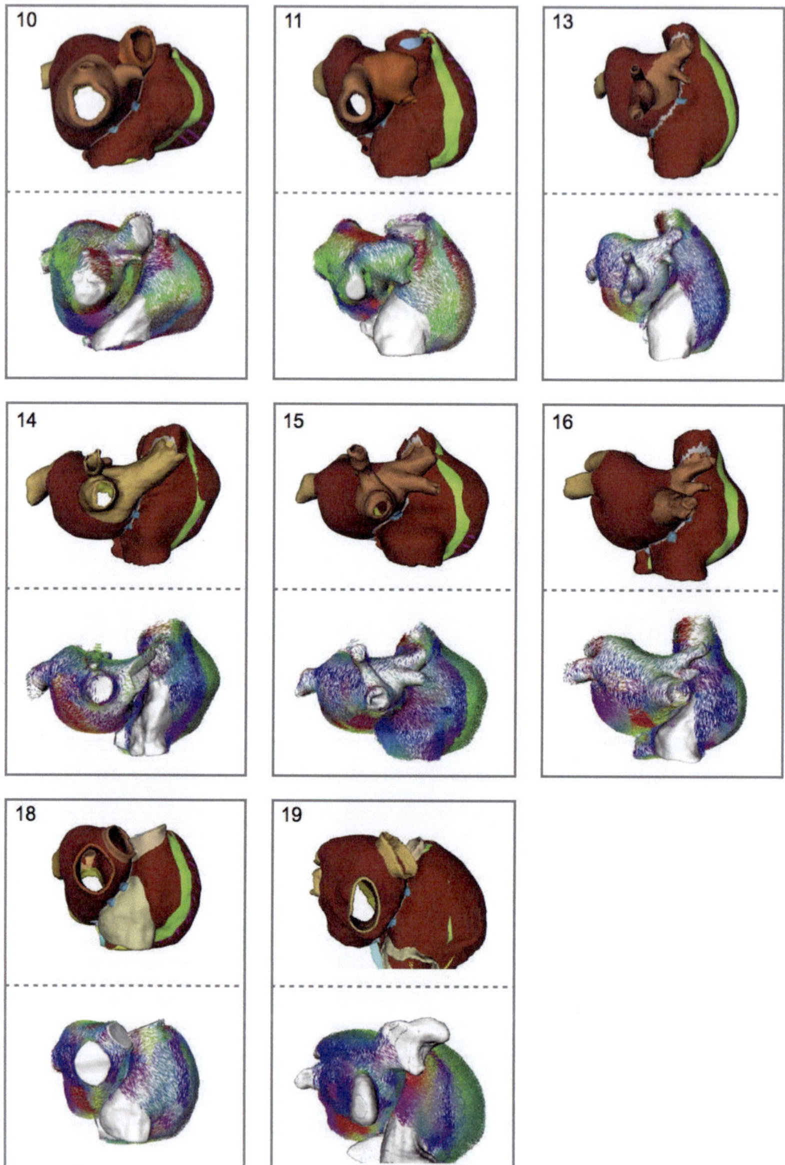

Fig. 7.16. Posterior view of atria models 10–19 augmented with fiber architecture (Tab. 4.1).

Personalization of the Cellular Electrophysiology

In this section methods for the patient-group and patient-specific personalization of models of the atrial electrophysiology are presented.

8.1 Patient-Group Models

Measuring cellular electrophysiological properties *in-vivo* is not yet feasible due to technical, health and ethical constraints. The adaptation therefore focused on the personalization of the electrophysiological models by choosing the most suitable existing electrophysiological model for each patient. This patient-group specific personalization is commonly done using models of AF-remodeled electrophysiology for AF patients. This was also done in the course of this work (subjects 8 – 11, Tab. 4.1). In a next step, this personalization process could be refined by choosing the appropriate model for the stage of AF (paroxysmal, persistent, long term persistent, permanent). Regional adaptations could also cover the use of different electrophysiological models with one patient geometry. E.g. a model for the electrophysiology of fibroblasts [236, 294, 397] could be used for fibrotic regions extracted from LGE-MRI (Sec. 7.1.1). The patient-group personalization approach relies on the diagnosis of the patient's disease and thus on the expertise of the clinician.

8.2 Patient-Specific Adaptations

A patient-specific adaptation of the electrophysiological model was achieved by setting the extracellular ion concentration parameters of the CRN model to mea-

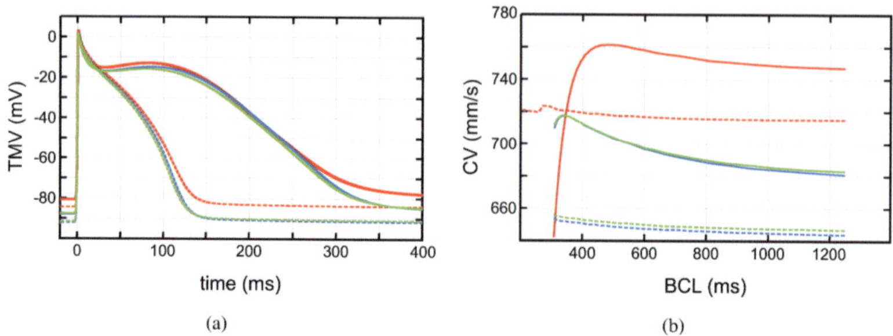

Fig. 8.1. Influence of altered extracellular ion concentrations to patient blood electrolyte concentrations. Red: CRN model, blue: patient 8, green: patient 9, dashed: AF-remodeling. Figure adapted from [225].

sured values of blood electrolyte concentrations of the same patient [225]. Changes in the concentration of electrolytes in the blood will wash in/out into the extracellular space of the cardiomyocytes [398, 399] and thus such personalization becomes feasible. For subjects 8 and 9 (Tab. 4.1), blood electrolyte concentrations were measured (Tab. 8.1). The CRN model was augmented to be able to also reproduce the effects of altered extracellular potassium concentrations (Sec. 5.1.1) and subsequently, patient-specific CRN models of the electrophysiology of AF patients 8 and 9 were created.

Figure 8.1 shows the impact of the altered extracellular ion concentrations on the regular CRN model and the AF-remodeled CRN model for both patients. In all configurations, the resting membrane potential was reduced by the electrolyte alterations and the conduction velocity was slowed [225]. Similar effects were observed in patients undergoing hemodialysis therapy to compensate for lost renal function (Sec. 15) [399]. Table 8.2 provides results from a quantitative analysis of the changes in action potential morphology in the healthy CRN model in four atrial regions described in [92]. The decrease in upstroke velocity dV_m/dt was the cause for the reduced conduction velocity observed in the tissue simulations (Fig. 8.1(b)).

A further patient-specific adaptation of the model was performed by adjusting the basic cycle length (BCL) in the simulations to the patient's heart rate. The heart rates were extracted either from BSPM measurements or from ECGs recorded from the electroanatomical mapping systems (Tab. 4.1).

Table 8.1. Values for extracellular / blood substance concentrations in the CRN model and AF patients 8 and 9. Creatinine and urea measurements were not used for the model personalization and are provided only for completeness. CRN values from [72].

	$[K^+]_o$ mmol/L	$[Na^+]_o$ mmol/L	creatinine mmol/L	urea mmol/L
CRN model	5.4	140	–	–
subject 8	4.0	141	108	–
subject 9	4.1	137	66	6.9

8.3 Discussion

In this chapter methods for the adjustment of the general model of atrial electrophysiology to the patient and the patient's disease were introduced. The impact of AF-remodeling outruled the effects of changes in extracellular ion concentrations on the action potential morphology and thus on APD and ERP in two AF patients. Nevertheless, the alterations in extracellular ion concentrations significantly reduced the excitation conduction velocity in the personalized models, caused by the decrease in $[K^+]_o$ (Sec. 15). For the conduction velocity, the effects of AF remodeling and hypokalemia augmented each other and may provide a better substrate for the initiation and maintenance of AF, as conduction velocity, APD and ERP were reduced.

Electrical remodeling due to persistent AF [73] has a significant effect on the APD, ERP and conduction velocity (Fig. 8.1, Sec. 5.2.2). The changes in electrophysiology in patients with paroxysmal AF will most certainly be less pronounced. As the cellular experimental database is sparse with respect to paroxysmal AF, the persistent AF-remodeled CRN model was used in this work to personalize the electrophysiological model to all AF patients, neglecting the stage of disease progression. As soon as more data become available, further parameter sets or new models for paroxysmal or permanent AF may allow for a personalization to the stage of the disease (patient subgroup).

The personalization of the model to the measured blood electrolyte concentrations poses a simple way for a patient-specific model adaptation, as blood samples are regularly analyzed in clinical environments. Attention should be paid to a sampling shortly before or after the situation which should be simulated, as electrolyte concentrations may show short-term variations.

Table 8.2. Impact of patient's extracellular ion concentration on the CRN model for four right atrial regions [92]. amp: amplitude.

region	$\Delta dV_m/dt$ V/s	ΔAPD_{90} ms	Δpeak mV	Δamp mV
CAM	-4.40	4.35	-1.23	6.03
CT	-3.70	1.02	-1.09	6.13
AVR	-5.00	-7.54	-1.36	5.94
APG	-4.40	1.33	-1.21	6.03

The effects on the action potential and tissue properties were minimal, as all simulations were performed in sinus rhythm. At BCLs of sinus rhythm, the restitution slopes of the tissue properties are flat (Fig. 8.1, Sec. 5.2.2) and therefore the model behavior does not significantly change.

In the future, the adaptation of electrophysiological models could base on the adaptation of model parameters and channel kinetics to a-priori knowledge of the patient. Known genetic mutations in patients could be recreated *in-vitro* and the experimental findings could be used to investigate consequences of the genetic mutation on the atrial electrophysiology [400–402]. The presented methods pose the first steps towards the creation of such patient-specific electrophysiological models. Further experience with the electrophysiological model personalization will speed up the engineering of new personalization methods in the future.

Personalization of the Excitation Propagation

The atrial excitation propagation results from the atrial anatomy as well as global and local electrical tissue properties (Fig. 6.1). Therefore, model personalization must consider these properties. In this chapter, two methods for the personalization of the sinus rhythm simulations in patient-specific atrial models are presented and evaluated. First, a method to personalize the inter-atrial conduction routes from intracardiac LAT maps and BSPM measurements is described. Second, a method to adjust the monodomain tissue conductivity to fit the sinus rhythm simulation to the measured P-wave duration is presented. The methods complemented each other and were combined in a workflow for the personalization of the excitation propagation in patient-specific atrial models (Fig. 9.1). The workflow produces a patient-specific atrial excitation model and can provide a model-based prediction of the LAT and conduction velocity direction and magnitude.

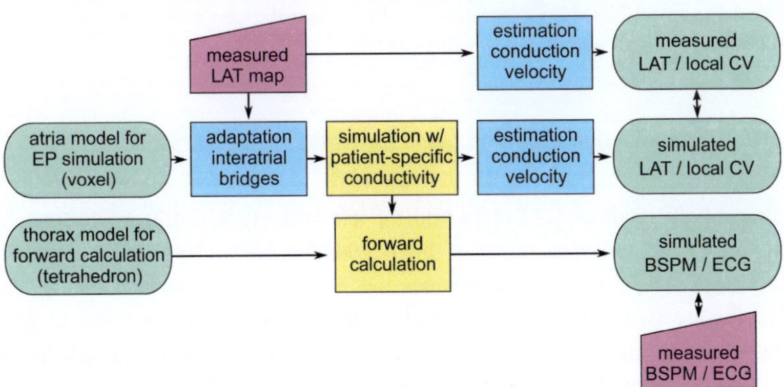

Fig. 9.1. Workflow for patient-specific simulation of the atrial sinus rhythm and ECG / BSPM.

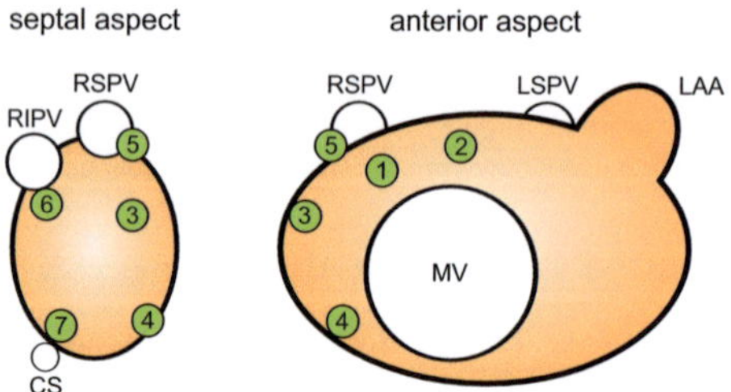

Fig. 9.2. Expected points of early left atrial activation. 1: Short Bachmann bundle. 2: Long Bachmann bundle. 3: Limbus of fossa ovalis. 4: Low anterior bridge. 5: High anterior bridge. 6: High posterior bridge. 7: Musculature of the coronary sinus.

9.1 Personalization of Interatrial Conduction Routes

The human atria are electrically isolated by the interatrial septum (Sec. 2.1.1.1). Various muscular bridges cross the septum and provide electrical connections between the atria. The most common bridges are semi-automatically introduced into the geometrical atria models along with the semi-automatic placement of atrial fiber orientation (Sec. 5.4). Bachmann's bundle is the most prominent interatrial connection. In contrast to the other bridges, it extends into the right atrium and LA, has well aligned fiber orientation and allows a fast excitation conduction. With the fiber-placement method, Bachmann's bundle, a low anterior bridge, two posterior bridges and a bridge connecting the ostium of the coronary sinus to the LA were modeled. Bachmann's bundle was thereby electrically isolated from the LA in about one fifth of its extent into the LA. In the simulations with these models, the first LA activation usually occurred over BB.

9.1.1 Personalization Based on Local Activation Time Maps

Intracardially recorded LAT maps from the left atrium (Sec. 2.4.2) provide insights into the LA activation sequence of the individual patient. The creation process of LAT maps usually lasts several minutes and involves input from the clinician. The data quality is therefore variable, making an automatic analysis of these maps not usable at this point. In the course of this thesis, a manual approach for the personalization of the interatrial conduction was developed.

Table 9.1. Sites of early left atrial activation observed in the LAT maps and altered interatrial bridges. Subject numbers correspond to patients in Tab. 4.1 and models in Tab. 7.1. Site numbers correspond to Fig. 9.2.

subject	early LA activation	BB (CV) 1 / 2	FO 3	low anterior 4	high anterior 5	posterior 6	CS 7
8	site 2, site 7	long (-50%)	no	yes	no	yes	yes
9	size 3, site 6	no (–)	yes	yes	no	yes	yes
10	site 1 / site 5	short (–)	no	yes	yes	yes	yes
11	site 1, site 3, site 4	short (-50%)	yes	yes	no	yes	yes

Personalization Approach

First, the user analysed the LAT map of a patient visually. Based on this analysis, the user determined which bridges from the semi-automatic segmentation approach should be included and which should be removed for the simulation. Additionally, the length and the conduction velocity of BB could be varied. The anaylsis of the LAT map was guided by the scheme shown in Figure 9.2. The scheme shows seven locations for left atrial activation, which correspond to the locations of the interatrial bridges. The user needs to determine in which of these locations a left atrial activation was to be seen in the measured LAT map. The corresponding bridges were set for the subsequent simulation, whereas all other bridges were removed from the model.

If an inferior location and a BB location in the measured LAT map yielded an activation, the conduction velocity in BB was reduced by -50%. This needed to be done, as a fast conduction via BB will activate the left atrium so fast, that a further inferior wavefront from the right atrium may not enter the left atrium, because the left atrium is acitvated in the corresponding region already.

Personalization Results

The described course of actions was applied to four AF patient datasets (8–11, Tab. 4.1, Tab. 7.1). The results of the LAT map analysis are listed in Table 9.1.

Simulation Results

Two simulations were performed for models 8 – 11. One simulation was conducted with the standard configuration of interatrial bridges. A subsequent simulation was

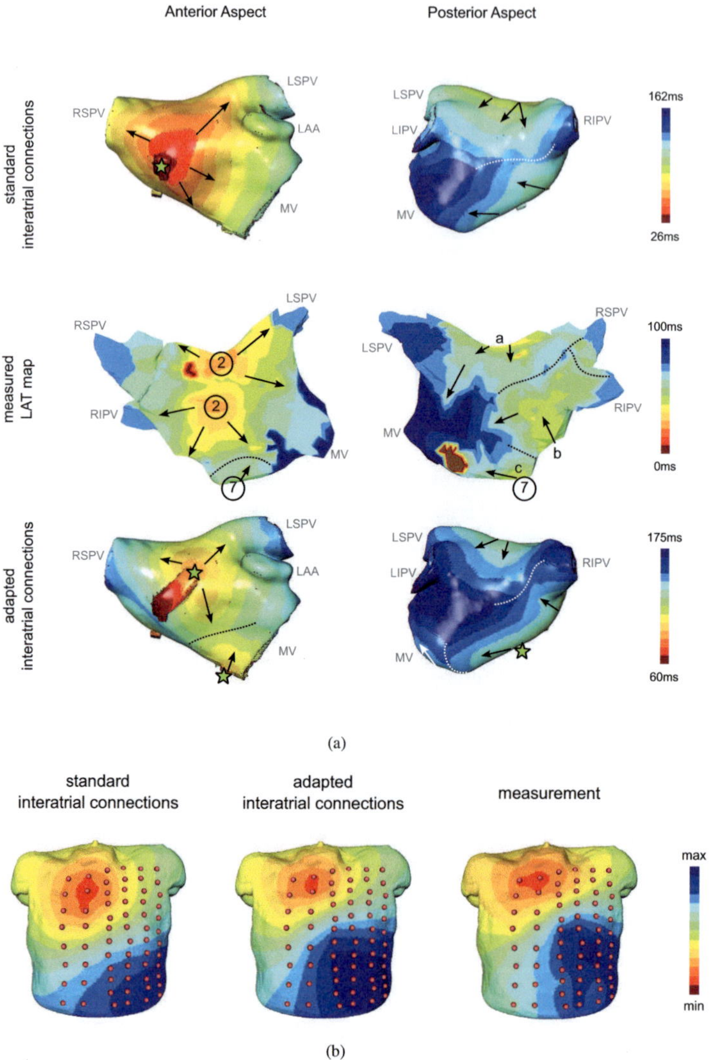

Fig. 9.3. a) Simulated and measured LATs in the left atrium of subject 8. Green stars indicate sites of early activation in the simulations. Encircled numbers correspond to sites of early activation in the measured LAT maps (Fig. 9.2). Arrows indicate major direction of propagation. Dotted lines mark the collision of two wavefronts. b) Integrated BSPMs over the P-wave for an initial simulation with interatrial conduction via BB and a simulation which had been adjusted to the earliest LA activation near the region of the FO. On the right side the measured integrated BSPM is shown. BSPMs were generated by interpolation of the integrated simulated or measured signals in the electrodes onto the body surface.

conducted after adapting interatrial bridges specifically to each patient (Tab. 9.1). Figures 9.3 – 9.6 show the simulated and measured LAT in the left atrium of the patients. Complete activation time span cannot be determined accurately from the LAT maps, as measurement errors at PVs and MV obscure the precise determination of the latest activation. Scale extremes were set for each simulation / measurement such that the colors at sites of early activation and colors at late activation near the LPVs correspond between the setups.

Simulations with the standard bridge configuration showed an early activation at the left anterior side of the LA. Two wave fronts activated the posterior wall of the LA. One wavefront came from the LA roof and the other from the right inferior. The wavefronts merged in the low posterior LA wall. The latest region to depolarize lay left inferior of the ostium of the LIPV.

In subject 8, the measured LAT map showed two sites of early activation depolarizing the anterior LA (sites 2, Fig. 9.3 middle row). Another wavefront arose from the region of the low posterior septum (site 7, Fig. 9.3 middle row). The activation from site 2 (BB) was faster and did not allow a wide spread of the wavefront arising from site 7. The posterior side showed three wavefronts (a, b, c, Fig. 9.3 middle row right). Wavefronts a and b merged between the RSPV and the low left posterior LA (dashed line, Fig. 9.3(a) middle row). In the simulation with adapted bridges, the LA was activated early at one site centered on the anterior LA. As in the measurements, another wavefront swapped over the anterior LA from the inferior septum (CS musculature). Three wavefronts activated the posterior LA. Two of these merged between the RIPV and the region inferior to the LIPV.

For subject 8 the simulated transmembrane voltages were forward calculated to retrieve BSPMs for both simulations. Figure 9.3(b) shows the simulated and measured time integrated BSPMs in this subject. The adjustment of the interatrial bridges to the LAT map of the patient led to an upward shift of the isopotential line. Additionally, the line was bent at the center of the thorax front in contrast to the standard simulation. The BSPM of the adjusted simulation showed a better correspondence to the measured BSPM pattern (RMSE adjusted bridges: 0.028 mVs vs. regular bridges: 0.040 mVs).

In the measured LAT map of subject 9, first LA activation occurred inferior of the RSPV at two adjacent sites (site 3 and site 6, Fig. 9.4 middle row). Three wavefronts activated the posterior LA from the LA roof, the right PVs and the

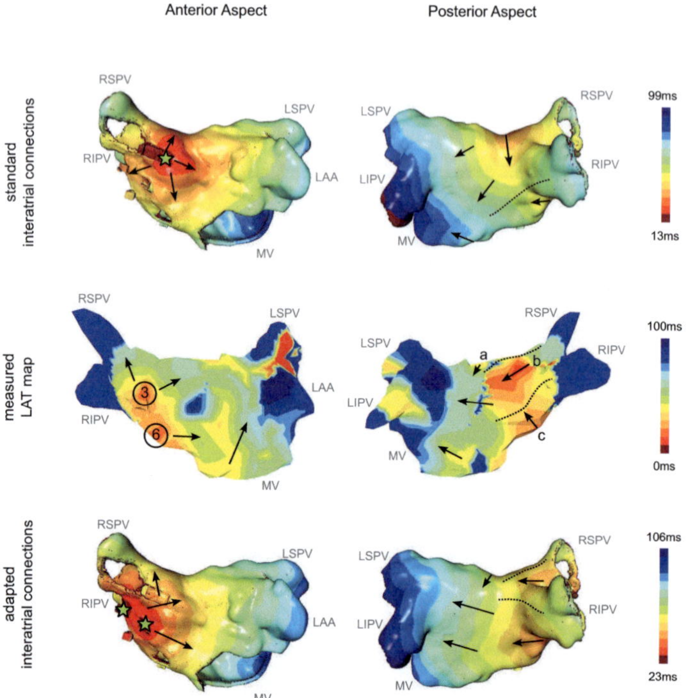

Fig. 9.4. Simulated and measured LATs in the left atrium of subject 9. Green stars indicate sites of early activation in the simulations. Encircled numbers correspond to sites of early activation in the measured LAT maps (Fig. 9.2). Arrows indicate major direction of propagation. Dotted lines mark the collision of two wavefronts.

right inferior LA (a, b, c, Fig. 9.4 middle row right). The wavefronts collided left of the RSPV and left of the RIPV. The latest activation was measured on a line connecting the MV and the left PVs. The simulation with the adjusted interatrial bridges recreated the sites of early LA activation (green stars, Fig. 9.4 bottom row). Additionally, the adjusted simulation showed three activation wavefronts on the posterior side and a similar line of latest activation as in the measurements (Fig. 9.4 bottom row).

In the measured LAT map of subject 10, the LA was activated early right inferior of the RSPV with a fast activation of the high septal region (site 1, Fig. 9.5 middle row). The LAT map showed a number of artifacts on the posterior LA (areas of

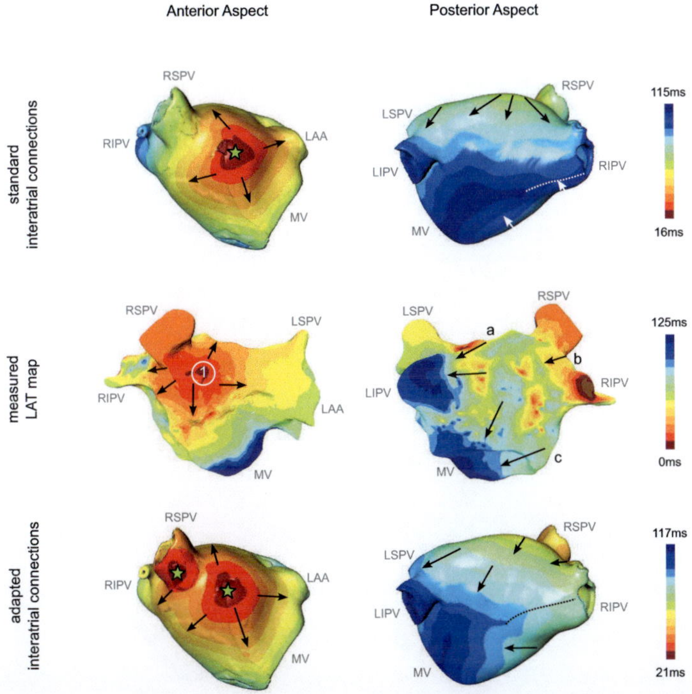

Fig. 9.5. Simulated and measured LATs in the left atrium of subject 10. Green stars indicate sites of early activation in the simulations. Encircled numbers correspond to sites of early activation in the measured LAT maps (Fig. 9.2). Arrows indicate major direction of propagation. Dotted lines mark the collision of two wavefronts.

early activation in the free wall) (red regions, Fig. 9.5 middle row right). Nevertheless, three wavefronts could be observed which activated the posterior LA (a, b, c, Fig. 9.5 middle row right). All wavefronts ran towards the left and left inferior LA. In the adjusted simulation, an early activation of the high septum was achieved over two superior anterior bridges (green stars, Fig. 9.5 bottom row). Two early and one later wavefront activated the posterior LA towards the left and the left inferior region similar to the activation pattern observed in the measurement (Fig 9.5 bottom row).

In subject 11, the measured LAT map showed four sites near the atrial septum activating a large area of the anterior LA nearly simultaneously (sites 1, 3, 4,

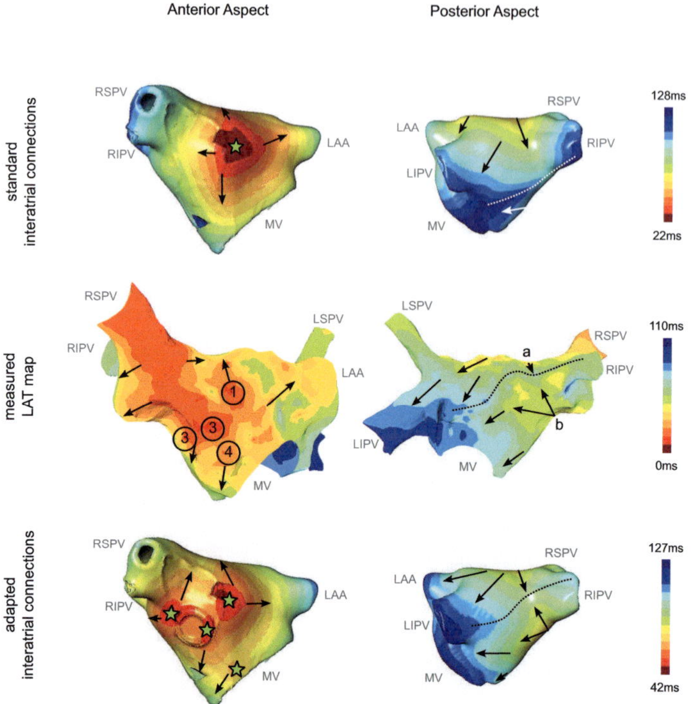

Fig. 9.6. Simulated and measured LATs in the left atrium of subject 11. Green stars indicate sites of early activation in the simulations. Encircled numbers correspond to sites of early activation in the measured LAT maps (Fig. 9.2). Arrows indicate major direction of propagation. Dotted lines mark the collision of two wavefronts.

Fig. 9.6 middle row). Two major wavefronts were observed on the posterior LA (a, b, Fig. 9.6 middle row right). These merged on a line between the LIPV and the RIPV. Wavefront a progressed towards the left inferior LA. Wavefront b activated the inferior LA from the right to the left side. In the adjusted simulation, the introduction of the fossa ovalis bridge and slowing of conduction along BB, led to simulated isochrones comparable to the measured activation sequence (Fig. 9.6, bottom row).

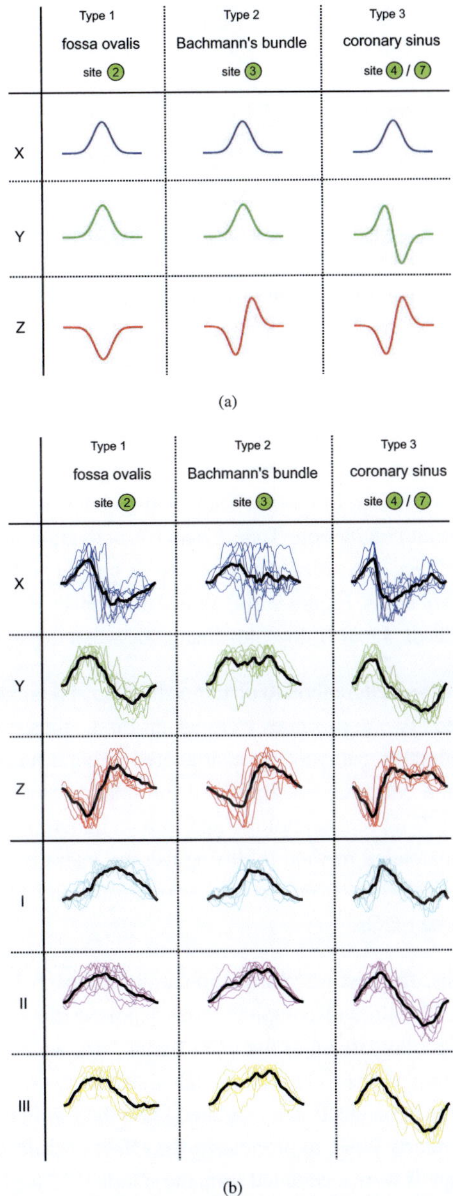

Fig. 9.7. a) P-wave morphologies identified by Holmqvist et al. for different interatrial conduction routes [33, 117, 403]. b) P-wave morphologies of simulated VCG signals under different interatrial conduction routes and P-wave morphologies of simulated Einthoven ECG signals. Black curves represent the mean of the simulated signals. The site numbers correspond to Figure 9.2.

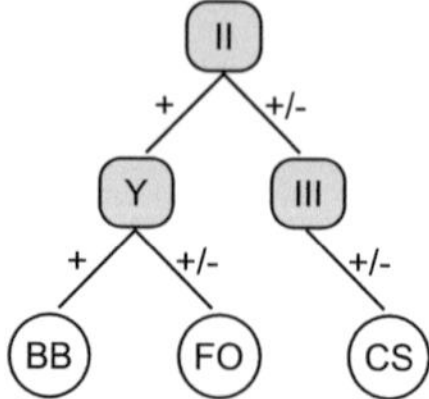

Fig. 9.8. Proposed differentiation method based on VCG and Einhoven ECG leads.

9.1.2 Impact of Interatrial Conduction on the ECG & VCG

The manual adjustment of the interatrial connections in the previous section led
to an altered left atrial depolarization sequence. This also caused a variation in the
computed BSPM pattern, as was to be expected. Holmqvist et al. investigated the
morphology of the P-wave in the three VCG leads in patients with different interatrial conduction routes [33, 117, 403]. They distinguished between three types of
preferential interatrial conduction. Type 1 was LA activation in the region of the
fossa ovalis. Type 2 was activation via Bachmann's bundle and type 3 was activation via the coronary sinus. Figure 9.7(a) shows the typical P-wave morphologies
in the three VCG leads. Lead X does not change between the types.

In this section, an extended method to distinguish interatrial conduction from ECG
signals is presented. Two approaches were taken. First, sinus rhythm with different interatrial conduction pathways was simulated in eight models (model 8–11,
Sec. 7.3.1) to gain a better understanding of the relationship between interatrial
conduction and the P-wave morphology and to test the hypothesis of Holmqvist
et al.. Second, an extended method to distinguish the interatrial conduction route
from the ECG was extracted from the simulation results and applied to the simulations and measured ECGs.

For the simulations, the fast-marching simulation approach introduced in Section 5.7 was used. The time of complete atrial depolarization was normalized to
100 ms for an easier comparison of the ECG signals afterwards. The results from
the sinus rhythm simulations were forward calculated to retrieve ECG signals and
BSPMs. BB was set to be electrically isolated from the LA in 60% of its LA extent, instead of regularly 20%, to pronounce the effect of different LA activation
locations. VCG signals were calculated from the standard 12 lead using the inverse
Dower method (Sec. 2.4.1.1).

Table 9.2. Analysis of the simulation results. Grey shaded cell indicate correctly identified simulated interatrial conduction.

	Holmqvist et al.				proposed			
	BB	FO	CS	not clear	BB	FO	CS	not clear
simulated BB	7	0	1	0	7	1	0	0
simulated FO	3	0	5	0	2	5	0	1
simulated CS	0	0	7	1	0	0	8	0

Figure 9.7(b) shows the simulated VCG and Einthoven ECG signals for the eight models and the averaged signal for each lead. The VCG signals of the Bachmann's bundle connection confirmed the findings from Holmqvist et al.. For conduction via fossa ovalis, the simulated VCG signals were biphasic, whereas the proposed scheme showed only monophasic signals (Fig. 9.7(b)). These correspond to the first half of the simulated VCG signals. For conduction via the coronary sinus musculature, leads Y and Z in the simulations and scheme correspond well, but lead X showed low amplitudes and a multiphasic signal, from which the overall shape was difficult to generalize. Einthoven ECG signals showed less variation between the eight simulations than VCG signals did. Einthoven signals from fossa ovalis and Bachmann bundle conduction could not be distinguished (all signals monophasic positive in all leads), but conduction via the coronary sinus showed significantly different signals in all leads (biphasic vs. monophasic) (Fig. 9.7(b)).

Based on the averaged simulation results, a method to differentiate between the three major interatrial conduction routes was developed (Fig. 9.8). Einthoven ECG and VCG signals which showed a great inter-model variance (e.g. lead X), were not taken into consideration for the differentiation method. In the proposed method, signals are distinguished as either being positive, but not necessarily having only one maximum, or as having positive and negative signals parts.

The differentiation method was applied on the simulated ECG / VCG signals (Tab. 9.2). The proposed method showed better results than the method proposed by Holmqvist et al., especially with respect to FO conduction (Tab. 9.2). With both methods, conduction via CS could be clearly identified. The method proposed by Holmqvist et al. could not predict FO conduction correctly in any cases, whereas the proposed method detected this in 5 of 8 simulations correctly. Conduction via BB was correctly identified with both methods in 7 of 8 simulations.

9.1.3 Discussion

In this section two methods to personalize interatrial conduction pathways from LAT maps and ECG data were presented. Manual adjustment of the rule-based interatrial bridges led to a better match of the simulated to the measured left atrial activation sequence in four AF patients. The data indicated that the interatrial conduction is highly variable between individuals. The simulation results highlight the importance to incorporate an isolating layer in the interatrial septum in the models, to allow for distinct interatrial activation breakthroughs.

LAT maps were created by manually selecting the time of local activation in catheter signals. Such approach is prone to user caused errors. In all LAT maps, such errors in terms of very late and very early activation times were found. Such area can be identified, as they show not gradual increase in activation time into the surrounding tissue and thus no wave propagation. These errors need to the removed prior to an adaptation of the interatrial conduction in the model to the measurement data. Similarly, in the area of the atrial septum, the far field from right atrial activation might be interpreted as left atrial excitation sequence. Special caution needs to be taken during the LAT recording process to avoid such misinterpretation. For future studies an automatic detection of LATs is desirable to avoid such errors from the beginning. This would also enable an automatic detection of sites of early activation and thus an automatic model personalization.

So far only locations of early left atrial activation were extracted from LAT map data. In a next step also regions of slow conduction or conduction block could be identified from the data and the model could be adapted accordingly. The mean left atrial conduction velocity determined from the LAT measurements could be used to tune the left atrial conduction velocity in the models prior to the personalization of the global conduction velocity to the models (Sec. 9.2). This could provide a more precise match of the simulations and measurements. Technical advances of the EAMS allow for a registration of MRI morphology data to the catheter measurements. This needs to be done prior to data recording to avoid misregistration afterwards. LAT maps based on MRI morphology data will have a more precise geometry and will thus provide more reliable conduction velocity data. Additionally, a quantitative comparison between locally simulated and measured activation times as well as a direct adaptation of local conduction velocity will be possible. A combination of LAT recording and a pacing protocol covering multiple cycle length could also provide information about local conduction velocity restitution [84], which could be directly incorporated into the model [223].

Using a cohort of eight atria and thorax models it was possible to investigate the influence of the three major interatrial connections (BB, FO, CS) on the ECG. Previous studies proposed a diagnosis of interatrial conduction based on VCG signals [33, 117, 403]. These studies showed only moderate sensitivity and specificity. The simulated ECG and VCG signals indicated different typical VCG patterns for the interatrial conduction via fossa ovalis. The VCG, although calculated from the ECG signals and thus not containing more information than these, seems to reveal important information about the interatrial conduction. The proposed differentiation method based on VCG signals was enhanced to use signals from VCG lead and ECG leads. This combination of signals yielded a clearer differentiation between the types of interatrial conduction. In a future study, the proposed differentiation method could be applied to measured ECG signals from patients, in whom the interatrial conduction pathways are known, to clinically validate the methodology.

The use of a cohort of anatomical models could rule out anatomy-specific effects in the simulated ECG signals. The approach additionally provides information about the variance of simulated ECG signals under the different conditions and may thus provide a measure of confidence of the averaged signals. The limb ECG leads showed a smaller variance in signals compared to the simulated VCG signals. This was most likely caused by a variance in ECG electrode positions between models (Fig. 7.7 – 7.14). Although limb lead positions also show a variance, the signals in the area further away from the cardiac electrical sources show less variation and thus the variance in electrode positions does not cause great variance in the signals.

The measured LAT maps indicate a combination of interatrial conduction routes rather than having only a single left atrial breakthrough site. This will cause a distortion of the ECG / VCG signals. Nevertheless, the study from Holmqvist et al. indicates a dominance of the preferential conduction routes in the ECG signals. Future simulation studies could investigate the ECG signals of simultaneous conduction over multiple bridges using the presented simulation framework.

The fast-marching simulation approach allowed a high throughput of simulations in a cohort of eight models. This procedure ruled-out results specific to an individual anatomy, which could lead to misinterpretation of the results. In the future, the influence of the conduction velocity of BB or the influence of a variation in sinus node location on the ECG could be investigated similarly to gain a further in-depth understanding of the atrial excitation sequence and P-wave morphology.

Both proposed methods to personalize interatrial conduction routes could be coupled to the automatic segmentation of the atria from MRI data (Sec. 7.1). Thereby, the location of myocardial structures not visible in clinical images data such as fast conducting bundles and interatrial conduction pathways could be retrieved from LAT data. These could subsequently be used to adjust the location of these structures in the anatomical segmentation model [60] for the individual patient.

9.2 Personalization of the Global Conduction Velocity

The adaptation of the conduction velocity in simulations using the bidomain or monodomain model for the excitation spread cannot be performed straight forward (Sec. 5.6). The local conduction velocity in such simulations depends on the local tissue conductivity, fiber orientation, model geometry, wavefront curvature and wavefront incidence angle. In this section, a two step approach is presented to personalize the global monodomain conductivity in sinus rhythm simulations on patient-specific geometries. First an initial simulation with literature values for the conductivity is performed. Then a correction formulation is applied to the conductivity to receive the patient-specific global conductivity for the sinus rhythm simulation. Figure 9.9 depicts the workflow. In the end of the section, the results of the personalization of the global conduction velocity are shown and resulting ECG and BSPM signals are compared to the measurements.

Formulation for Tissue Conductivity Adjustment

The monodomain tissue conductivity can be described by a quadratic function of the simulated excitation conduction velocity [351, 352]:

$$\sigma = a \cdot CV^2 + b \cdot CV + c. \tag{9.1}$$

Figure 5.30 provides an example of this relationship. If two simulations with different conductivities are performed, a relation between the simulated conduction

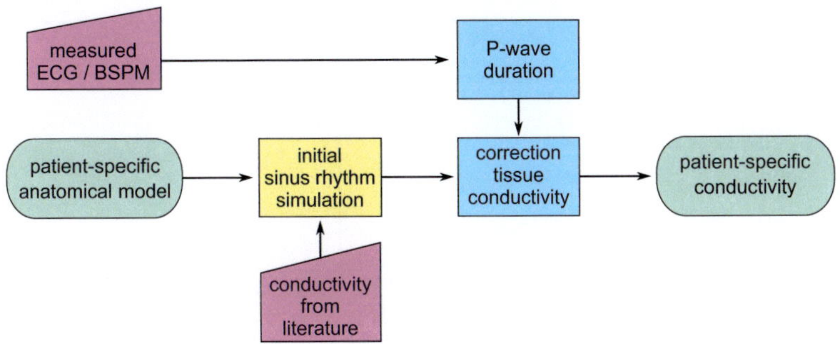

Fig. 9.9. Workflow to determine the patient-specific tissue conductivity.

velocities can be formed by

$$\frac{\sigma_2}{\sigma_1} = \frac{a \cdot CV_2^2 + b \cdot CV_1 + c}{a \cdot CV_1^2 + b \cdot CV_1 + c}. \tag{9.2}$$

Let simulation 1 be an initial simulation with literature values for the conductivity σ_{init} and conduction velocity CV_{init}. For a second simulation, the conductivity should be adjusted such that another, e.g. measured, conduction velocity CV_{meas} is reproduced by the simulation. For the second simulation, the value of the unknown conductivity σ_{adapt} is required. As the quadratic term in equation 9.1 dominates the others [351] (Sec. 5.6), equation 9.2 can be reduced and approximated to a relationship of the conductivities to the square of the conduction velocities.

$$\frac{\sigma_{adapt}}{\sigma_{init}} = \frac{a \cdot CV_{meas}^2 + O(CV_{meas})}{a \cdot CV_{init}^2 + O(CV_{init})}$$

$$\frac{\sigma_{adapt}}{\sigma_{init}} \approx \frac{CV_{meas}^2}{CV_{sim,init}^2} \tag{9.3}$$

The conduction velocity can be rewritten as the ratio of the longest possible path in the model s and the time for complete atrial depolarization T_{cAD}. In the case of the initial simulation, the latter is the time $T_{cAD,init}$ where the last model element depolarizes. For the measured conduction velocity, this corresponds to the P-wave duration PWd.

$$\frac{\sigma_{adapt}}{\sigma_{init}} \approx \frac{\left(\frac{s}{PWd}\right)^2}{\left(\frac{s}{T_{cAD,init}}\right)^2} \tag{9.4}$$

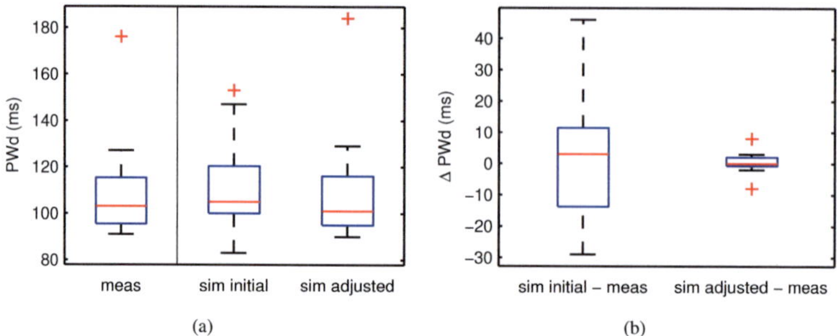

Fig. 9.10. a) Measured *PWd* and simulated T_{cAD} in subjects 1–8. b) Absolute difference between measured and simulated time of complete atrial depolarization. Adjustment of the conductivity using equation 9.5 led to a precise match of PWd. Outliers are the two simulations with greatest offset to the measured PWd in the initial simulations.

As the size of the atrial model corresponds to the size of the atria of the patient from whom the ECG was taken, the equation can be reduced and rearranged to receive a formula for the adapted tissue conductivity

$$\sigma_{adapt} \approx \sigma_{init} \times \left(\frac{T_{cAD,init}}{PWd} \right)^2 . \tag{9.5}$$

This approach to adapt the monodomain tissue conductivity does not influence the anisotropy ratio between the longitudinal and transversal conductivity, as both are scaled by the same factor.

Simulations with Personalized Global Conduction Velocity

The formula was applied in sinus rhythm simulations using eleven patient models (model 1–11, Tab. 7.1) and the respective ECG measurements. The measured and simulated PWds are summarized in Fig. 9.10. A precise match of the time of complete atrial depolarization to the PWd was achieved with a single adaptation step. The mean absolute difference between the simulated and measured PWd was 2.8 ms.

Comparison of Simulated and Measured ECG Signals

The method presented in this section aimed at adapting the global conduction velocity to fit the simulated and measured PWd, and not the P-wave morphology. It

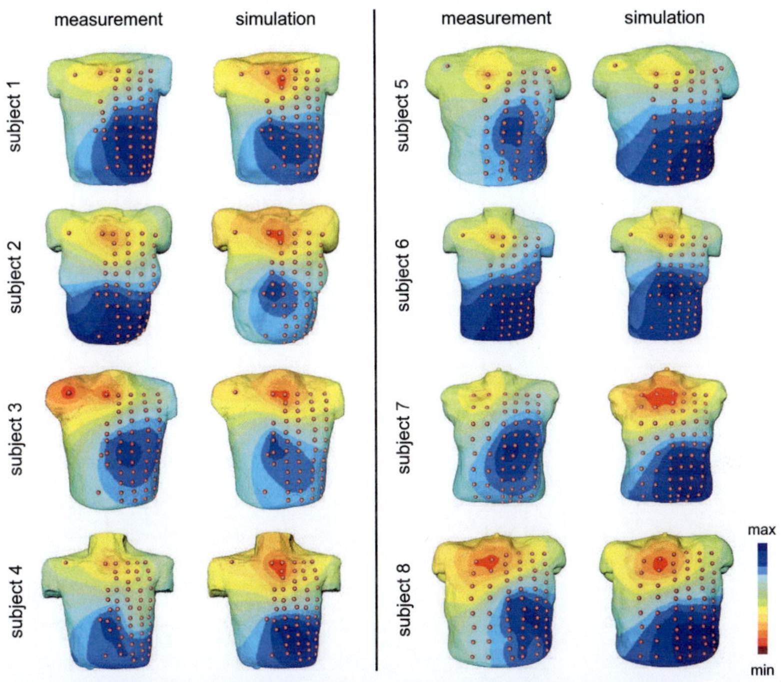

Fig. 9.11. Measured and simulated time-integrated BSPM during the P-wave.

is nevertheless interesting to see a comparison of these. Figures 9.11 shows the measured and simulated integral BSPMs for subjects 1 – 8. Figures 9.12 and 9.13 show a comparison of the simulated and measured ECG signals. The length of the signals are the same, as this was the aim of the personalization process. The BSPMs corresponded well overall (RMSE 0.267 ± 0.097 mVs). A precise match of BSPM patterns was not achieved. The mean of the median correlation coefficients between the personalized simulated ECG signals and the measured signals was 0.577 ± 0.206. The polarity and phase of the limb lead ECG signals were reproduced in the majority of the cases, whereas VCG signals showed a stronger deviation from the measurements.

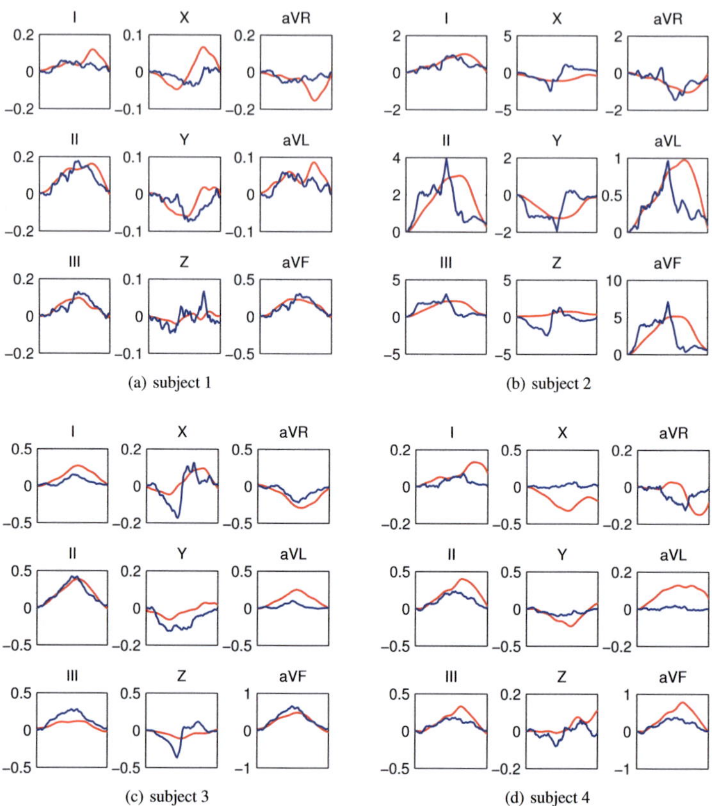

Fig. 9.12. Measured (red) and simulated (blue) P-wave signals in various ECG leads. X-axis limits correspond to the individual P-wave duration, which was matched before (Tab. 4.1). Y-axis is the ECG signal in mV. Figures continues in Fig. 9.13.

9.2.1 Discussion

In this section a fast method to adjust the global conduction velocity in patient-specific anatomical models to the measured PWd was presented. The approach only required one additional simulation prior to the personalized simulation. Literature values for the tissue conductivity were adjusted using a quadratic term derived from the relationship between monodomain tissue conductivity and conduction velocity. In simulations in which the conductivity adjustment factor was

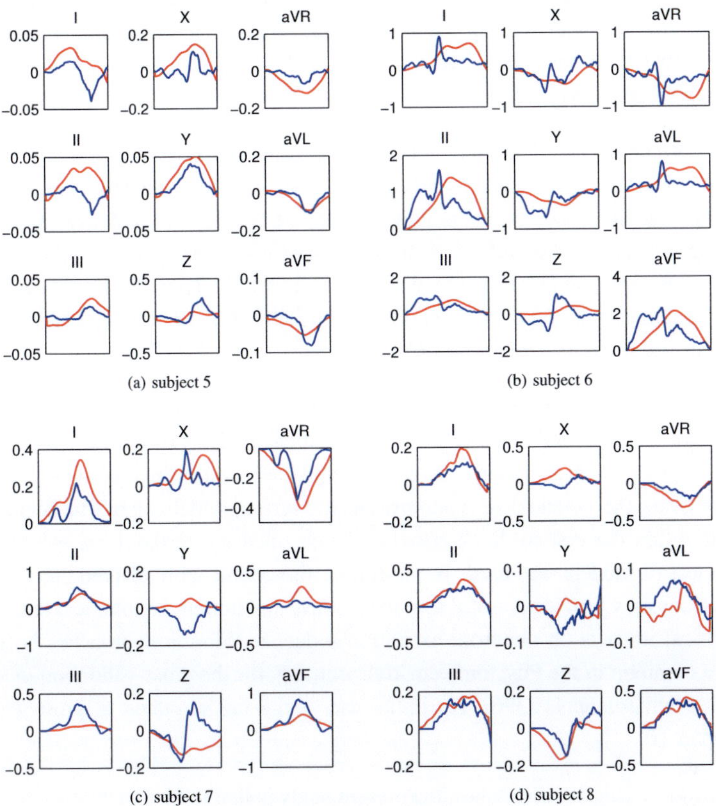

(a) subject 5

(b) subject 6

(c) subject 7

(d) subject 8

Fig. 9.13. Continuation of Figure 9.12.

considerably large, the time increment of the Euler scheme needed to be adjusted to allow for numerical stability of the model (e.g. $10\mu s$ instead of $20\mu s$). This prolonged the duration of the simulation.

Heterogeneous conduction velocities in the models were adjusted by the same factor. The atria contain different fast conducting bundles and slow conducting zones. It is not clear whether these structures show the similar inter-individual variation of conduction velocity as regular left and right atrial myocardium [84]. Nevertheless, the adjustment of all model conduction velocities in the same way led to a precise match of the simulated and measured P-wave duration. Adaptation of regional variations in conduction velocity, e.g. based on LAT maps (Sec. 9.1.1) could further provide a better match of integral BSPMs and ECG signals. Similar approaches have been presented recently based on signals recorded with circular mapping catheters [223].

The correlation coefficients of the simulated and measured ECG signals were rather low. This was caused by electrode signals near the isopotential line of the BSPM. Slight variations of the isopotential line inverted the polarity of signals in adjacent electrodes and led to negative correlation coefficients, although the general pattern and ECG limb lead signals looked well. To minimize this artifact, the median correlation coefficient (CC) of all ECG channels for each patient was determined instead of the mean CC.

The use of a cohort of model and patient datasets allowed a validation of the general methodology and enabled a simple statistical analysis of the variation of the results of the method under different prerequisites. Using singular anatomies always bears the potential of anatomy-specific errors, which cannot be identified from within the system. Such errors could be ruled out in the mean solution as a cohort of models was used. In the future, the cohort with adapted conduction velocities may provide a useful tool for the investigation of pathological and therapeutical impacts on the atrial excitation sequence. Similar approaches become more common in the Physiome environment, e.g. for the cross-validation of simulation environments [396]. An initial use of the model cohort is presented in Chapter 10.

To the knowledge of the author, the present study is first to quantitatively compare simulated and measured ECG signals from the same individuals. Previous studies may have shown better correspondence of measured and simulated signals [230], but in these studies the atria model, the thorax model and the ECG data were derived from three different individuals. The predictive power of such adapted simulations is therefore limited. In the future, the adjusted global conduction velocity could be used as an initial estimate for ECG imaging (ECGI) methods [287]. Additionally, ECGI could provide information about atrial conduction pathways which could be fed into the model personalization process. Analysis of the atrial activation sequence derived from the adjusted models can provide non-invasive insights into the patient's conduction velocity. These results are shown and discussed in the next Section.

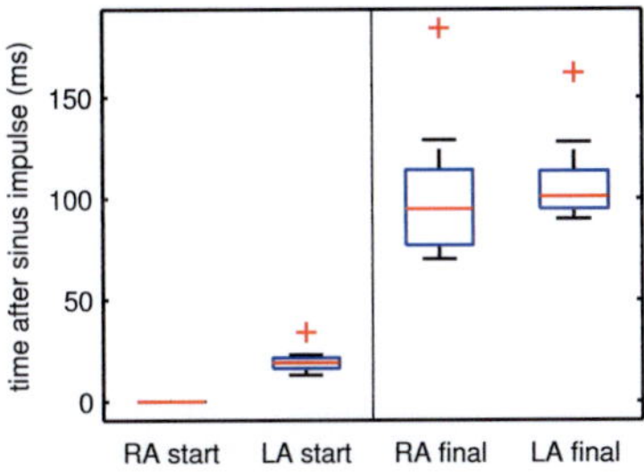

Fig. 9.14. Time of first and final activation of the left and right atrium in eleven models with personalized global conduction velocity.

9.3 Analysis of Conduction Velocity in Personalized Simulations

Simulations with personalized global conduction velocity allowed an insight into the bi-atrial depolarization sequence of the individual patient without the need of invasive measurements. In this section, local conduction velocities in personalized simulations and respective measured LAT maps are analyzed and compared. Local conduction velocity was calculated from the LATs generated from the PWd-adapted simulations for subjects 1–7 and from adapted simulations and measured left atrial LATs in subjects 8–11. Conduction velocity was calculated as described in Section 4.4.

Analysis of Simulated Local Activation Times

In average, LA activation started 19.9 ± 5.5 ms after the sinus node impulse. Right atrial activation was completed after 101.7 ± 33.0 ms. LA activation was completed 88.5 ± 17.0 ms after first LA activation respectively 108.4 ± 21.0 ms after the sinus node impulse. Figure 9.14 shows box plots of the distributions of the first and final activation of both atria in the 11 models.

Analysis of Simulated Conduction Velocity

Conduction velocities simulated with standard values for the tissue conductivity varied between 0.60 m/s and 0.79 m/s. The adjusted conduction velocity showed a greater range between 0.56 m/s and 0.98 m/s. Table 9.3 summarizes the conduction velocities in the 11 models. Conduction velocities in the adjusted simulations

Table 9.3. Mean conduction velocities in the models of eleven individuals (data from right and left atrium). Boxplots of the conduction velocity distribution in each adjusted simulation are shown in Fig. 9.15.

Subject	CV_{init} m/s	$CV_{adjusted}$ m/s
1	0.67 ± 0.15	0.62 ± 0.14
2	0.69 ± 0.17	0.86 ± 0.21
3	0.76 ± 0.16	0.82 ± 0.17
4	0.63 ± 0.12	0.70 ± 0.14
5	0.79 ± 0.15	0.64 ± 0.12
6	0.63 ± 0.16	0.61 ± 0.15
7	0.78 ± 0.16	0.89 ± 0.18
8	0.60 ± 0.12	0.56 ± 0.11
9	0.58 ± 0.12	0.98 ± 0.19
10	0.67 ± 0.13	0.80 ± 0.16
11	0.66 ± 0.13	0.69 ± 0.14

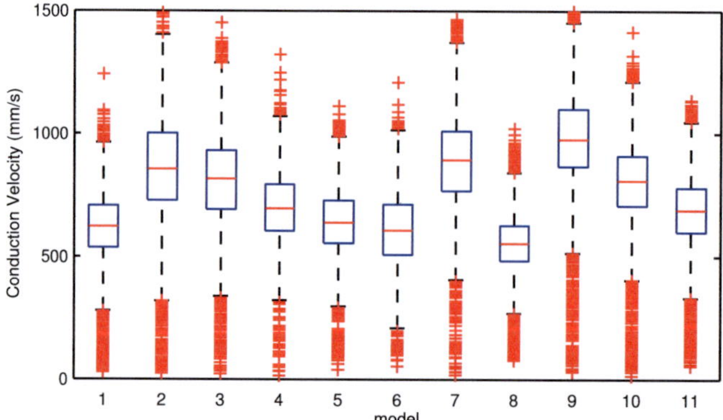

Fig. 9.15. Estimated conduction velocities in the models of eleven individuals. The global conduction velocity was adapted such that the measured PWd was met by the simulation. Mean conduction velocity values are listed in Tab. 9.3.

showed a significant variance. Figure 9.15 provides box-plots of this intra-model variation of conduction velocity. The conduction velocity magnitudes were distributed Gaussian like around the mean conduction velocity in the simulations.

Comparison of Measured and Simulated Conduction Velocity

For subjects 8–11, also the conduction velocities in the LAT maps (Sec. 9.1.1) were analyzed. Conduction velocities showed a great variance within each LAT

Table 9.4. Simulated and measured conduction velocities in the left atrium of four AF patients. Boxplots of the conduction velocity distribution are shown in Fig. 9.17.

Subject	$CV_{LA,adjusted}$ m/s	$CV_{LA,meas}$ m/s
8	0.52 ± 0.11	0.78 ± 0.46
9	0.97 ± 0.20	0.77 ± 0.50
10	0.82 ± 0.16	0.64 ± 0.45
11	0.70 ± 0.13	0.84 ± 0.51

map (Fig. 9.16). The mean values of the conduction velocity determined from the activation times are listed in Table 9.4. Histogram plots of the conduction velocity magnitude revealed a single sided exponential distribution for the measured conduction velocities. The mean measured conduction velocities correspond only to a limited extent to the mean simulated conduction velocities, which showed a Gaussian distribution. Nevertheless, mean simulated conduction velocities lay within one standard deviation of the measured mean conduction velocities. Boxplots of the measured and simulated conduction velocities highlight the great variance of the measured conduction velocities (Fig 9.17). Simulated and measured conduction velocities in subject 11 fit well, whereas median simulated conduction velocities in the other subjects only lay within the 25th to 75th percentiles of the measured conduction velocities.

9.3.1 Discussion

In this section, LATs and local conduction velocities from eleven models and respective measurements were analyzed. This revealed some interesting findings.

First, conduction velocity in simulations with literature values for the tissue conductivity fit into the range of measured atrial conduction velocities (Tab. 2.2). This shows, that for atrial models, for which no ECG data is available, the literature values for the tissue conductivity provide a good estimate of the atrial conduction velocity. Subjects 8 and 9 showed the lowest mean conduction velocity in the literature value simulations. In these subjects, the CRN model was adjusted to reflect electrical persistent-AF remodeling and the extracellular ion concentrations in the model were adjusted to respective measurements (Sec. 8.2). This accounts for the slow conduction velocity in these simulations. Similar low conduction ve-

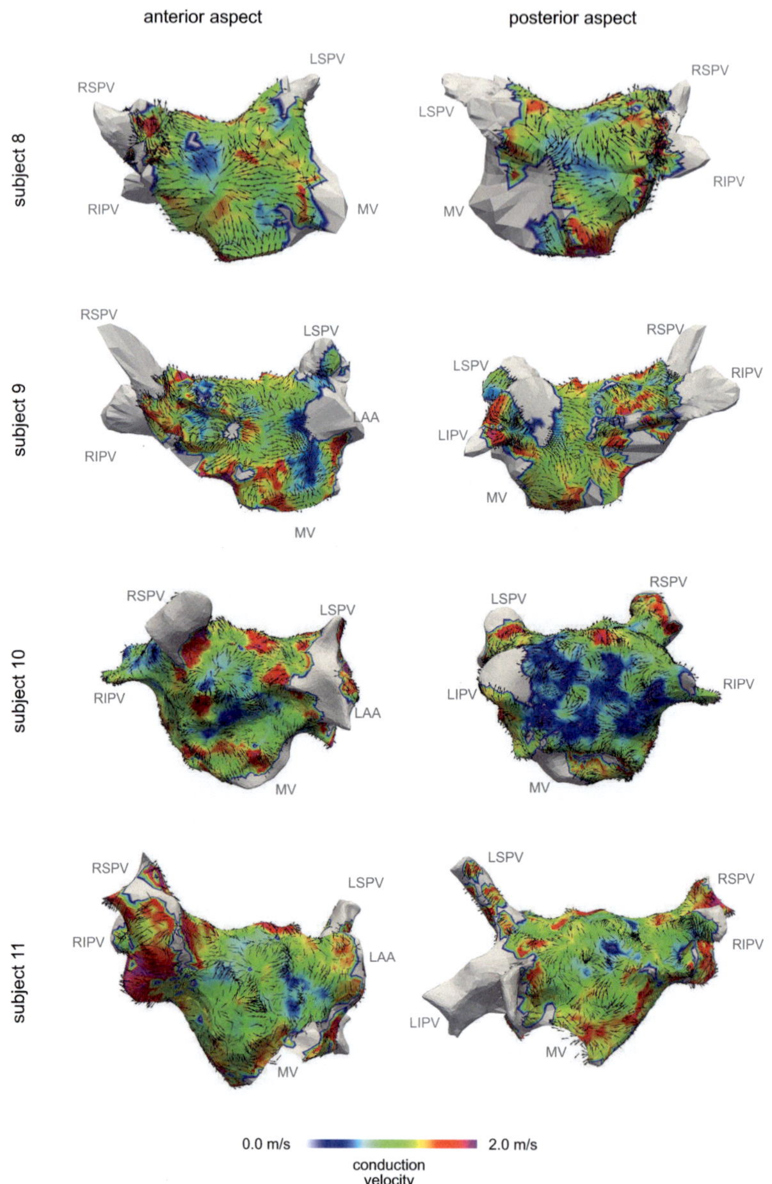

Fig. 9.16. Local conduction velocity magnitude and direction derived from left atrial LAT maps of patients 8–11 (Tab. 4.1). The conduction velocity magnitude is shown on the endocardial surface extracted from the EAMS. The excitation direction is shown as black arrows overlayed on the endocardial surface.

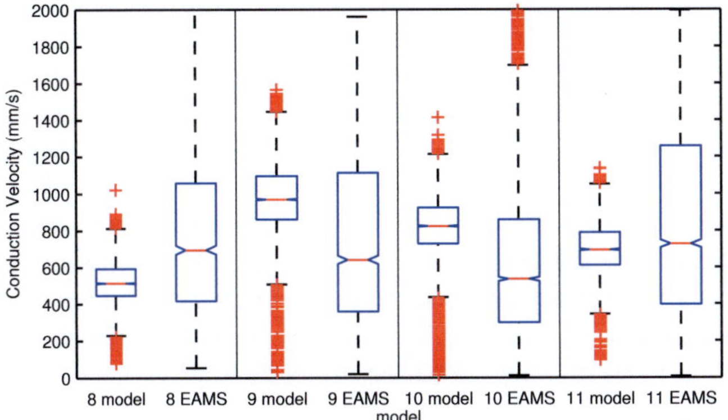

Fig. 9.17. Estimated and measured conduction velocities in the left atrial models of four AF patients. Mean conduction velocity values are listed in Tab. 9.4.

locities are commonly observed in patients with atrial flutter and atrial fibrillation (Tab. 2.2).

Second, simulated conduction velocities showed variations within each simulation. These variations were caused by differences in extent and size of fast conduction pathways, variable local fiber orientation and wavefront incidence angle as well as differences in local surface curvature. E.g. in region where fibers were well aligned, conduction velocity was higher than in regions with frequent change in fiber orientation, although the tissue conductivity was the same in both regions. Similar variation in conduction velocity was also observed in the measurement data. This shows that an adaptation of the global conduction velocity in the simulations adjusts the global activation time on the one hand. On the other hand, the adjusted simulations might provide an insight into abnormal regional conduction velocities.

Third, conduction velocity calculations from measured LAT maps were prone to errors. For one, errors in location activation time (Sec. 9.2.1) led to very high local conduction velocities. Furthermore, in areas where activation time was constant, e.g. in PVs were measurement data were missing, conduction velocity was determined to be zero. The majority of these artifacts were canceled out prior to the conduction velocity computation by removing the according LAT / coordinates pairs from the calculation. Nevertheless, some erroneous measurement values remained present in the data. Secondly, the geometry of the LAT maps might be distorted. Besides the rough outer shape of the LAT maps, in some cases the parameters for a correct 3D localization of the surface were not set in the EAMS (field-scaling, e.g. subject 9). This flattens the geometry between the anterior and posterior side which results in larger conduction velocities at the left atrial roof, as this area is squeezed. For future LAT map acquisitions, the EAMS geometry should be merged with CT or MRI morphological segmentations of the atria and a high density of sampling points should be used. Furthermore, an automated detection of local tissue activation would remove user-introduced errors. If these steps are taken, a direct parameterization of the model with measured local conduction velocities could be performed. This would create a tool for the model-based analysis of local tissue structure and the planning of RFA interventions.

Part IV

Applications

Essentially, all models are wrong, but some models are useful.

George E.P. Box

Study of Atrial Repolarization in the ECG

The ECG signal of the atrial repolarization sequence is usually obscured by the signal of the ventricular depolarization (QRS complex). In patients with AV-block, the atrial repolarization signal can be recorded in the ECG (Ta-wave, Sec. 2.4.1). Computational modeling provides a tool to get a clearer look at the atrial repolarization sequence in the ECG and thus gain a deeper understanding of the repolarization sequence in healthy and diseased individuals. In this chapter, results of the simulation of the atrial depolarization and repolarization sequence in a cohort of models are presented.

10.1 Simulation Setup

Sinus rhythm activity (BCL=800 ms) was simulated in eight atrial models and then forward computed in thorax models of the same individuals to retrieve ECG and BSPM signals (models 1–8, Tab. 7.1). For the simulations, the heterogeneous electrophysiological model of Courtemanche et al. described in Section 5.2 was used. Cell coupling in the atria was calculated using the monodomain equation. Simulated transmembrane voltages were forward computed on finite element thorax models as described in Section 3.4.

To validate the simulation outcome, ECG data from four AV-block patients were recorded at the Department of Cardiology, Lund University, Lund, Sweden. The patients were awaiting implantation of permanent cardiac pacemakers and had no history of other heart disease and were not taking anti-arrhythmic drugs [404]. ECG signals over at least 10 seconds were recorded with a standard 12-lead ECG system. The signals were high-pass and low-pass filtered afterwards and then a

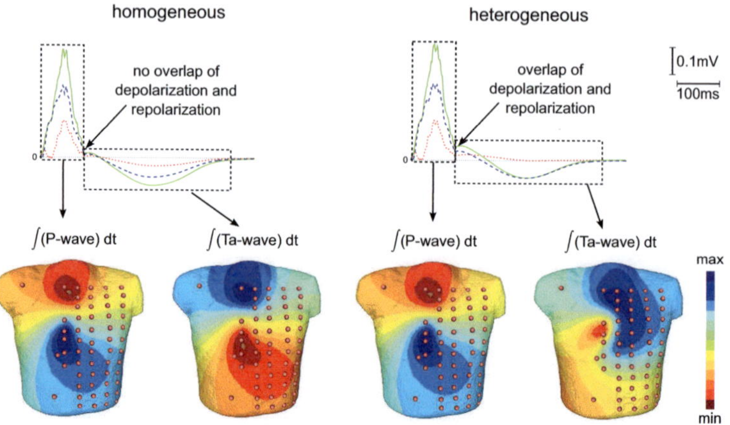

Fig. 10.1. Representative example of atrial ECG signal and BSPM in a healthy model with and without electrophysiological heterogeneities. Figure adapted from [295].

signal averaging was performed [154]. P-wave and Ta-wave onset and offset were marked manually by the clinical experts. The data was provided by Professor Pyotr Platonov and colleagues (Department of Cardiology and Center for Integrative Electrocardiology, Lund University, Sweden).

10.2 Results

10.2.1 Homogeneous and Heterogeneous Electrophysiology

In a first set of simulations (2x8), the influence of the modeled electrophysiological heterogeneities on the BSPM and ECG were investigated. In the homogeneous simulations, the Ta-wave started right after the P-wave reached the isopotential line again. This indicates that the repolarization starts after both atria were completely activated. From the standard leads, Einthoven II had the strongest signal during the Ta-wave. The BSPM of the Ta-wave was precisely the inverted pattern of the P-wave (Fig. 10.1).

The P-wave signal and BSPM during the P-wave was not altered by the introduction of the heterogeneities, which was expected, as mostly repolarizing currents were altered. In the heterogeneous simulations, the dipolar pattern of the BSPM

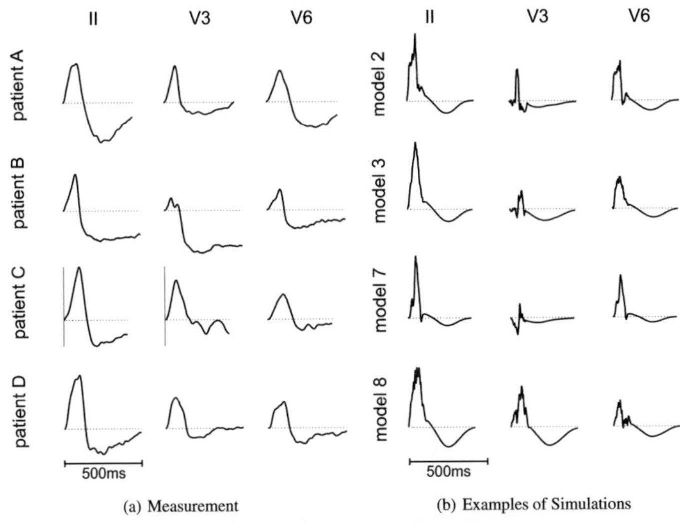

(a) Measurement											(b) Examples of Simulations

Fig. 10.2. Measured and simulated atrial ECG signals.

during the Ta-wave was twisted clockwise on the thorax front. The signal polarity in the standard ECG leads was inverted during the Ta-wave compared to the P-wave. The signals of Einthoven II and III showed a similar morphology and amplitude. The signal of Einthoven I had a significantly smaller amplitude compared to the other leads and also compared to the homogeneous simulations. The P-wave ending and Ta-wave beginning overlapped each other. This led to an offset from the isopotential line at the end of the P-wave. The amplitude of the Ta-wave in the signal of Einthoven II was smaller in the heterogeneous simulations compared to the homogeneous simulations (-0.047±0.025 mV vs. -0.062±0.035 mV), but the time of the maximum amplitude was not altered (125±50 ms vs. 123±50 ms).

ECG signals from the four AV-block patients showed inverted polarity of the Ta-wave compared to the P-wave in the vast majority of the signals. The P-wave and the Ta-wave signal were strongest in Einthoven II lead in three out of four patients. The simulated ECG signals had a morphology similar to the measured ECG signals (Fig. 10.2), although models and ECG measurements were taken from different individuals. P-wave and Ta-wave duration were longer in the measurements than in the simulations. For the simulations the monodomain tissue conductivity was chosen such that conduction velocity was similar to those used in stan-

Table 10.1. Impact of heterogeneities and electrical remodeling in the CRN model on the 3D excitation sequence and ECG. Corrected PTa (PTac) calculated as proposed by [150]. PWd: P-wave duration, TaWd: Ta-wave duration, RSb: begin of the repolarization sequence, RSe: time when last cell starts repolarization.

Model / Measurement	Heart Rate (bpm)	PWd (ms)	PTa (ms)	PTac (ms)	TaWd (ms)	RSb (ms)	RSe (ms)
healthy	75	101 ± 13	393 ± 20	441 ± 20	293 ± 18	75 ± 11	254 ± 16
[150]	75		435	483	316		
[152]			386	420	276		
[151]	70	124 ± 16	449 ± 55	512 ± 60	323 ± 56		
[404]	86 ± 15	129 ± 17	451 ± 53	519 ± 57			
persistent AF	75	109 ± 14	229 ± 15	277 ± 15	121 ± 5	67 ± 3	151 ± 17
paroxysmal AF [404]	73 ± 17	126 ± 15	408 ± 47	441 ± 68			

dard atrial models. Simulated Ta-wave duration and PTa interval, which is the sum of P-wave and Ta-wave duration and reflects the temporal summation of all atrial electrical events [150], fit well into the duration ranges reported in literature (Tab. 10.1).

10.2.2 Influence of Persistent AF Remodeling

Under persistent AF-remodeled conditions, the P-wave duration remained nearly unchanged in all eight models. The Ta-wave was significantly shorter than in the healthy case and its amplitude was larger (-0.074±0.048 mV vs. -0.062±0.035 mV). Besides the drastic shortening of the Ta-wave, the morphology was comparable to the healthy Ta-wave, e.g. Einthoven II and III signals were similar and Einthoven I had a near-zero amplitude. The observed overlap of the P-wave and Ta-wave was negated and strongly pronounced in the remodeling simulations (for an example see Fig. 10.5).

This strong overlap of P-wave and Ta-wave was caused by the early onset of the repolarization, which occurred while the atria were still depolarizing (Fig. 10.6, RSb<PWd in Tab. 10.1). Additionally, the intra-model APD variation was siginificantly reduced compared to the healthy case (Fig. 10.6(b)).

The BSPM during the Ta-wave showed a dipolar pattern between the upper chest and the bowel, centered at the median axis (Fig. 10.5). The area of negative potential spread out over the entire frontal hip in most simulations. P-wave pattern remained unchanged compared to the healthy simulations.

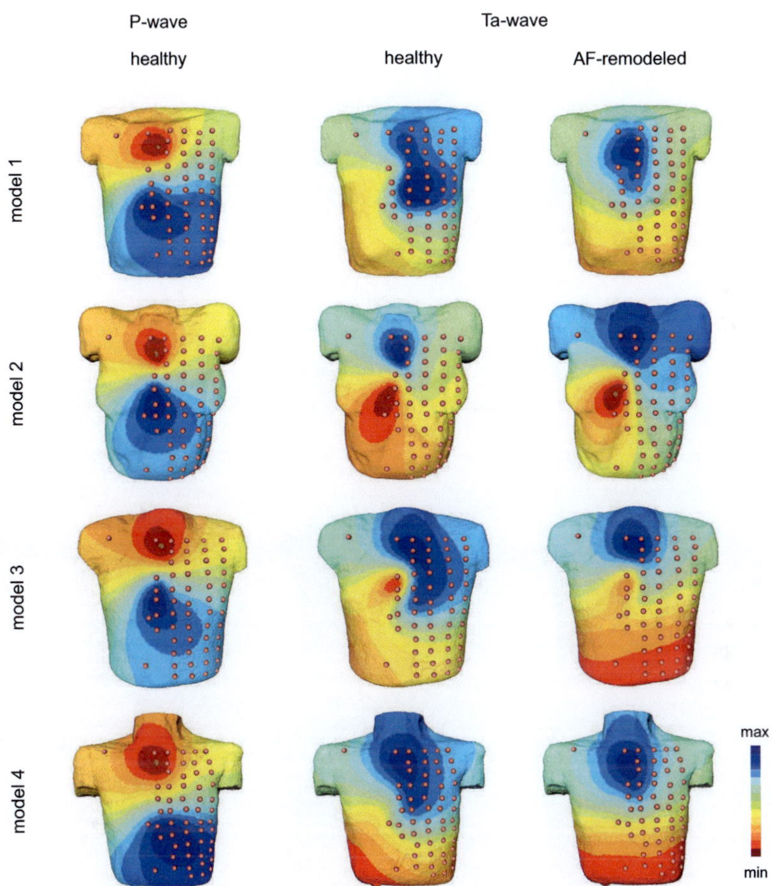

Fig. 10.3. BSPM of integrated P-wave and Ta-wave of models 1-4.

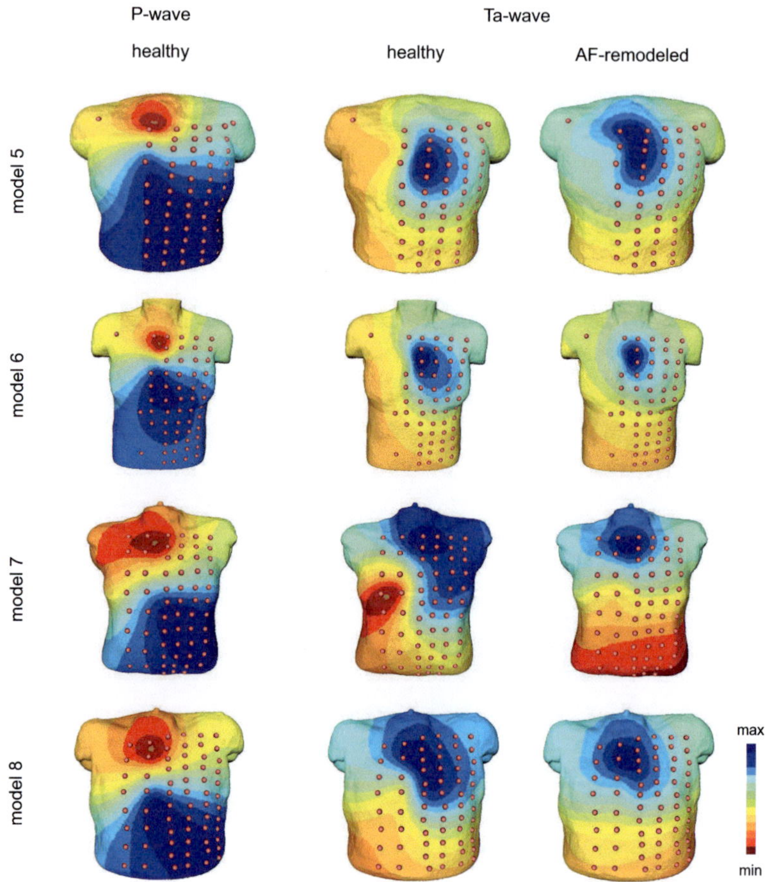

Fig. 10.4. BSPM of integrated P-wave and Ta-wave of models 5-8.

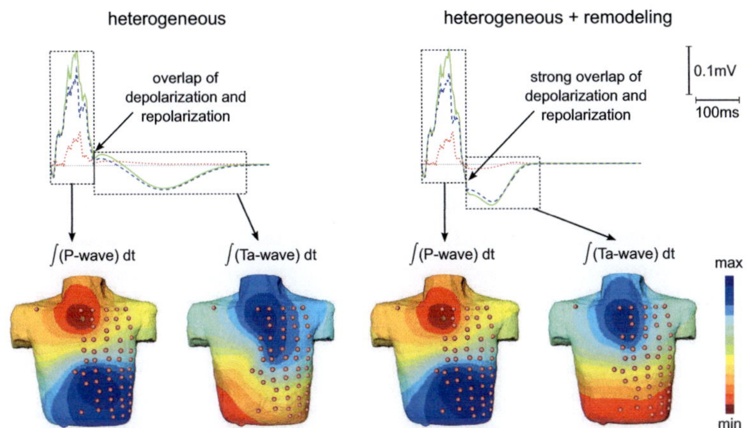

Fig. 10.5. Representative example of atrial ECG signal and BSPM in a healthy model and a model reflecting persistent AF electrical remodeling. Figure adapted from [295].

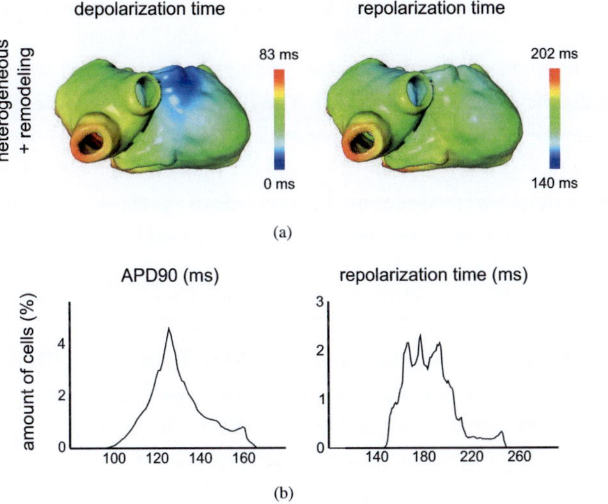

Fig. 10.6. a) Representative distribution of depolarization and repolarization time in a bi-atrial simulation (model 5). b) Representative histogram plots of the distribution of APD and repolarisation time (model 2). The variation in APD (width of histogram curve) is significantly reduced compared to the healthy simulations (Fig. 5.8).

10.3 Discussion

In this chapter simulations of the atrial depolarization and repolarization sequence and the corresponding ECG signals under healthy and persistent AF-remodeled conditions were presented. For this purpose a cohort of eight atria and thorax models were used. Modeling studies commonly use only one atrial model coupled with a generic thorax model or a thorax model from a different individual [218, 230]. This approach may introduce model-specific uncertainties in the results. The workflow described in Section 7 allowed for the creation of a cohort of volunteer-/patient-specific models in a reasonable time and thus enabled the multi-model simulation study. The variation in simulated ECG fiducial times (Tab. 10.1) of up to $\pm 13\%$ highlights that the use of only one model could have resulted in misleading simulation outcome.

The polarity and morphology of the simulated P-waves and Ta-waves were in agreement with the measurement data from four AV-block patients and to the findings reported in literature [151, 152, 404]: The polarity of the Ta-wave was inverted compared to the P-waves in the standard leads and the signal in Einthoven II had the largest amplitude for both the P-wave and Ta-wave. Lead Einthoven II covered the greatest potential variance during the P-wave and was thus aligned with the dipolar pattern of the BSPM. The BSPM pattern over the P-wave agreed with previously published measurement data [144]. The BSPM pattern over the Ta-wave indicated that a bipolar ECG lead between the suprasternal notch and the right hip, or unipolar electrodes on the right side of the chest could record Ta-wave signals with a larger amplitude than the regular 12 lead ECG in AV-block patients.

The simulated P-wave duration was shorter than the P-wave duration reported in the Ta-wave ECG studies (Tab. 10.1). For the simulations, conduction velocities were adapted to reflect velocities used in established models [91]. A decrease of the model conduction velocity will prolong the P-wave and thus provide a better match to the measurement data. This is likely to also resolve the difference in PTa duration. The simulated Ta-wave durations fitted into the range of measured values.

In the measurement data, Ta-wave duration decreased with faster heart rates ($323 \pm 56\,$ms (BCL $714\,$ms) [151], $316\,$ms (BCL $800\,$ms) [150], $276\,$ms (BCL $856\,$ms) [152]). Debbas et al. reported a linear relationship between the Ta-wave

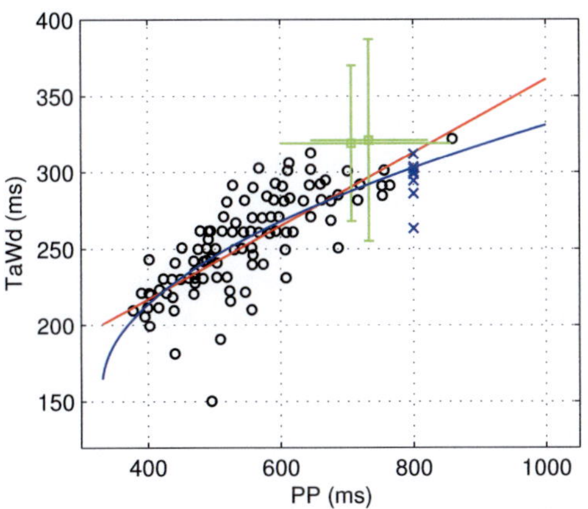

Fig. 10.7. Relationship of TaWd to PP interval. Black circles indicate data from Debbas et al. [150], green squares with error bars show data from Holmqvist et al. [151] and blue crosses mark the new simulation data. The red curve represents the linear fit of Debbas et al. [150] to the respective measurement data. The blue curve is a square root fit to the data from Debbas et al. and the simulation data.

duration and the pacing interval [150] which fits to previous findings from Hayashi et al. [152]. The non-linear restitution curves of APD and conduction velocity (Sec. 5.2) indicate a more complex relationship between the two measures. E.g. a square root equation in the form of

$$TaWd = \left(\sqrt{(PP - a)} * b \right) + c \tag{10.1}$$

with the unknown coefficients a, b and c was fitted to the measurement data from Debbas et al. and the simulation results using a least squares optimization. These coefficients were determined to be $a = 331.1$ ms, $b = 6.676$ ms and $c = 158.4$ ms. The RMSE between the fit and the data was 20.87 ms compared to an RMSE of the original linear fit of 21.15 ms. Figure 10.7 shows the data and the fits. The major difference between both assumptions can be observed at very high (>150 bpm) and regular sinus rhythm heart rates (<75 bpm). A shorter repolarization duration at high frequencies, as predicted by using equation 10.1, might contribute to the development or maintenance of AF.

In the heterogeneous simulations, a zero-offset at the end of the P-wave was observed due to a temporal overlap of the atrial depolarization and repolarization sequence, which concurs with *in-vivo* findings [150]. The homogeneous simulations were not able to reproduce this behavior.

Holmqvist et al. reported a decreased Ta-wave duration in patients suffering from paroxysmal AF [404]. The simulations under persistent AF-remodeled conditions showed a more pronounced shortening of the Ta-wave duration. As AF is a progressing disease [99], these findings accompany each other. The simulations provided the insight, that the shortened Ta-wave duration was caused by a combination of i) a more dense repolarization sequence (shorter APD in all regions) and ii) early repolarization while the depolarization was still ongoing.

The presented methodology to simulate the atrial repolarization sequence in the ECG using a cohort of anatomical models and the heterogeneous CRN model could be used for further investigations of the atrial repolarization sequence under healthy and pathological conditions in the future. E.g. the relationship between PP and TWd could be further explored and the influence of a reduced cell coupling (gap junctional remodeling) on the repolarization sequence could be investigated. The knowledge gained from the simulations could be used to better understand the onset and maintenance of atrial arrhythmias due to repolarization disturbances.

Simulation of Atrial Fibrillation

Atrial models have been used to investigate the mechanisms leading to atrial arrhythmias and atrial fibrillation for a long time [2]. The vast majority of the utilized models were 3D surface or shell models with a rather small number of computational nodes. Additionally, most often the models were tuned to artificial conditions to initiate wave-breaks, reentries and rotors (very slow conduction velocity, shorter refractory period, sudden regional variations in electrophysiology etc.). Initiating atrial fibrillation by imitating the pathophysiological initiation process, e.g. by ectopic firing in the pulmonary veins [100], is becoming more popular since the models are better understood (Tab. 11.1). Surface models have a reduced demand in computational power and allow longer simulations. With the constant increase in available computational power, also the simulation of AF in atrial models with a finite wall thickness is becoming feasible. Beforehand, such simulations could only be performed using rule-based simulation approaches (e.g. [257]). These are usually unsuitable, as the discretization artifacts coming along with this approach play a significant role at high excitation frequencies and might produce false results.

Table 11.1. Initiation of AF in 3D atrial models. ES: ectopic stimulus.

study	year	EP model	AF-init	comments
Gong et al. [206]	2007	CRN	ES-train (10) at PVs	
Plank et al. [94]	2008	CRN	ES at RIPV	LA only
Reumann et al. [187]	2008	CRN	2x overlapping ES at PVs	
Deng et al. [405]	2010	CRN	ES and cross-field	
Tobon et al. [93]	2010	Nygren	ES in 6 different locations	
Uldry et al. [406, 407]	2010	Luo-Rudy		
Lu et al. [217]	2011	Nygren	ES at RSPV and IVC	
Aslanidi et al. [218]	2011	CRN	ES between CT and PM	
Zhao et al. [96]	2012	Fenton-Karma	ES at PVs	

Fiber orientation plays an important role in the maintenance of AF. Especially in regions with high anisotropy, e.g. fibrotic tissue, the underlying fiber orientation contributes significantly to the development and maintenance of the arrhythmia [94, 96]. Fiber architecture is organized in complex, multi-layer patterns in the human atria (Sec. 2.1.1.3). This structure cannot be reproduced by 3D surface models. Furthermore, Allessie and co-workers recently presented evidence for transmural reentry [408] caused by endo-epicardial dissociation [409–411]. Such phenomena may not be reproduced in shell models. In this chapter, an example of the simulation of atrial fibrillation in a volumetric atrial model under pathophysiological initiation conditions is presented.

11.1 Simulation Setup

The model of subject 4 (Tab. 7.1) was chosen for this simulation for no particular reason, except for the availability of a thorax model from the same individual. At the time of the study, the data of subject 8, an AF patient, was not processed yet, otherwise this model would have been chosen. The model contained complete rule-based fiber orientation (Sec. 5.4) and heterogeneous electrophysiological properties based on the CRN model (Sec. 5.2). Furthermore the CRN model was adjusted to reflect persistent-AF electrical remodeling (Sec. 3.2.2). The monodomain anisotropy ratio was set to 10:1 (CV anisotropy 3:1) in accordance with previous volumetric modeling studies [94, 96]. Such anisotropy ratio is thought to reflect fibrotic atrial myocardium. The longitudinal tissue conductivity was reduced to 40% of the regular value [216]. A similar approach has been chosen before in other AF simulation studies [412]. The resulting reduction of conduction velocity was approximately 65%. Simulations were carried out on a Mac OS X 10.6 computer with 8 x 2.8GHz Intel Xeon processors and 32 GB RAM using a C++ simulation framework based on PETSc [413]. The problem was discretized using finite elements and solved using the forward Euler approach with a time step of $20\,\mu$s. Prior to the simulations, the nodes in the model were initialized with 10 beats of single cell pre-calculation at regular sinus rhythm frequency 75 bpm (BCL=800 ms). Ten seconds of atrial activity were simulated and subsequently forward-calculated every millisecond to the body surface ECG. Computation took approximately five days for the simulation of the atrial excitation and another seven days for the ECG computation.

Table 11.2. Regular and sequence of ectopic stimuli during the simulation.

stimulus	time	site	tissue under stimulus	excitation outcome
0	0.000	sinus node	excitable tissue	sinus rhythm excitation
1	0.227	junction RSPV/LA	partly refractory tissue	reentry (center moves from origin to left anterior LA)
2	2.027	junction RSPV/LA	partly refractory tissue	reentry (center meanders around stimulus location)
3	3.827	junction RSPV/LA	partly refractory tissue	reentry (same location as 2, then moves into RSPV, center moves out of PV after 1.1s)
4	5.627	junction RSPV/LA	depolarized tissue	none
5	7.427	junction RSPV/LA	excitable tissue	none
6	9.227	junction RSPV/LA	refractory tissue	figure of eight reentry / two rotors (one in RSPV, one at junction of RSPV and LA

At first, regular excitation was initiated at the site of the sinus node in the superior right atrium. Then, ectopic activity was initiated next to the RSPV with a BCL of 1.8 s. The location of the stimulus was chosen based on the excitation pattern and the anatomical structures. The location of the ectopic activity lay on the border of the PV and the left atrium. Furthermore, two wavefronts collided in this region during sinus rhythm activation providing a repolarization wavefront, which allowed excitation propagation in only one direction. The time of the first ectopic activity was determined as the time where the underlying tissue repolarized to -75 mV. The ectopic stimulus was spherical with a radius of 3 mm.

11.2 Results

Right atrial activation was dominated during regular excitation and during the arrhythmia phase by fast conduction along the major muscular bundles (BB, CT, PM). The first ectopic beat initiated a stable rotor in the left atrium, which moved from the initiation location towards the anterior left side of the left atrium. The other ectopic stimuli party caused a second rotor. These rotors meandered around the location of the ectopic activity until the rotor center moved out of the PV tissue. No ectopic stimulus initiated a third rotor. Stimuli 4 and 5 did not cause a rotor at all, but stimulus 6 resulted in a figure of eight reentry. Table 11.2 summarizes the findings. Figure 11.1 shows the transmembrane voltage distribution at the time of and after each ectopic stimulus.

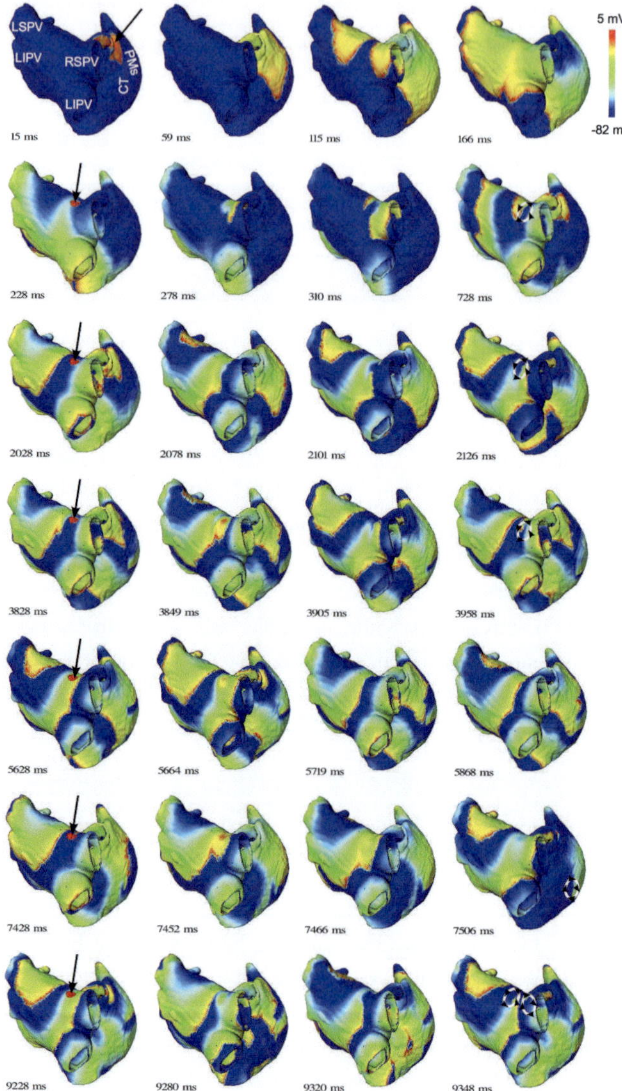

Fig. 11.1. Example of ectopic foci triggering atrial fibrillation. After one sinus node activation, ectopic activity was stimulated at the RSPV (BCL 1800 ms, e.g. 242 ms). Multiple rotors developed. The most stable and largest rotor was created from the first ectopic beat and manifested at the anterior left atrial roof. The left atrial rotor therefore drove the right atrial activation.

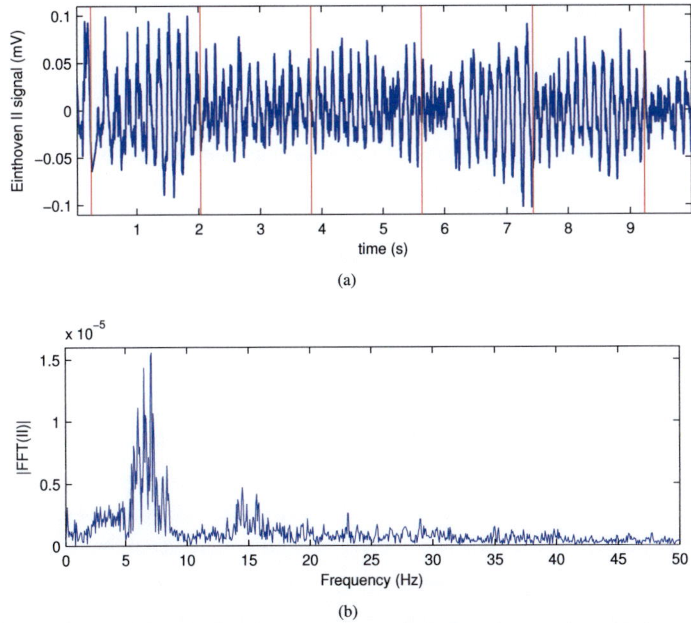

Fig. 11.2. a) Computed Einthoven II ECG signal from the AF simulation. Vertical red lines indicate times of ectopic stimuli at the RSPV. b) Power density spectrum of the fast Fourier transform of the ECG signal.

During the arrhythmia phase additional reentries developed spontaneously in the right atrium temporarily. One reentry (2.5 - 3.0 s) moved around the border of the CT and the PMs. Another one developed along the inferior end of the CT (6.3 - 8.1 s). Further stationary rotors developed and diminished at the anterior junction of pectinate muscles and the TVR, where fiber orientation changes abruptly.

The forward-calculated Einthoven II ECG signal is shown in Figure 11.2 along with the power-density spectrum calculated from the fast Fourier transform of the signal. The frequency with the maximal amplitude was 7.14 Hz, which corresponds to 428 bpm or a cycle length of 140 ms. The amplitude of the signal was reduced after the second ectopic stimulus and further reduced after stimuli 4 and 6.

11.3 Discussion

In this chapter, an example of the initiation of atrial fibrillation by single ectopic activity in a volumetric 3D biatrial model was presented. The reduction in tissue conductivity was greater than the conductivity reduction commonly used to model electrical remodeling ($\sigma_{remod} = 0.7 \times \sigma_{healthy}$ e.g. [73]). With tissue conductivity reduced to 70% it was not possible to induce reentry in the model. The conduction velocity was similar to the conduction velocity observed in AF patient 8 (Fig. 9.17). The dominant frequency observed in the body surface ECG was in the range of clinically observed frequencies during persistent AF [414].

Notably, rotor centers of stable rotors manifested at borders of adjacent tissue regions (LAA–LA, TVR–RA, CT–PM). In these areas an abrupt change of fiber orientation and cellular electrophysiology was present. These findings coincide with findings from simulations studies, which induced AF more easily in heterogeneous models [68, 69]. The use of gradients to ensure a smooth transition of electrophysiological properties between neighboring regions counteracted the manifestation of rotor centers in the present study.

The simulation took a considerable amount of time. The present approach does therefore not allow for a systematic investigation of the influence of different model parameters. The use of unstructured grid models would reduce the computational load, as bath nodes may not need to be included into the computation. The use of simplified excitation propagation models which can still reproduce diffusion and repolarization [262–264] as well as the use of phenomenological electrophysiology models [240] would further speed up the simulations in the future, once all relevant model parameters are understood and a reduction of the models becomes feasible.

The cells in the model were initialized at a regular sinus rhythm frequency. Initialization with higher pacing rates could yield an easier initiation of reentries, as the ERP and conduction velocity will be reduced (Sec. 5.2). Similar approaches are also undertaken by other groups (BCL=230 ms [218], BCL=280 ms [406], BCL=375 ms [94]). Also other stimulus location could be investigated, e.g. other PVs [94, 187], the caval veins [217], between the CT and PM [218] and at various locations across the atria [93, 415].

Besides fiber orientation, also a method to introduce sheets of myocardial tissue was introduced in Section 5.4. Up-to-now, conduction properties were set isorotational along and across fiber orientation. Using the information about local sheet orientation enableds the use of variations in conduction properties within a sheet and between sheets. This could e.g. be used to further investigate transmural dissociation and reentries as proposed in the literature [408–411].

In the future, heterogeneous tissue properties reflecting regions of increased fibrosis could be introduced based on LGE-MRI measurements from AF patients (Sec. 13.2.3). This could favor wavebreaks and produce more stable arrhythmia. Also, all prerequisites are set to model the atrial deformation during AF (Chap. 14). This could provide new insights into mechanisms leading to thrombus development during AF.

The performance of the AF simulation framework can be increased in various aspects. A decrease in computation time in conjunction with the existing anisotropic heterogeneous atrial models will allow a systematic investigation of anatomical and electrophysiological parameters with respect to the initiation and maintenance of AF in the future. This will provide new insights into the mechanisms leading to AF.

12

Investigation of Conduction Block in Ablation Lesion Gaps

Radio-frequency ablation of atrial fibrillation shows only moderate long-term success rates after the first RFA intervention [1]. Most often gaps between ablation lesions develop over time and reverse the initial termination of the arrhythmia. Gaps in circumferential PVI can occur nearly anywhere, but are slightly more commonly present at the LPVs compared to the RPVs [111]. The isthmus area between the LIPV and the mitral valve was shown to be especially prone to gaps [111]. During the initial lesion formation process, ablation lesions are wider compared to the chronic state of the lesions, as explained in Section 2.3. Initially, necrotic, edema and inflammatory regions are caused by the application of the radio-frequency current. Over a period of several weeks, the damaged non-necrotic tissue partly regenerates and partly forms scar tissue. Ablation lesions, which are initially electrically isolating can form electrical pathways during this healing period. Additionally, differences in cell electrophysiology caused by RFA, e.g. shorter APD

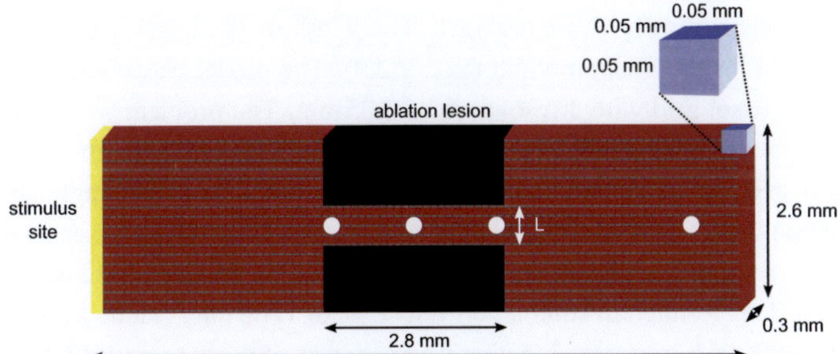

Fig. 12.1. Setup of simulation environment to investigate lesion gap conduction under different electrophysiological and tissue conditions. L is the variable lesion gap width. White circles indicate the position of voltage sensors. Dashed grey lines indicate longitudinal fiber direction. Yellow tissue indicates the stimulation zone.

Table 12.1. Simulations settings for the investigation of lesion gap conduction.

	tissue	EP model	fiber orientation w.r.t. to gap direction
A1	RA	physiological	longitudinal
A2	LA	physiological	longitudinal
A3	PV	physiological	longitudinal
A4	RA	physiological	transversal
A5	LA	physiological	transversal
A6	PV	physiological	transversal
B1	RA	physiological	isotropic
B2	LA	physiological	isotropic
B3	PV	physiological	isotropic
B4	RA	AF-remodeled	isotropic
B5	LA	AF-remodeled	isotropic
B6	PV	AF-remodeled	isotropic

near the edge of a lesion [126], may play a role in the recurrence of AF. In this chapter, a simulation study investigating the conduction behavior of lesion gaps under different anatomical and electrophysiological conditions is presented. The effects of fiber orientation underlying the lesion gap, the atrial region (Sec. 5.2) and electrical remodeling due to AF (Sec. 3.2.2) were investigated.

12.1 Simulation Setup

A geometrical model representing a plain patch of atrial tissue with an idealized ablation scar (zero conductivity, sharp edges) was created. Four sensors were placed in the atrial tissue at which the action potential was recorded. The setup is shown schematically in Figure 12.1. The width of the lesion gap was within the range of ablation lesion width (Sec. 2.3.3). The model was constructed on a structured voxel grid with a resolution of 0.05 mm. The problem was discretized using finite differences and solved using a semi-implicit Euler scheme [416] with a timestep of $20\,\mu$s. Voxels were initialized with 50 beats of single-cell pre-calculation before the tissue simulation was started. Monodomain tissue conductivity was set to 0.176 S/m in fiber direction and 0.062 S/m perpendicular to fiber direction. This resulted in conduction velocities of 1200 mm/s along fiber orientation and 680 mm/s transversal to fiber orientation at sinus rhythm (75 bpm) in right atrial tissue. The tissue patch was stimulated at one side with 240 pA/pF for 1 ms. The value was determined as twice the rheobase of the right atrial tissue patch setup.

Atrial electrophysiology was simulated using the CRN model for right atrial, left atrial and pulmonary vein tissue (Sec. 5.2) in regular and persistent-AF remodeled state (Sec. 3.2.2). Reduced tissue conductivity from electrical remodeling [73] was not considered to isolate the effect of altered cell electrophysiology. Simulations were performed with 14 different pacing cycle lengths ranging from 50 ms to 1000 ms. For each setup, gaps of varying width in the ablation lesion were introduced. 25 gap widths were investigated (0.05–2.0 mm) for 12 different anatomical situations (Tab. 12.1). Overall 2520 simulations were performed. Afterwards, the transmembrane voltage signals of the sensors were evaluated to determine the cross-gap conduction and the conduction velocity in the gap. Voxel activation was thereby determined as the time when the transmembrane voltage during the action potential upstroke increased above -20 mV. The threshold was chosen in a way that changes in transmembrane voltage caused purely by diffusion did not lead to an activation detection.

12.2 Results

Comparison of the behavior of different atrial tissue types revealed that right and left atrial tissue were slightly less prone for cross-gap conduction compared to pulmonary vein tissue (Fig. 12.2(a)). Fiber orientation had a greater influence on the gap conduction. Gaps with transversal fiber orientation showed a cross-gap conduction at gaps of width of 0.55–0.6 mm compared to 0.35–0.4 mm with longitudinal fiber orientation. Electrical activity with pacing cycle length (PCL) below 400 ms showed a better gap isolation in all cases. Fiber orientation also influenced the wave-front curvature of the wave exiting the lesion gap (Fig. 12.3).

Electrical remodeling of the cell electrophysiology increased gap isolation width by approximately 0.1 mm (Fig. 12.2(b)). Additionally, the edge of the PCL dependent isolation variation was shifted towards shorter PCLs under the influence of electrical remodeling (200 ms vs. 400 ms) due to shorter APD and ERP in remodeled tissue (Sec. 5.2). The differences between tissue types with respect to the lesion gap conduction remained present under electrical remodeling. Ablation lesion width and thus the length of the gap corridor did not alter the results. Pacing at cycle lengths shorter than the tissue ERP did not result in an excitation wave, but diffusion raised the membrane potential slightly in the proximity to the stimulus location.

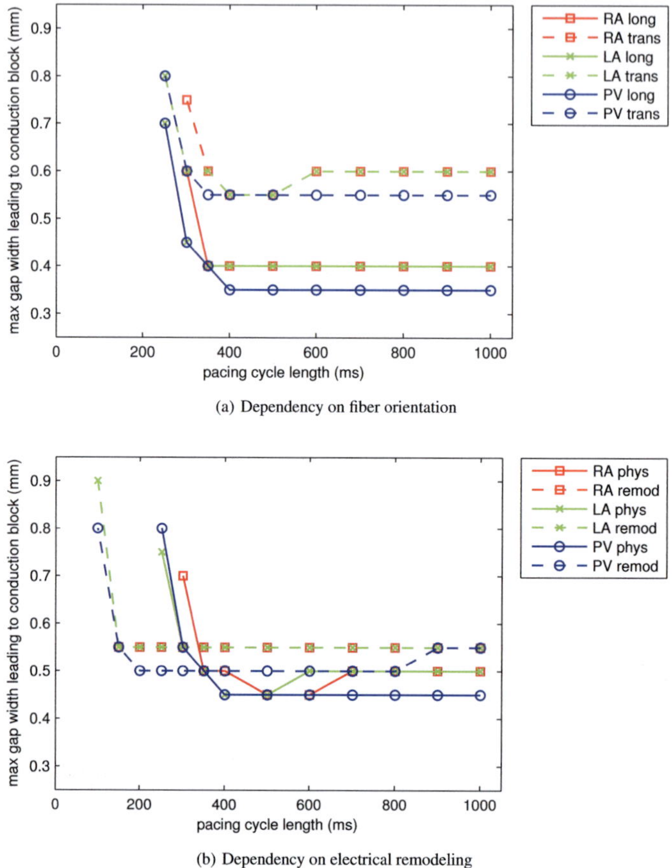

(a) Dependency on fiber orientation

(b) Dependency on electrical remodeling

Fig. 12.2. Dependency of conduction over a gap in an ablation lesion on fiber orientation (a) as well as on electrophysiological tissue properties (b). The area under each curve indicates configurations which results in conduction block across the lesion gap. The area above each curve represents configurations which results in cross-gap conduction. Red: right atrium, green: left atrium, blue: PVs. a) Solid: longitudinal, dashed: tranversal fiber orientation. b) Solid: healthy, dashed: persistent AF remodeling.

In cases where the excitation wave did not reach the tissue behind the incomplete lesion, the tissue within the lesion gap was always activated: The excitation block occurred at the outlet of the lesion (Fig. 12.1, left). In cases where the lesion gap was wide enough to allow an activation of the tissue behind the lesion, the con-

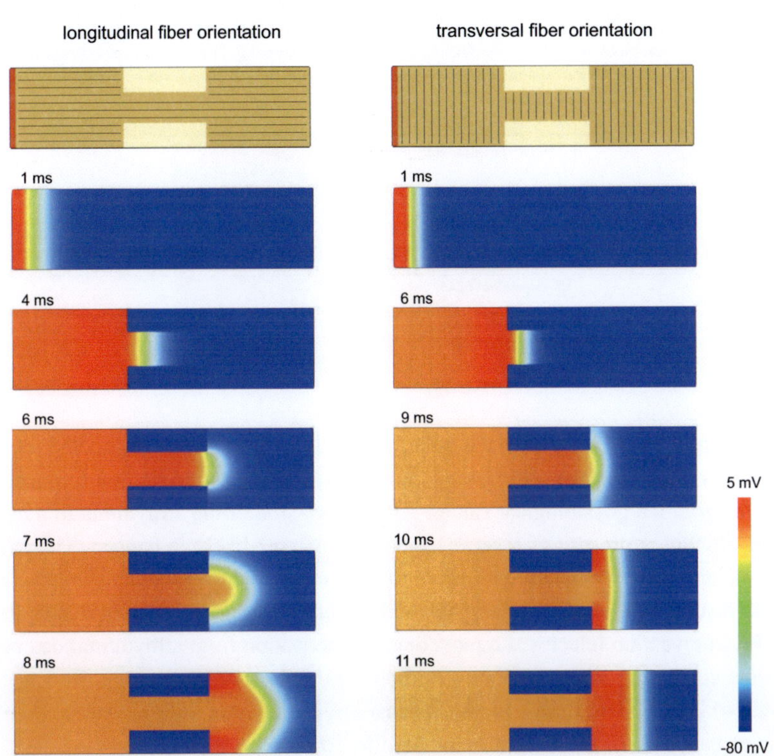

Fig. 12.3. Example of the effects of fiber orientation in the tissue patch (pulmonary vein tissue, 1000 ms BCL, 1 mm gap). Shown are transmembrane voltages at different time instances during the tissue patch activation. Wave-front curvature after exiting the gap differs significantly between the two cases. Please note the different time instances between the two cases caused by different conduction velocity in longitudinal and transversal fiber direction.

duction velocity over the lesion gap was slowed down in narrow lesion gaps. The main findings are summarized in Table 12.2.

12.3 Discussion

In this chapter an analysis of the effects of various factors on the conduction across gaps in ablation lesions was presented. The results might serve as one explanation for the recurrence of AF after RFA therapy and thus for the moderate long-term success rates in contrast to short term success [1].

Table 12.2. Main findings of the simulation study. Red parameters increased the maximum gap width at which conduction block was observed, thus more gaps were conducting. Green parameters decreased the maximum gap with at which conduction block was observed, thus the setup was less vulnerable for lesion gap conduction. EP: electrophysiology.

region-dependent block	pulmonary vein
	right atrium
	left atrium
fiber-dependent block	longitudinal (high conductivity)
	transversal (low conductivity)
EP-dependent block	physiological / normal
	electrical remodeling
rate-dependent block	sinus rhythm
	arrhythmia

At the time of RFA therapy, the atrial myocardium is in an AF-remodeled state. Thereby, the cell electrophysiology, tissue conductivity and excitation frequency differ from the physiological values. After conversion of the arrhythmia to a regular rhythm, a process of reverse remodeling is most likely to happen. Thereby, tissue and cell parameters will return to physiological values again. The simulation results indicate that lesion gaps which were electrically isolating during the RFA intervention (electrical remodeling, fast activation from arrhythmia) can become conductive under healthy conditions (normal electrophysiology, slow acitvation) independent from structural changes of the ablation lesions. This could be an explanation for the recurrence of AF after RFA procedures.

Ranjan and colleagues [110] proposed a recovery of gap conduction based purely on an increase in tissue conductivity during the reverse remodeling process. This could be caused by a recovery of gap junction coupling and / or by healing of inflammatory tissue. In the monodomain model, transversal fiber orientation is realized as a reduced tissue conductivity. The transversal, low conductivity setup showed in the present simulations also a better gap conduction block, which supports the hypothesis from Ranjan et al..

Thomas et al. [417] showed an effect, that with isthmus cross sectional areas below $0.7\,\text{mm}^2$ ($0.5\,\text{mm} \times 1.4\,\text{mm}$) there is no conduction across lesion gaps at BCL shorter than 400 ms. Gaps having $0.9\,\text{mm}^2$ ($0.9\,\text{mm} \times 1.0\,\text{mm}$) cross sectional area allowed conduction at all tested BCLs. The measurements were done at the area of Crista Terminalis, a fast conducting muscle bundle in the right atrium with well aligned fiber orientation. The present findings agree with these measurements, as

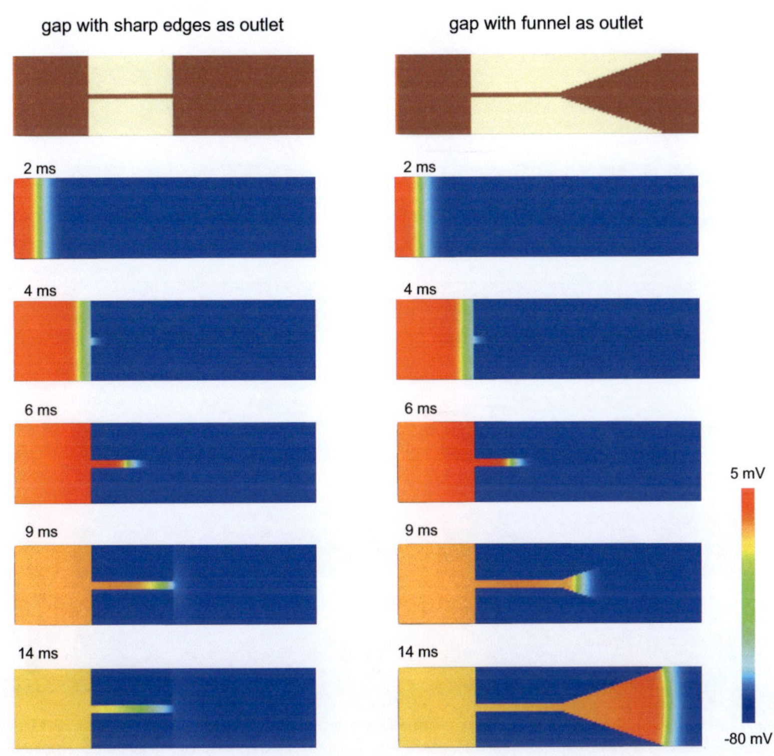

Fig. 12.4. Simulation of a non-conducting sharp edged outlet and of a conducting funnel shaped outlet. Left atrial electrophysiology, transversal fiber orientation, 1000 ms BCL, 0.2 mm gap.

for all BCLs that allowed conduction, the 0.9 mm gap was always conducting in case of longitudinal fibers.

Pérez and co-workers [418] measured lesion gap conduction in more complex gap geometries and with different fiber directions in a rabbit right ventricle. They proposed that conduction through gaps in RF lesions is associated with gap geometry and cross-sectional area and that complex geometry of gaps could lead to unidirectional blocks and/or rate-dependent blocks. The findings of the simulation study agree with these measurements. In particular rate-dependent and fiber-dependent block were observed. On the other hand, activation always spread within the lesion gap and block occurred at the outlet of the lesion. This is most likely to be caused by a source-sink mismatch at this site, such that the large amount of cells behind

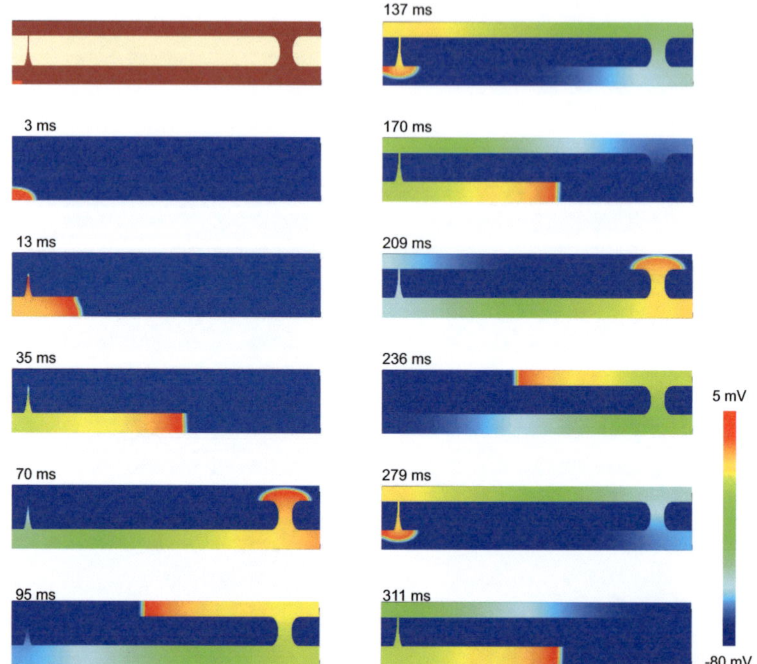

Fig. 12.5. Reentry in a model with an incomplete ablation lesion in left atrial myocardium (Δx=0.33 mm, smallest width of funnel gap 0.2 mm, electrically remodeled electrophysiology and tissue conductivity [73]). The lesion contained a large gap at the right side and a small, funnel shaped gap at the left. The latter gap caused single sided conduction and thus acts like a diode (35 & 137 ms). In this chronic state the border zone diminished and makro-reentry was observed between the two gaps (137 ms ff).

the gap cannot be activated from a small amount of cells at the gap outlet. Simulation of other outlet geometries which slowly equal out the source-sink mismatch, e.g. geometrically seen a funnel, confirmed this hypothesis (Fig. 12.4), although a systematical analysis of e.g. the opening angle was not performed. Both findings rather point to a significance of the outlet size and geometry than to a significance of the complete gap geometry. The funnel shaped outlet led to a gap conduction in one direction only (diode like behavior), which may cause makro-reentry circuits if another conductive gap in the lesion was present (Fig. 12.5). The reentry pathway was closed when a border zone was set around the ablation lesions (Fig. 12.6). This reflects the case acutely after RFA therapy (Sec. 2.3).

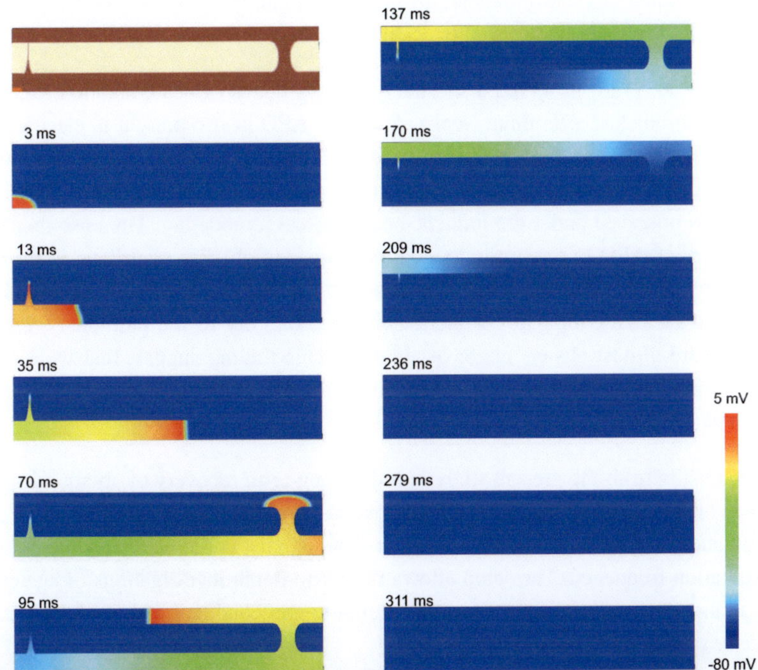

Fig. 12.6. Reentry in a model with an incomplete ablation lesion in left atrial myocardium (same setup as in Fig. 12.5). In the acute phase after ablation, the lesion was surrounded by a low-conductivity border zone of 0.1 mm width ($\sigma_{border} = 0.01 \times \sigma_{myocardium}$). The border zone blocked conduction in both ways through the funnel gap (170 ms). In the chronic state the border zone diminished and makro-reentry was observed between the two gaps (137 ms ff).

The fiber orientation dependent wave-front curvatures observed after the wave exited the lesion gap (Fig. 12.3) confirmed findings from a combined simulation and measurement study [419]. The findings of slowed conduction across narrow but conductive lesion gaps also agreed with the findings from this study by Cabo et al. [419].

In the present study, conduction across lesion gaps was achieved for rather small lesion width compared to the Ranjan study which observed conduction block in gaps between 1.5 to 4 mm [110]. This could be caused by differences in simulation setup (monodomain vs. bidomain, 3D patch vs. 2D sheet) and by differences in tissue conductivity. Nevertheless, the principle findings agree between the present study and the study undertaken by Ranjan et al..

Wood et al. [126] observed a shorter APD near edges of acute ablation lesions, which diminished after three weeks. A shorter APD is also present in electrically remodeled tissue as used in the present simulations. It is likely that the ablation-induced APD shortening might lead to the conduction-block across small lesion gaps, as observed under the influence of electrical remodeling. The post-ablation recovery of APD in proximity to ablation lesions might also contribute to the recurrence of gap conduction and thus possibly AF. The present study also revealed an increased risk for gap conduction in the proximity to the pulmonary veins, which are almost always target of ablation. This finding suggest that wide area PVI might be less sensitive for conduction across the ablation lesions compared to narrow PVI.

The results from the present study hint that long-term recovery of ablation lesion gap conductivity and thus possible recurrence of AF, could be explained by the simultaneous reverse remodeling of tissue conductivity, cell electrophysiology and excitation frequency. The setup allows further in-depth investigations of anatomical and electrophysiological factors leading to gap conduction in the future, e.g. tissue conductivity variations.

Model-Based Evaluation of Ablation Therapy

Patient-specific atrial models may enable personalized model-based therapy evaluation and planning in a clinical environment. Such approach requires fast computational methods to not unnecessarily lengthen the clinical procedure. On the other hand, image data is commonly acquired at least one day prior to a major clinical intervention providing time for performing simulations over night. In this section a clinical application of patient-specific atrial models is described. The goal was to evaluate the outcome of radio-frequency ablation (RFA) interventions using atrial models as support to the image data. Two scenarios were considered for this. First, the success of pulmonary vein isolation (PVI) in two AF patients (models 8 & 9, Tab. 7.1) was evaluated from image data acquired acutely after the intervention (< 24 h after RFA). Second, the chronic development of RFA lesions after PVI was investigated in three AF patients (models 9c, 10 & 11 Tab. 7.1) based on image data acquired at least 3 month post ablation.

13.1 Simulation Setup

First, volumetric, anisotropic atrial models were generated for each case as described in Section 7.1. Afterwards, structural information about the RFA lesions were extracted from late Gadolinium enhancement (LGE) MRI (Sec. 7.1.1). The LGE intensity is variable between patients and scans. The relative LGE-MRI intensity threshold to determine scar regions was therefore varied between 0 and 100%. For each threshold a bi-atrial model with scar tissue in regions where LGE-intensity was above the threshold was created. Next, sinus rhythm excitation was simulated in each model utilizing the monodomain (acute RFA lesions) and the anisotropic fast-marching level set approach (Sec. 5.7, (acute and chronic lesions)). Monodomain simulations were run for thresholds provided by expert clinicians from King's College London, UK. The computational load did not permit a complete sensitivity analysis using the monodomain approach. Analysis of the sensitivity of PV isolation on LGE-MRI threshold was done using the fast-marching approach. For each simulation the isolation and activation of tissue within each PV sleeve was analyzed for each threshold value. Two ratios were determined.

$$R_{ia,threshold} = \frac{\text{inactive PV sleeve myocardium}}{\text{regular PV sleeve myocardium}} \tag{13.1}$$

$$R_{s,threshold} = \frac{\text{PV sleeve scar tissue}}{\text{regular PV sleeve myocardium}} \tag{13.2}$$

If the complete sleeve tissue of a PV was set to be scar tissue at a certain threshold (e.g. 0%), R_s will be 1.0 and R_{ia} will be 1.0. If a PV was successfully isolated from the left atrial body, R_{ia} will be 1.0, but R_s will be smaller than 1.0 for thresholds > 0%. The difference between both ratios described the relative portion of PV sleeve tissue which was not ablated, but isolated.

$$R_{i,threshold} = \frac{\text{inactivate PV sleeve myocardium} - \text{PV sleeve scar tissue}}{\text{regular PV sleeve myocardium}} \tag{13.3}$$

PVI was considered successful for $R_{ia} \to 1.0$ with $R_s \to min$. Additionally, the change in left atrial activation sequence caused by the scar tissue was assessed visually.

13.2 Results

13.2.1 Acute LGE-MRI and Virtual Ablation

In the left atrial model of subject 8, the weakest isolation from the left atrial body occurred in the LSPV, whereas the LIPV showed large portions of ablation scar already at high threshold values. The right PVs showed circumferential isolation from threshold below 62%. At 61.2%, the excitation pattern in the left atrium revealed two gaps in the lesion around the RIPV (Fig. 13.1(a)). Two additional ablation lesions were set *in-silico* at the spots of conduction break-through (virtual ablation, Fig. 13.1(a), encircled). These isolated the RIPV electrically from the left

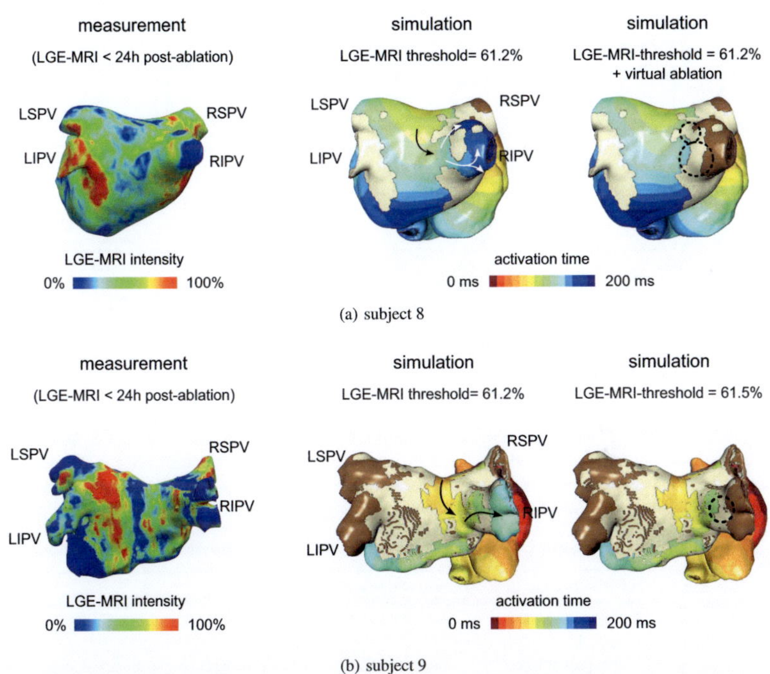

(a) subject 8

(b) subject 9

Fig. 13.1. Monodomain simulation of atrial excitation on models with RFA scar based on LGE-MRI acquired <24 h (acutely) post-ablation. Arrows indicate activation wavefronts. Dashed encircled are additional lesions placed virtually (subject 8) or caused by an increase in LGE-MRI threshold (subject 9). In both cases, this led to an isolation of the RIPV.

atrium. The overall left atrial excitation sequence was not altered at these LGE-MRI threshold values.

In subject 9, the RIPV was the weakest isolated PV. Simulation of the sinus rhythm at a threshold of 61.2% revealed a lesion gap at the posterior side of the RIPV. An increase of the threshold by 0.3% closed this gap such that all PVs were isolated from the left atrial body (Fig. 13.1(b)). In contrast to the ablation lesions of subject 8, tissue acutely affected from RFA was distributed more equally over the left atrium in subject 9. This led to a significantly reduced portion of active left atrial myocardium in the simulations. Small corridors between both pairs of PVs and along the mitral valve annulus allowed a regular sinus rhythm left atrial activation sequence.

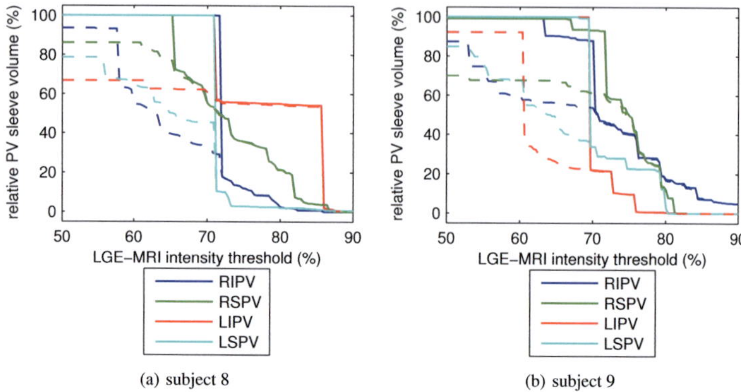

(a) subject 8   (b) subject 9

Fig. 13.2. Results of sensitivity analysis of acute LGE-MRI threshold and PV isolation. Solid curves: relative amount of inactivate PV sleeve tissue R_i, dashed curves: relative amount of scar in PV sleeves R_s. The difference between both curves marks non-scar isolated amount of PV sleeve tissue. PV are considered to be isolated at the threshold value at which both curves of the PV split.

Table 13.1. Overview of LGE-MRI acquisition, clinical findings and simulation observations.

subject	LGE-MRI acquisition	clinical observations	*in-silico* findings
8	24h post-ablation	successful PVI	all PVs reconnect nearly at same threshold
9	24h post-ablation	successful PVI	all PVs reconnect nearly at same threshold
	3 month post-ablation	successful PVI	all PVs reconnect nearly at similar thresholds
10	pre-redo-ablation	all PVs reconnected	RIPV most prone to reconnection
11	pre-redo-ablation	RSPV reconnected	RIPV most prone to reconnection

The sensitivity analysis showed in both cases that all PVs were isolated from the left atrium in a very narrow range of threshold values (Fig. 13.2), thus not providing a reliable indicator for the reconnection probability of the PVs. From a clinical perspective, PVI was successful in both patients (no AF recurrence three month post-ablation). This could correlate to the sensitivity analysis findings, as no single PV was significantly more prone to reconnection as the others.

13.2.2 Chronic LGE-MRI / Redo procedures

Three AF patients underwent LGE-MRI after at least 3 month post-ablation. Subject 9 had a successful first RFA procedure, but subject 10 and 11 needed to un-

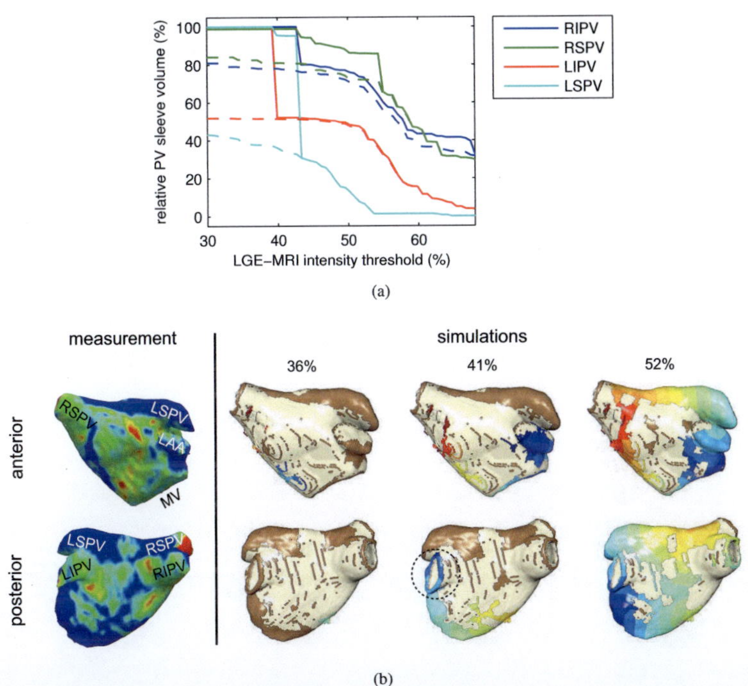

Fig. 13.3. a) Activation of the four pulmonary veins depending on the threshold of LGE-MRI for the transfer of scars into the model of patient 9c. Dashed lines: scar tissue, solid lines: tissue which is not activated during sinus rhythm. Differences between the two curves of each PV denotes tissue, which is isolated from regular activation. 0% threshold means that all regions which have a LGE-MRI value are considered to be scar tissue. 100% means that no region is considered to be scar. b) Endocardial shell with LGE-MRI intensity distribution and simulated activation times using different LGE-MRI thresholds. The dashed circle denotes the PV most likely to reconnect to the left atrial myocardium.

dergo a second RFA intervention (redo-procedure, Tab. 13.1). LGE-MRI was acquired prior to the redo-procedures.

In subject 9c the right PVs showed a significant portion of scar tissue at high threshold values. The RSPV was the first to be partly isolated from the left atrial body. Complete PV isolation was achieved in all PVs at similar threshold values (Fig. 13.3(a)). This corresponds to the findings from the acute post-ablation sensitivity analysis of the same subject (Fig. 13.2(b)). The LIPV was the PV to be isolated from the left atrium at the lowest threshold and thus had potentially the weakest isolation (Fig. 13.3(b)).

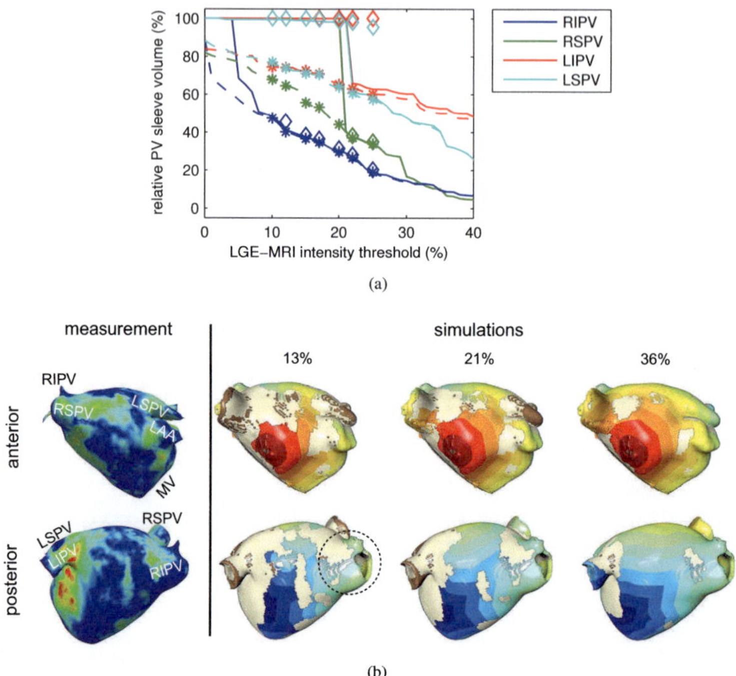

Fig. 13.4. a) Activation of the four pulmonary veins depending on the threshold of LGE-MRI for the transfer of scars into the model of patient 10. Dashed lines: scar tissue, solid lines: tissue which is not activated during sinus rhythm. Differences between the two curves of each PV denotes tissue, which is isolated from regular activation. 0% threshold means that all regions which have a LGE-MRI value are considered to be scar tissue. 100% means that no region is considered to be scar. b) Endocardial shell with LGE-MRI intensity distribution and simulated activation times using different LGE-MRI thresholds. The dashed circle denotes the PV most likely to reconnect to the left atrial myocardium.

In subject 11, the sensitivity analysis revealed that the RIPV has the weakest isolation (Fig. 13.4(a)). The portion of scar tissue in each PV was in this case nearly linearly proportional to the LGE-MRI threshold value. Electroanatomical mapping during the second RFA procedure in this patient revealed that all PVs were reconnected (Tab. 13.1). This would correspond to LGE-MRI thresholds above 23%, as shown by the sensitivity analysis.

In subject 12, the simulations showed that the RIPV was most likely to reconnect to the left atrium, although the threshold values for all PVs were closer together than in subject 11 (Fig. 13.4(a)). Intracardiac measurements during the redo proce-

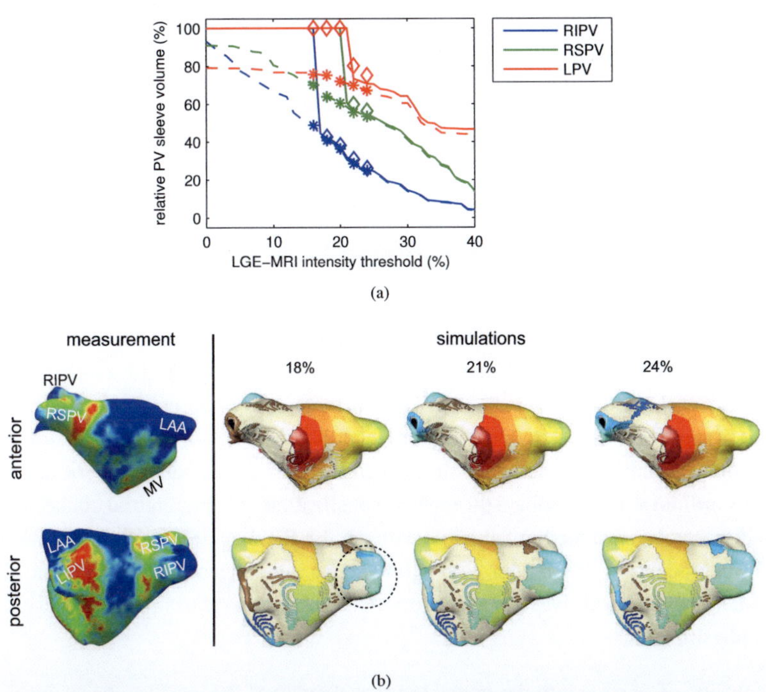

Fig. 13.5. a) Activation of the three pulmonary veins depending on the threshold of LGE-MRI for the transfer of scars into the model of patient 11. Dashed lines: scar tissue, solid lines: tissue which is not activated during sinus rhythm. Differences between the two curves of each PV denotes tissue, which is isolated from regular activation. 0% threshold means that all regions which have a LGE-MRI value are considered to be scar tissue. 100% means that no region is considered to be scar. b) Endocardial shell with LGE-MRI intensity distribution and simulated activation times using different LGE-MRI thresholds. The dashed circle denotes the PV most likely to reconnect to the left atrial myocardium.

dure revealed that the RSPV was reconnected to the left atrium (Tab. 13.1). In the simulations, the region around both right PVs showed small LGE-MRI intensities (Fig. 13.4(b)).

13.2.3 Fibrosis from pre-ablation LGE-MRI

For subject 9, the intensity of LGE-MRI prior to the first RFA procedure was also mapped onto a 3D shell of the left atrium (Tab. 4.1). Strong LGE-MRI intensity in images acquired prior to an RFA procedure might reveal pre-existing scars.

Table 13.2. Tissue properties for three degrees of fibrosis transferred from pre-ablation LGE-MRI.

fibrosis degree	LGE-MRI threshold	CV longitudinal	CV anisotropy	σ_{tissue} longitudinal	σ_{tissue} anisotropy
none	100%	100%	1.9:1	100%	3.75:1
mild	56%	80%	4:1	64%	16:1
moderate	60%	60%	8:1	36%	64:1
severe	64%	40%	12:1	16%	144:1

Additionally, there's strong evidence that such approach can also identify regions of tissue fibrosis [272, 274].

Using the same approach as to mark RFA scar tissue in volumetric left atrial models (Sec. 7.1.1), fibrotic regions were marked in the model (Fig. 13.6 top 2 rows). Three thresholds were considered to mark three degrees of fibrosis (Tab. 13.2). Tissue properties were altered such that myocardial dissociation increased and conduction velocity decreased with stronger fibrosis. Tissue conductivity for healthy myocardium was determined through personalization of the interatrial connections and the global conduction velocity to match the PWd as described in Chapter 9. The remodeling CRN model with extracellular ion concentration adapted to the patients blood electrolyte concentrations (Chap. 8) was used to simulate atrial electrophysiology.

Inclusion of fibrosis into the model led to a slowing of the wavefront propagation in the fibrotic regions compared to the regular simulation (Fig. 9.4). Furthermore, excitation did not propagate into the RSPV, which sleeve tissue was made up of severe fibrosis. The left atrial activation sequence was altered on the anterior left atrium. Late activation occurred near the left atrial roof and the LAA (Fig. 13.6, encircled). A line of severe conduction slowing could be observed between the right and left PVs on the posterior side (Fig. 13.6, dashed line). Both effects could also be seen in the measured LAT map from the same patient, although the degree of conduction slowing was less. Figure 13.6 provides a comparison of the fibrotic model and the simulation outcome as well as the measured LAT map.

13.3 Discussion

In this chapter results from LGE-MRI based model augmentation in six cases were presented. LGE-MRI data collected shortly after an ablation procedure (within the

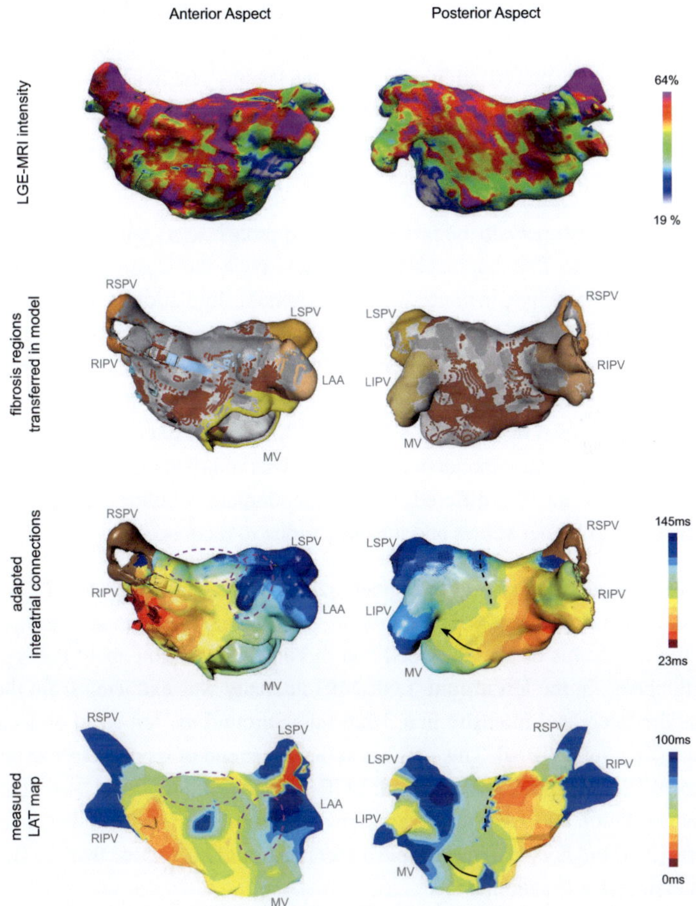

Fig. 13.6. LGE-MRI intensity based fibrosis model in subject 9. Fibrotic regions can be seen in the second row as grey zones. The darker the grey the more severe the fibrosis. Fibrosis regions might in part be an explanation for late activation on spots on the anterior left atrium (encircled in magenta) and conduction block on the left atrial roof (dashed lines). The phenomena could not be reproduced with the regular model (Fig. 9.4).

first 24 h) showed large signal intensity in regions of scar, edema and inflammatory tissue. In order to reliably image scar formation after RFA in the left atrium, LGE-MRI data needs to be recorded at least 3 month after the ablation procedure [138]. Such was done for subject 9c, 10 and 11.

The transfer of scar tissue into the models based on LGE-MRI intensity showed great variability, especially using LGE-MRI data acquired acutely after RFA. Thereby, virtually placed additional ablation lesions resulted the same simulation outcome as slight variations in LGE-MRI intensity threshold. To provide a quantitative analysis of PV reconnection probability, sensitivity analyses of the PV isolation with varying LGE threshold were performed.

For the sensitivity analysis the fast-marching approach was used to simulate sinus rhythm activation. This approach had the draw-back that small gaps in ablation lesions were conductive, in contrast to monodomain simulations and experimental findings (Chap. 12). For subjects 10 and 11 a reduced number of monodomain simulations with thresholds around the crucial PV isolation thresholds were performed to validate the simplified simulation approach. The outcome of the monodomain simulations is shown in Figures 13.4(a) and 13.5(a) with stars and diamonds. For subject 12, a good correspondence could be observed, whereas in subject 11 results of the superior PVs differed. In the monodomain solution the superior PVs were isolated at higher values and thus less prone to reconnection.

LGE-MR images are difficult to segment [274]. This is caused by low LGE-MRI resolution, contrast-agent accumulation in anatomical structures next to the atria (e.g. the esophagus or great vessels) and missing information about the regional wall thickness in the left atrium. LGE-MRI intensity was extracted from the image as the integrated intensity in a 3 mm range around the left atrial endocardial surface (Sec. 7.1.1) [270]. This range is at the upper end of reported left atrial wall thickness (Sec. 2.1.1.4), to capture all LGE-signal within the atrial wall. The low image resolution and significant variations in manual scar segmentation between experts [274] did not permit a reliable direct transfer of scar tissue from LGE-MRI into the model at the moment.

Tissue conductivity was set to zero in scar regions. Similar approach is commonly done in ablation modeling studies [93, 185, 187, 420]. Ablation lesions might be more complex in their microscructure (Sec. 2.3). The modeling approach could be enhanced by adding border zones of slow conduction and isotropic tissue properties in scar and surrounding regions to model acute lesions and the lesion healing over time. In this presented cases, this was not necessary, as the outcome of chronic scar formation was regarded [138].

Minor problems with the isolation of the LSPV were observed. The automatic image segmentation approach only provided the left atrial endocardial surface, which was dilated afterwards. This caused a connection between the LSPV and the LAA epicardial surface in some of the models. Therefore, the LSPV may be activated from the LAA, although it was isolated at its orifice from the left atrial body, thus distort the sensitivity analysis results. This was observed in subject 10. Segmentation techniques using volumetric statistical shape models could prevent such errors in future studies [60, 361].

The results of the sensitivity analysis of subjects 10 and 11 did not correspond to the equivalent clinical findings. This might be caused by various reasons. (i) The applied fast-marching level set method to simulation atrial excitation is not able to reproduce diffusion processes and wavefront curvature effects yet and thus allows excitation through any kind of small gaps. (ii) The model-based analysis of the image data relies on the segmentation of LGE-MRI data, which was shown to be an unreliable task yet [274]. It might therefore be necessary to tune LGE-MR image acquisition, morphological image acquisition (wall thickness) and to use fast simulation methods with diffusion [262] to receive reliable results from the sensitivity analysis. Nevertheless, the presented framework has the potential to support diagnosis of AF patients awaiting redo-procedures. The simulations provide a visual impression of the excitation pathways in the left atrium and eventual propagation into reconnected pulmonary veins and can thus help to plan RFA interventions.

LGE-MRI data acquired prior to an EP intervention might be able to reveal regions of fibrotic tissue in the atria [269, 272]. Based on such data, fibrotic left atrial myocardium was introduced into the model of subject 9. This simulation study was quite speculative and was meant to serve as a proof of concept.

Three degrees of fibrosis were set depending on the LGE-MRI intensity. The conduction velocity magnitude and anisotropy were varied with the LGE intensity. The values for anisotropy were quite strong, to pronounce the effects of the fibrotic regions. Plank et al. [94] modeled severe fibrosis with a tissue conductivity ratio of 12:1. Zhao et al. [96] on the other hand assumed regular atrial conductivity anisotropy as 10:1 and Tobon et al. set remodeled atrial conductivity ratio to 4:1 [412]. Measured conduction velocity anisotropy ratios in healthy atrial tissue range from 8.5:1 (70:1 for tissue conductivity) in human pectinate muscles [79], 6.55 (43:1) [421] in healthy canine Bachmann bundle to 5.5 (30:1) [422] in healthy canine right atrial free wall. The assumed values for mild and moderate fibrosis lay within this range. In future studies more degrees of fibrosis or a continuous transfer function between degree of fibrosis depending on the LGE-MRI intensity could be used.

In the fibrosis model only atrial myocyte cells were considered. Other cell types, primarily fibroblasts, might be present in fibrotic regions. Fibrosis could therefore also be modeled as a mixture of myocytes and fibroblasts [236, 397]. Modeling studies investigating the electrotonic coupling between such cell types and other microstructural tissue heterogeneities [423] provided explanations for the genesis of CFAEs and other signal fractionation. Microfibrosis can also be modeled by the introduction of conduction barriers between adjacent computational nodes [424]. In the present case, the modeling approach needed to be more macroscopic, as the aim was to investigate the influence of fibrotic patches on the biatrial excitation sequence. The other approaches require a finer model resolution, which was not sensible for the present aim.

Fibrosis in the model led to a slowing of the global and local excitation wave spread. This resulted in zones of slowed conduction or near conduction block. These could serve as an explanation for similar zones in the measured LAT map. Nevertheless, the thresholds for the transfer of fibrosis from LGE-MRI to the model were chosen based on an educated guess and thus the findings are to be taken cautiously. A quantitative measure for the determination of the threshold would provide more confidence in the *in-silico* findings.

The established modeling and simulation framework was able to provide new insights into the impact of ablation scars and fibrosis on the left atrial activation. Once a more quantitative imaging and segmentation of LGE-MRI data is available, the framework might be used to predict long-term success of RFA therapy and for the planning of redo procedure. It thereby might help to understand reentry pathways and to localize gaps in ablations lesions prior to the EP intervention. Modeling of local fibrosis based on LGE-MRI might help to guide ablation procedures and could contribute to the understanding of abnormal electrical signals observed in the left atrium during the EP procedure.

14

Influence of Ablation Lesions on the Cardiac Function

Radio-frequency ablation is performed in AF patients to restore cardiac function by isolating arrhythmic triggers and substrate from the left atrial myocardium (sec. 2.3). The main goal is to reduce the risk of clot formation and thus for ischemic stroke. Additionally, the quality of life of the patient should increase. It is well known that in general the success of an RFA procedure (AF termination) increases with the amount of scar tissue created. On the other hand, it is not clear how the ablation scars influence the cardiac pump function. The effect is hard to study in clinical trials as the ablation scars change their material properties over time, but simultaneously reverse remodeling of the atria and ventricles after RFA

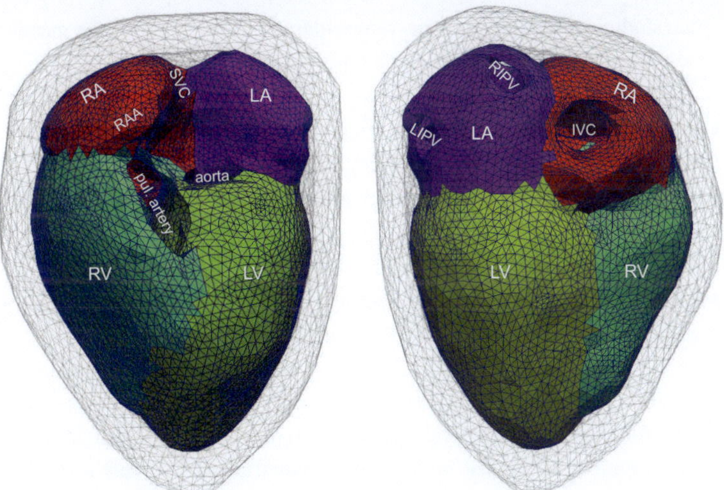

Fig. 14.1. Four chamber tetrahedron model of subject 19 (Tab. 4.1 & 7.1).

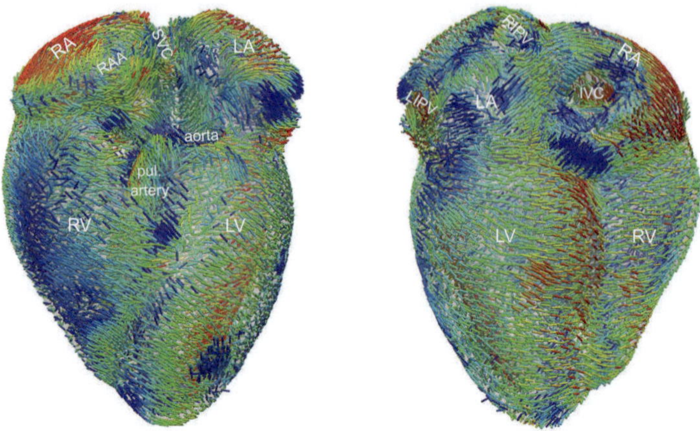

Fig. 14.2. Fiber orientation in the four compartment model of subject 19. The y value of the fiber orientation vector is color coded using an HSV color map. Perspectives as in Figure 14.1.

therapy might overlap the effect of the introduced ablation scars. Computational models bear the potential of examining the effects separately. In this section, a study is presented which investigates the influence of various ablation lesion patterns on the left atrial and left ventricular stroke volume.

In-silico studies examining the atria-ventricular relationship during the mechanical cardiac cycle comprise several requirements. Among these are:

- Coupled geometrical atrial and ventricular model with fiber orientation.
- Electrophysiological and cell coupling model for the atria and for the ventricles.
- Tension development and elastomechanical model of the atrial and ventricular myocardium.
- Elastomechanical model of ablation scars.
- Measurement data of the cardiac activity for model validation, preferably from the same subject as the geometry was derived from.

As the ventricles have been subject of electrophysiological and elastomechanical modeling for a long time, the prerequisites with respect to the ventricles are given in many research institutions. In contrast, volumetric atrial models with fiber orientation are given in only very few groups (Tab. 3.1) and models of tension development and elastomechanics of the atria have not been developed yet. In this thesis, solutions for the majority of the prerequisites concerning the atria were presented which now allow for combined atria-ventricular electrophysiology and eleastomechanical simulations.

In this chapter, the impact of different ablation lesion patterns on the cardiac deformation is investigated. The simulation study was done in collaboration with Dipl.-Phys. Thomas Fritz, who was responsible for the biomechanical simulation framework. The fundamentals of the elastomechanical simulation of cardiac tissue are therefore not described in detail in this thesis.

14.1 Simulation Setup

A whole heart model was derived from a healthy volunteer (subject 19, Tab. 4.1 & 7.1). Ventricular fiber orientation was approximated according to the rules described by Streeter et al. [167]. Atrial fiber orientation was modeled as described in Section 5.4. Figure 14.2 depicts the fiber orientation in the whole heart model. Electrophysiology and tension development were simulated on a voxel grid (Δx = 0.33 mm). Cardiac deformation was computed in a tetrahedron model of the four cardiac compartments within an elastic shell representing the pericardium [425]. The four chambers were not directly connected to the pericardium but could slide against each other. The model contained approximately 25.000 elements. Model fixation was derived from multi-slice cine-MRI from subjects 10, 11 and 19 (Tab. 4.1). In this data, the orifices of the caval veins and PVs as well as the apex did not move significantly and were thus assumed to be stationary. The model was fixated around the openings of the caval veins, around the orifices of the four PVs and at the apex. The AV plane was not fixated. Figure 14.1 shows the four compartment tetrahedral model used for the biomechanical simulations.

The ventricular electrophysiology was simulated with the electrophysiological model from ten Tusscher et al. [426]. The ventricles were activated endocardially with a realistic stimulation profile [427] and excitation propagation was computed with the monodomain model. More detail on the simulation of the ventricular electrophysiology can be found in [386]. Atrial electrophysiology was simulated using the gradual heterogeneous CRN model (Sec. 5.2). Tension development in the ventricles was simulated using the HTD model [308, 311] and in the atria with the aHTD model (Sec. 5.3).

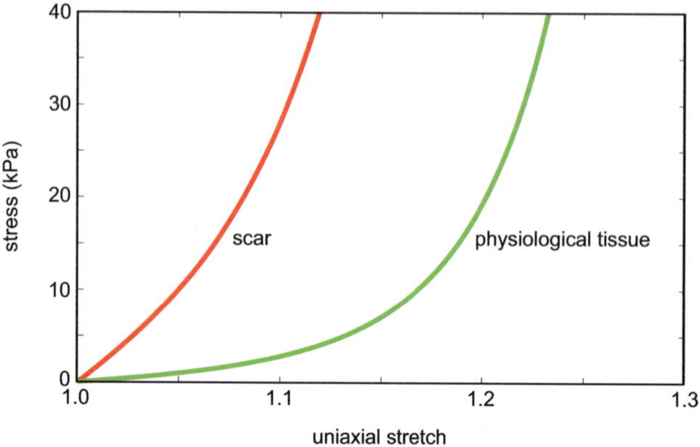

Fig. 14.3. Stress-strain relationship in atrial myocardium and atrial scar tissue as modeled in this study.

Ventricular and atrial elastomechanical properties were modeled using the model of Guccione et al. [428, 429]. The model was originally developed for ventricular myocardium. In the absence of respective measurement data for the atrial myocardium, it was assumed that the atria and ventricles behave similar. The model describes anisotropic hyperelastic and incompressible myocardial tissue. No information about the elastomechanical properties of RFA scars were found. However Takahashi et al. [430] suggest that RFA scars behave like myocardial infarction tissue in the long run. In a chronic state such tissue has a 6–8 fold higher stiffness than regular tissue [431]. In some simulations, the stiffness was increased by a factor of 10 [432, 433]. For the present study, the stiffness parameter C in the Guccione model was also increased by a factor of 10:

$$C_{scar} = 10 \cdot C_{regular}. \tag{14.1}$$

In an uniaxial stretch experiment, this altered the stress-strain relationship as shown in Figure 14.3.

Generic ablation lesion patterns were set in the left atrial model as introduced in Section 5.5. A subset of lesion patterns was chosen (no lesion, B, D, F). The other lesion patterns did not differ significantly from these patterns and would thus not provide additional insights.

Two sets of independent simulations were run. For one, the impact of ablation lesions on the active atrial deformation was investigated. Therefore, atrial depolar-

Table 14.1. Simulated and measured left atrial volume over the complete cardiac cycle.

		model	[434]	[435]	[18]	[19]
Maximum Volume	(ml)	66.4	97±27	90 (60–120)	–	–
Unloaded Volume	(ml)	52.0	–	–	–	62±22
Minimum Volume	(ml)	38.7	44±13	30 (35–40)	41.86	–
Stroke Volume	(ml)	27.7	53±21	–	–	–
Ejection Fraction	(%)	41.7	54±12	–	–	–

ization and repolarization and subsequently tension development and deformation were simulated for the regular model and five ablation lesion patterns. In a second set of simulations, the influence of ablation lesions on the passive atrial deformation was examined. Ventricular contraction following ventricular excitation was simulated six times with different ablation lesion patterns in the left atrium. In each simulation the left atrial and left ventricular volume curve were recorded.

14.2 Results

Simulation of the regular cardiac excitation in the atria started in the right atrium between the SVC and the RAA. The activation spread downwards the right atrium and coupled into the left atrium anterior via BB. The wavefront rolled over the roof of the left atrium and merged with a second wavefront coming from the inferior right atrium between the inferior PVs (Fig. 14.6, top row arrows). The tension development sequence followed the electrical activation is therefore not shown. The maximum force was greatest in the CT and showed a gradient from right to left and superior to inferior within the right atrium. The PVs and the atrio-ventricular rings showed the smallest tension development.

The deformation caused by the tension development (Fig. 14.4 led to a decrease in left atrial volume as shown in Figure 14.7(a). Ventricular contraction caused an increase in left atrial volume as shown in Figure 14.7(c). The extrema of the atrial volume are listed in Table 14.1. These lay within the range of experimental findings, although the left atrial volumes were at the lower end of measured ranges.

Introduction of ablation lesions did not alter the right atrial activation and tension development. Pattern B, D, F isolated the PVs from the electrical activation. Thus no force was developed in the PVs and the surrounding area in these patterns. Pattern D changed the direction of the activation wavefront in the left atrial region

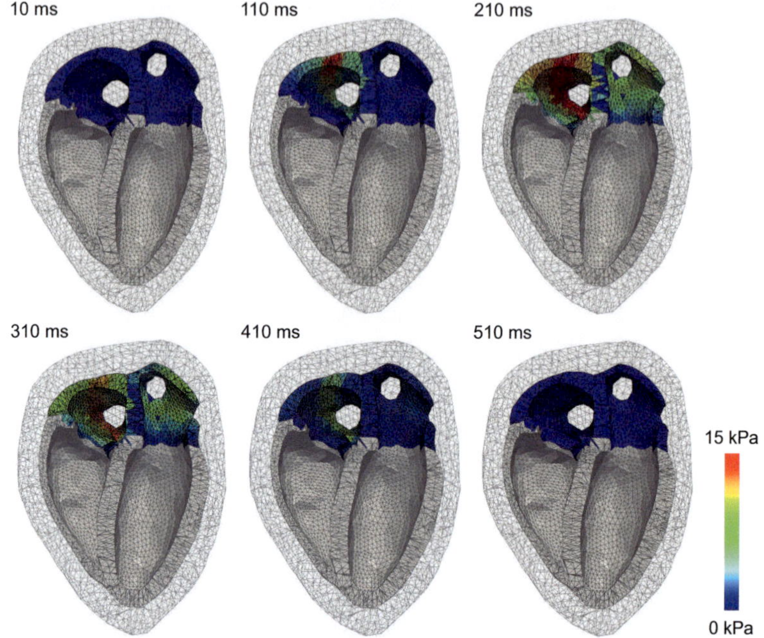

Fig. 14.4. Cardiac deformation during atrial contraction and relaxation. The active tension is color coded in the atria.

between the PVs. Pattern F electrically isolated a major portion of the left atrium (PVs and region in between). In this area force development was suppressed. There was no significant difference in tension development between lesions B and D. Figure 14.6 shows the electrical activation and maximum local tension for the regular case and the five ablation lesion models.

Introduction of ablation reduced the decrease in atrial volume during the atrial contraction and relaxation (Fig. 14.7(a)). Pattern F showed the greatest effect on the left atrial volume and had the smallest change from unloaded atrial volume. The atrial volume during ventricular contraction was affected only slightly by the ablation lesions (Fig. 14.7(c)). Ventricular volume during atrial and ventricular contraction was not significantly affected by the ablation lesions (Fig. 14.7(b) & 14.7(d)). Table 14.2 lists the maximum and minimum atrial volume in the healthy case and in the five ablation lesion cases over the cardiac cycle. The stroke

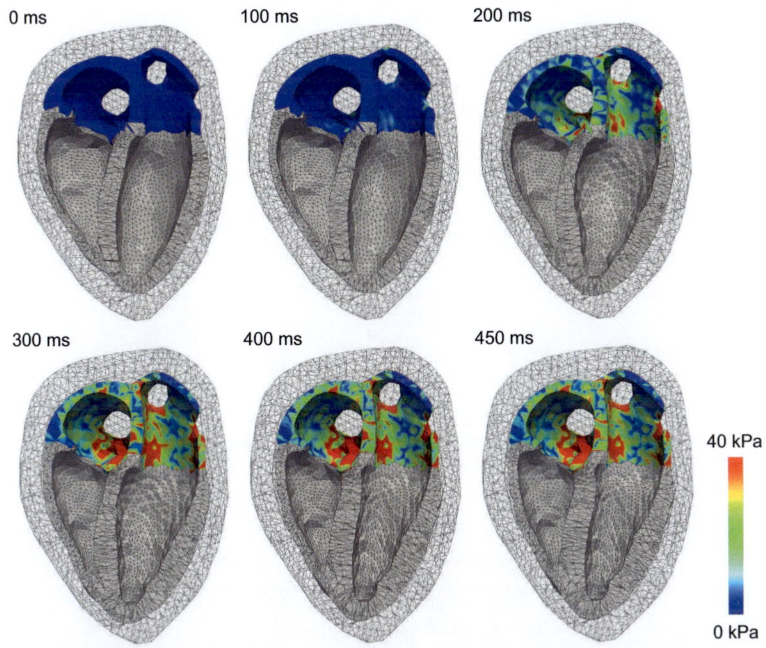

Fig. 14.5. Cardiac deformation during ventricular contraction. The L2 norm of the wall stress is color coded in the atria.

volume and ejection fraction were decreased in the ablation lesion models. Sole PVI (pattern B) had the least decrease of stroke volume and pattern F the greatest.

14.3 Discussion

In this section a combined electrophysiological and elastomechanical simulation study investigating the influences of standard ablation lesion patterns on the cardiac function was presented. To the knowledge of the author, this is the first study to simulate active atrial deformation. Previous computational studies of atrial elastomechanics concentrated on the left atrial deformation during ventricular contraction [318, 319] or during external deformation [317]. Further studies investigated the effects of stretch-sensitive channels on the initiation of AF in static atrial models [323–325].

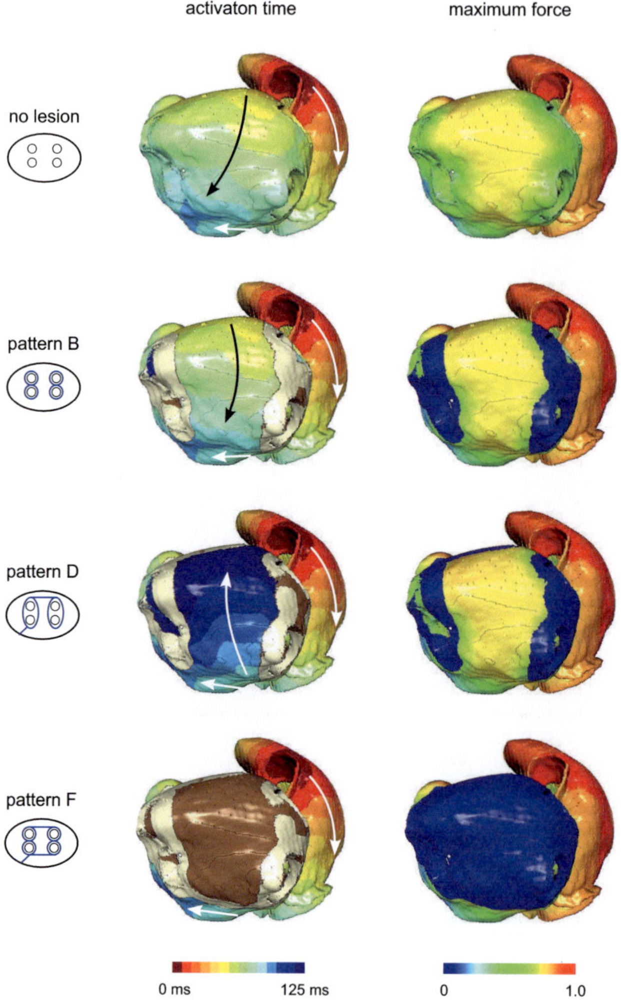

Fig. 14.6. Local activation time and maximum force in model 19 with different generic ablation lesion patterns during sinus rhythm.

Passive elastomechanical properties of the atria were approximated by a model of the ventricular elastomechanical properties due to the absence of a comparable model for the atria. This model was very recently adapted to reproduce the atrial elastomechanics [436] based on measured stress-strain curves from [317]. Simu-

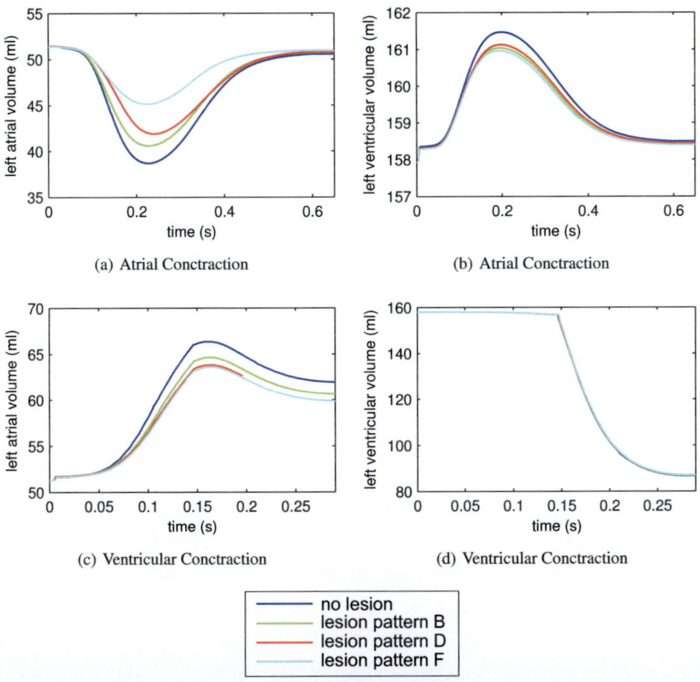

Fig. 14.7. Left atrial and left ventricular volumes during atrial and ventricular contraction. Lesion patterns described in Tab. 5.9 and Fig. 5.27.

lations using the same four chamber model did not reveal strong changes in the deformation sequence compared to the present simulations. Nevertheless, future studies should use the adapted model instead of the ventricular model. If simulations do not need to account for anisotropic elastomechanical properties in the atria, the Mooney-Rivlin model [437], which was also recently adapted to atrial properties [436], could be used.

The atrial wall thickness of model 19 was at the upper end of measured values of the wall thickness (Sec. 2.1.1.4). This did not significantly influence the electrical activation sequence, but led to an increased absolute stiffness during the elastomechanical simulations. In the simulations, the stiffness of the atria in the Guccione model was therefore adjusted qualitatively to allow for a regular contraction and deformation of the atrial chambers as observed in the cine-MRI (Fig. 14.8).

Table 14.2. Influence of different ablation patterns on the left atrial volume.

		no lesion	B	D	F	[430]
Maximum Volume	(ml)	66.4	64.6	63.8	63.6	–
Minimum Volume	(ml)	38.7	40.6	41.9	45.1	–
Stroke Volume	(ml)	27.7	24.1	21.9	18.5	5.75
Ejection Fraction	(%)	41.7	37.2	34.4	29.0	18±11

The elastomechanical properties of scar tissue were described by the Guccione model as well. This model describes the myocardium as an incompressible, anisotropic, hyperelastic material. Ablation scars either have isotropic tissue properties or fiber orientation in this region is chaotic [119], which can microscopically be described as isotropic tissue [130] (Sec. 2.3.1). It is not expected that modeling the scars as isotropic tissue, e.g. by using the Mooney-Rivlin model, will change the simulation results substantially, as the high stiffness of the scars does not permit a deformation of these anyway.

The stiffness of ablation scars was derived from the stiffness of ventricular chronic myocardial infarction tissue due to the lack of better experimental data. The stiffness of myocardial infarction tissue changes over the course of time after the infarction [431]. The tendency is that the scar becomes more stiff over time. A similar behavior could be assumed for ablation scars as well. Ablated tissue shows variations in tissue perfusion in LGE-MRI during the first three month postablation [138]. The simulation framework used for the present study could also be used to investigate the changing impact of ablation scars on the cardiac function over time.

During ventricular contraction, the atrial wall stress was strongly increased in elements near the fixation nodes around the orifices of the PVs and caval veins. The fixated structures were nearly motion-free in the cine-MRI available. Nevertheless, a little motion could be observed in these structures. It could therefore be useful to not apply strict boundary conditions in these structures, but allow for a reduced elastic deformation. This could be realized by attaching onsets of vessel tissue to the structures and fixating these at the end. This would also correspond better to the anatomical situation. Strong wall stress in the regions of the PVs could also be a factor influencing ectopic activitiy in this region through stretch-activated channels.

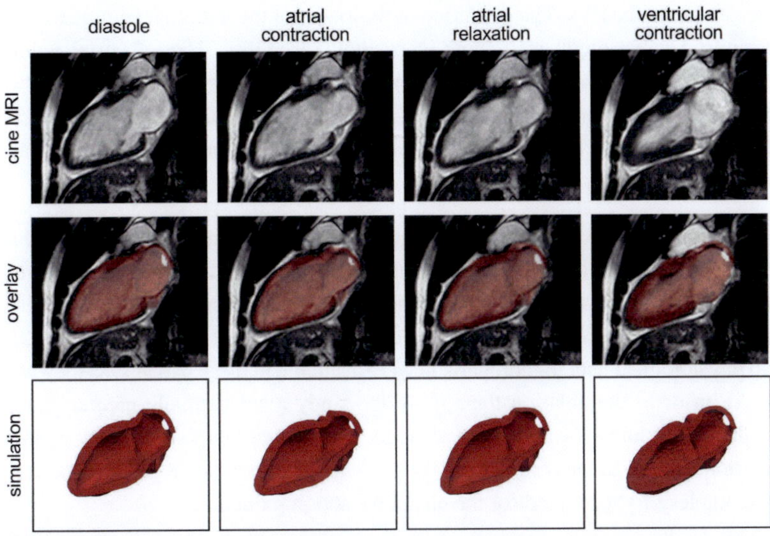

Fig. 14.8. Cardiac motion observed in the cine-MRI of subject 19 as well as in the corresponding four chamber model. Figure based on visualizations provided by Dipl.-Phys. Thomas Fritz, Institute of Biomedical Engineering, Karlsruhe Institute of Technology (KIT).

Maximum tension decreased from right to left and from superior to inferior. Weakest tension development was found in the PVs and valve rings. A strong contraction in these structures seems to be unrealistic, as it would work against the cardiac skeleton. These findings coincided with the single cell simulation results presented in Section 5.3. Right atrial contraction was significantly pronounced compared to the left atrium. This was caused by various factors. For one, tension development was stronger in the right atrium. Second, fiber orientation was well aligned along the CT and PMs in the right atrium. In the left atrium, two sheets of different fiber direction overlap another and thus may hinder contraction. Third, the left atrium was fixated at the PVs. The PVs and the left atrial wall in between lay within a plane. Therefore this part of the left atrial wall could not displace to a great extent. In the cine-MRI from the same patient, the right atrium also showed a stronger contraction than the left atrium.

The simulated left atrial volume time course as well as the ejection fraction and stroke volume fit into the range of measured values for these data (Tab. 14.1). The model was thereby at the lower end of the measured values. This was due to a small unloaded left atrial volume. The left atrial volume was determined without the volume in the PVs. The definition of the extent of the left atrial volume towards the mitral valve might also show inter-study variability. The left atrial volume curve also showed a similar behavior as measured curves [438, 439] except for the temporary volume decrease during ventricular contraction (Fig. 14.7(c), 150–300 ms). During this time period the ventricular contraction did not pull the valve plane towards the apex.

The main finding from the study was that ablation lesions in the left atrium reduce left atrial stroke volume and therefore may hinder left ventricular filling. Thereby the reduction went hand in hand with the amount of ablation scar set. Takahashi et al. also observed a reduced left atrial ejection fraction after ablation *in-vivo* [430]. No physiological data for comparison were available in these patients, as they were in AF prior to the ablation therapy. In the study a substantially greater portion of the left atrium was ablated in average, which could explain the low ejection fraction, similar to the results from the present study in which a greater amount of ablation lesions accounted for the ejection fraction reduction.

No relationship between the scar location and the comparment volumes could be observed. However, sole circumferential PVI resulted in the least decrease of atrial function. From the perspective of curing a patient from AF, usually more ablation scar results in a better long-term outcome of the ablation procedure. The simulation results indicate that extensive left atrial ablation may also reduce the quality of life of the patient in terms of cardiac function. Additionally, left atrial regions isolated from regular electrical activation did not show great deformation neither during atrial contraction nor during ventricular contraction. This may lead to an increased risk of blood clot formation in these regions post-ablation.

Atrial and ventricular activity were simulated independently from another. This approach was valid for regular sinus rhythm simulations, as atrial and ventricular contraction and deformation are happening sequentially under these circumstances. If faster sinus rhythm or atrial arrhythmia are to be simulated, combined whole heart simulation will most likely produce more reliable results. This approach would also allow for the investigation of atrio-ventricular diseases such as the Wolf-Parkinson-White syndrome. Whole heart simulations require significantly more computational resources (memory and time). It should therefore carefully be evaluated if separate simulations may be suitable for the given problem as well.

For the ventricles, only the contraction phase and not the relaxation phase was simulated. This was due to numerical problems during the relaxation phase. Ablation lesion patterns were not expected to have a significantly different influence on the atrial volume during this phase.

In a next step, atrial contraction and left atrial volume could be analyzed under the influence of AF-remodeling. A suitable tension development model is given in Section 5.3. For the elastomechanical properties, tissue anistropy should be increased, as observed during structural remodeling. Similarly, conductivity anisotropy should be increased and conduction velocity should be reduced in such simulations (Sec. 3.2.2).

In the present study, the electro-mechanical coupling [440] was simulated using the aHDT model (Sec. 5.3). Conversely, also the mechano-electrical feedback could be incorporated into the simulations [320]. The mechano-electrical feedback loop was shown to have significant impact on the initiation and perpetuation of AF [321, 322]. Computational modeling bears the potential to provide new insights in the processes underlying the initiation of AF by e.g. left atrial overload, dilatation or hypertension. Initial efforts in this area have been undertaken by Kuijpers et al. [323–325]. In these studies the atria were modeled as fixed structures. Therefore stretch-sensitive channels were artificially stimulated and only the impact on the electrophysiology could be observed. The modeling framework presented in this chapter opens up the possibility to investigate the more complex relationships between atrial motion under pathological constraints and atrial electrophysiology and vice-versa.

In this chapter an approach to model active and passive atrial deformation under physiological and post ablation therapy conditions was presented. The four chamber simulations revealed a reduction of left atrial ejection fraction with increasing portion of ablation lesions in the left atrium. The modeling framework enables a better understanding of post-therapy complications and may assess the risk of blood clot formation if coupled to computational fluid dynamics (CFD) software. In the future, simulations of atrial elastomechanics may allow for an improved planning of ablation lesions for the individual AF patient.

Impact of Hemodialysis on the Atrial Electrophysiology

Patients with end-stage renal disease (ESRD) have a greater prevalence for developing atrial fibrillation (AF) than the healthy population [441]. AF is also more

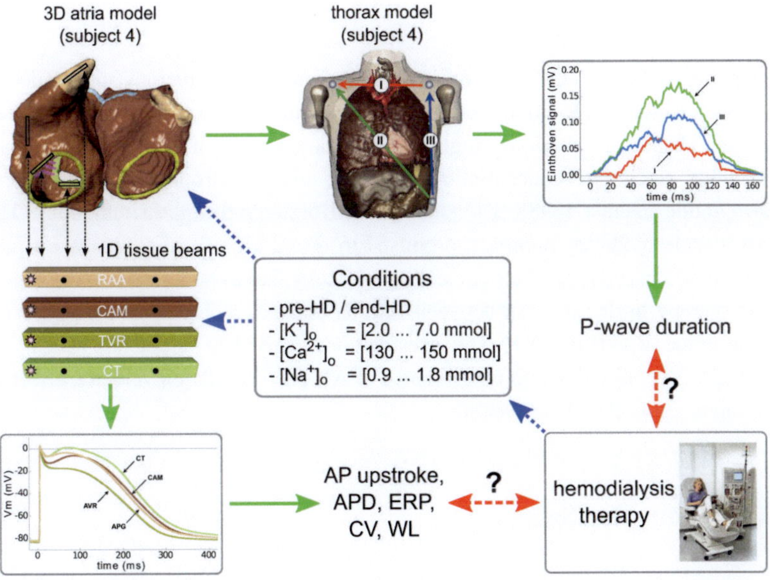

Fig. 15.1. Overview of the methodology used for the multi-scale modeling study. The goal was to understand the influence of hemodialysis therapy on the atrial electrophysiology on cellular and ECG level (red dashed arrows). Therefore a set of blood electrolyte conditions based on measurements from hemodialysis and ESRD patients (blue dotted arrows) was used to parameterize 1D and 3D simulations in atrial tissue (gree solid arrows).

Table 15.1. Measured plasma ion concentration before (pre) and after (post) hemodialysis (HD) sessions. Values given as mean $\pm$ standard deviation in mM. Data from the first six dialysate compositions from [446].

Dialysate K^+	Ca^{2+}	$[K^+]_o$ pre-HD	post-HD	$[Ca^{2+}]_o$ pre-HD	post-HD
2.0	1.25	5.0 ± 0.4	3.3 ± 0.2	1.17 ± 0.11	1.24 ± 0.12
2.0	1.50	5.0 ± 0.5	3.4 ± 0.2	1.16 ± 0.11	1.37 ± 0.06
2.0	1.75	5.2 ± 0.6	3.5 ± 0.3	1.18 ± 0.09	1.52 ± 0.07
3.0	1.25	5.1 ± 0.6	3.7 ± 0.3	1.17 ± 0.12	1.19 ± 0.05
3.0	1.50	5.1 ± 0.6	3.9 ± 0.4	1.16 ± 0.10	1.25 ± 0.12
3.0	1.75	5.1 ± 0.4	3.9 ± 0.3	1.21 ± 0.12	1.37 ± 0.07
[398]		4.9 ± 0.5	3.9 ± 0.4	1.18 ± 0.09	1.30 ± 0.07

common in patients undergoing hemodialysis therapy to compensate for renal failure [442, 443]. Additionally, during hemodialysis the probability for the occurrence of supraventricular extrasystoles is increased. In hemodialysis patients with AF morbidity and mortality are increased as a consequence [442, 444].

To better understand the relationship between hemodialysis therapy and AF development, a multi-scale simulation study was conducted [399, 445] and is presented in this chapter. A previous single-cell simulation study indicated that changes in blood electrolyte concentrations alter the atrial electrophysiology in a pro-arrhythmic manner [398]. The study lacked tissue and organ simulations. The multi-scale modeling approach promised to reveal a chain of evidence for pro-arrhythmic alterations of atrial electrophysiology due to hemodialysis therapy. It can thereby bridge the gap between cellular changes and macroscopic observations in the ECG (Fig. 15.1). The investigations focused on parameters which play a role for the initiation and maintenance of AF, e.g. conduction velocity, effective refractory period and wavelength.

15.1 Simulation Setup

Measurement data were retrieved from two studies [398, 446]. In the first study [446], 16 patients underwent six hemodialysis (HD) sessions each. In each session the composition of the dialysate bath was changed. Table 15.1 lists the dialysate bath compositions. Blood plasma samples were taken from each patient before and right after the HD session and potassium and calcium concentrations were deter-

Table 15.2. Extracellular ion concentration setups for the whole atria sensitivity analysis simulations. Values are in mM.

	Setup	$[K^+]_o$	$[Ca^{2+}]_o$	$[Na^+]_o$
	CRN model	5.4	1.8	140.0
1	hyperkalemia	5.2	–	–
2	hypokalemia	3.3	–	–
3	hypocalcaemia	–	1.16	–
4	hypercalcaemia	–	1.52	–
5	hyponatremia	–	–	135.0
6	hypernatremia	–	–	145.0
7	pre-HD	4.9	1.18	139.8
8	end-HD	3.9	1.30	141.6

mined (Tab. 15.1). In the second study [398], plasma electrolytes before and after HD therapy from 20 ESRD patients undergoing hemodialysis therapy with a standard dialysate bath for less than six months (mean 4 ± 2) were measured (Tab. 15.1, bottom row).

A hemodialysis session usually lasts two to five hours and patients with renal failure need to attend for therapy at two to four days a week. As the change in blood electrolyte concentrations progresses along with the therapy session duration, ion concentrations in the extracellular space of the myocardium are washed in /out according to the change in blood electrolytes. For the simulations, the CRN model of atrial electrophysiology was used. The model was adjusted to recreate the effects of variations in extracellular ion concentrations (Sec. 5.1.1).

Three simulation scenarios were considered. First, a 1D beam of coupled computational nodes was used to determine the change in upstroke duration and upstroke velocity between pre-HD and post-HD conditions. The upstroke duration AP_{up} was determined as the time between the beginning of the upstroke and its maximum. The upstroke velocity dV_m/dt was calculated as the positive maximum of the derivative of the transmembrane voltage over time during the upstroke.

The electrolyte configuration at the end of a hemodialysis session was composed of changes in all ion concentrations. To better understand the contribution of each ion, a sensitivity analysis of the influence of potassium, calcium and sodium on the electrophysiology was performed in the 1D beam setup as second scenario. Thereby, APD, ERP, conduction velocity and wavelength were determined as described in Section 5.2. The sensitivity analysis was done for each of the four right atrial regions (RA, CT, RAA, TVR). Five simulations with varying concentrations of each ion were performed for each atrial region. Maximum and minimum values of potassium and calcium were determined with a confidence interval from the measurements (Tab. 15.1). Maxima of sodium concentration were taken from [398].

Third, eight whole atria sinus rhythm simulations were conducted under minimum and maximum single ion concentrations as well as under pre-HD and end-HD conditions (Tab. 15.2) to determine the impact of the changes in cellular electrophysiology on the global repolarization sequence and the P-wave. For the simulations model 4 (Tab. 7.1) was used. Fiber orientation was set in the CT, PMs and BB, as described in [169]. All other regions had isotropic tissue properties. Heterogeneities in atrial electrophysiology were set as in [92]. The BCL was set to 844 ms in accordance to the heart rate observed in the ESRD patients [398].

15.2 Results

15.2.1 Action Potential Upstroke

Table 15.3 summarizes the change in the upstroke of the action potential in right atrial regions during a hemodialysis session. The upstroke duration was prolonged in all atrial regions after hemodialysis. Common atrial myocardium (CAM) showed the largest prolongation, AVR the smallest. This also caused a decrease in upstroke velocity. The decrease was in a similar magnitude in all tissues. The resting membrane voltage was hyperpolarized in all regions by -5.6 ± 0.1 mV at BCLs longer or equal to 475 ms (heart rates below 125 bpm). The general action potential morphology was preserved in all regions.

15.2.2 Extracellular Ion Concentrations

Table 15.4 summarizes the results from the sensitivity analysis in the 1D tissue beam setup and Figure 15.2 shows results for a sinus rhythm BCL. An increase in extracellular sodium increased APD, conduction velocity, ERP and thus wavelength in all right atrial cell types. An increase in extracellular calcium concentration caused a decrease in APD, ERP and wavelength in all cell types but did not alter the conduction velocity significantly. An increase in extracellular potassium

Table 15.3. Maximum upstroke velocity dV_m/dt_{max} (mV/ms) and upstroke duration AP_{up} (ms) at 844 ms BCL for different atrial tissues.

Tissue		pre-HD	end-HD	difference	
				absolute	relative
CAM	AP_{up}	4.94	5.67	+0.73	+14.7%
	dV_m/dt_{max}	119.3	115.4	-3.9	-3.3%
CT	AP_{up}	4.94	5.56	+0.62	+12.6%
	dV_m/dt_{max}	117.8	115.0	-2.8	-2.4%
TVR	AP_{up}	5.42	5.88	+0.46	+8.5%
	dV_m/dt_{max}	120.2	115.8	-4.4	-3.7%
RAA	AP_{up}	5.28	5.99	+0.71	+13.4%
	dV_m/dt_{max}	119.1	115.2	-3.9	-3.3%

Table 15.4. Results of sensitivity analysis. $\nearrow$ indicates a positive dependency on the rise in extracellular ion concentration; $\uparrow$ a strong positive dependency; $\searrow$ a negative dependency. $\simeq$ means nearly no change and $\bowtie$ indicates contradicting or a more complex dependency-pattern. See text for more information.

Res	Tissue	$[Na^+]_o \uparrow$	$[K^+]_o \downarrow$	$[Ca^{2+}]_o \uparrow$	end-HD
APD_{90}	CAM	$\nearrow$	$\bowtie$	$\searrow$	$\nearrow$
	TVR	$\nearrow$	$\bowtie$	$\searrow$	$\searrow$
	RAA	$\nearrow$	$\bowtie$	$\searrow$	$\simeq$
	CT	$\nearrow$	$\bowtie$	$\searrow$	$\simeq$
CV	CAM	$\nearrow$	$\downarrow$	$\simeq$	$\searrow$
	TVR	$\nearrow$	$\downarrow$	$\simeq$	$\searrow$
	RAA	$\nearrow$	$\downarrow$	$\simeq$	$\searrow$
	CT	$\nearrow$	$\downarrow$	$\simeq$	$\searrow$
ERP	CAM	$\nearrow$	$\searrow$	$\searrow$	$\searrow$
	TVR	$\nearrow$	$\searrow$	$\searrow$	$\searrow$
	RAA	$\nearrow$	$\searrow$	$\searrow$	$\searrow$
	CT	$\nearrow$	$\searrow$	$\searrow$	$\searrow$
WL	CAM	$\nearrow$	$\downarrow$	$\searrow$	$\searrow$
	TVR	$\nearrow$	$\downarrow$	$\searrow$	$\searrow$
	RAA	$\nearrow$	$\downarrow$	$\searrow$	$\searrow$
	CT	$\nearrow$	$\downarrow$	$\searrow$	$\searrow$

increased the conduction velocity and wavelength substantially and also increased ERP. The change in APD did not directly depend on the change in extracellular potassium. APD increased for $[K^+]_o$ decreasing below a concentration of 4.5 mM. APD decreased for decreased $[K^+]_o$ above 5.75 mM. APD did not change significantly for potassium concentrations between 4.5–5.75 mM. ERP was most sensitive to changes in extracellular potassium concentration.

The end-HD condition was a combination of hypokalemia, hypercalcaemia and hypernatremia (Tab. 15.1). Conduction velocity was decreased in all cell types by approximately -50 mm/s compared to the pre-HD case. This was mainly caused

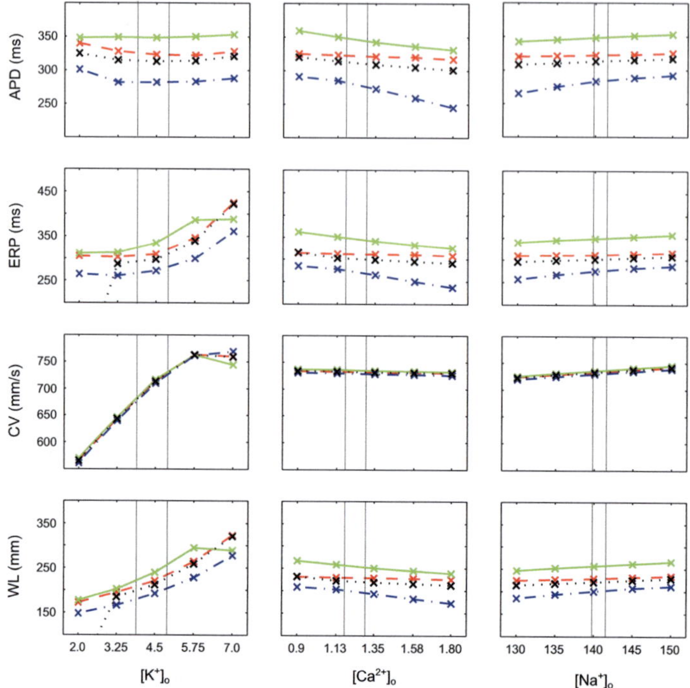

Fig. 15.2. Results of the sensitivity analysis for a BCL of 868 ms (69 bpm) for right atrial cell types (solid green: CT, dashed red: CAM, dotted black: APG, dash-dotted blue: AVR). Vertical lines indicate the bounds of pre-HD and end-HD concentrations as reported in [398]. The restitution behavior for the regular CRN model (RA) is depicted in Fig. 15.3.

by the hypokalemia, which also caused the decrease of ERP (Tab. 15.4). APD was increased in regular atrial myocardium, but decreased in the tricuspid valve ring and did not change in the appendage or the crista terminalis. The wavelength as an indicator for the development and maintenance of reentry circuits was decreased in all cell types after HD therapy (Fig. 15.4). TVR cells had a wavelength after hemodialysis 25% shorter than the wavelength before hemodialysis in regular atrial myocardium.

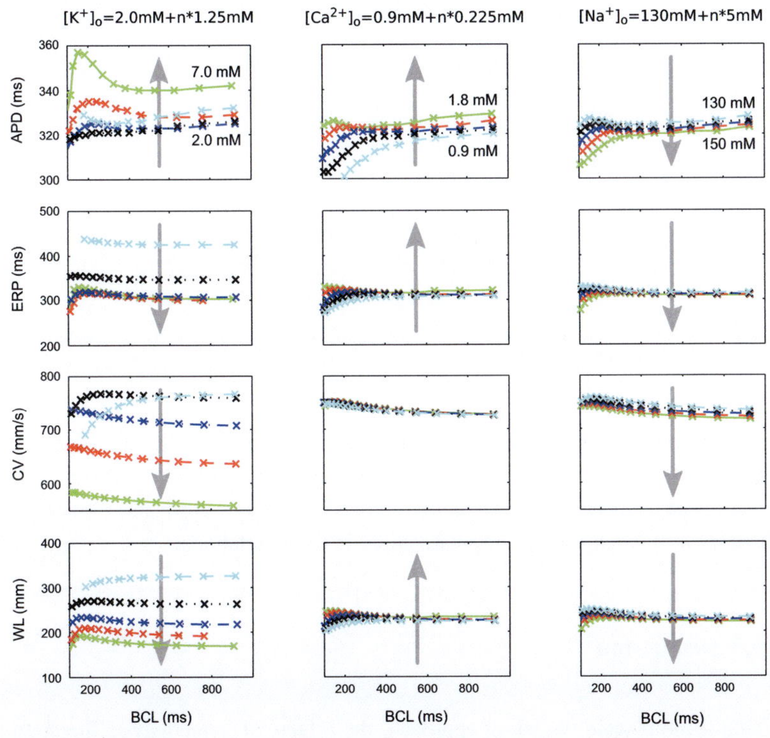

Fig. 15.3. Restitution curves for the CRN model (RA) for different concentrations of potassium, calcium and sodium ($n = 1 \ldots 5$). Green curves result from highest concentrations, cyan curves from smallest concentrations. Grey arrows indicate the impact of an increase in extracellular ion concentration on the restitution behavior.

15.2.3 Atrial Activation Sequence and P-wave

The time of complete depolarization of both atria was longer under end-HD conditions compared to pre-HD conditions (158 vs. 145 ms). Local APD was longer under end-HD conditions (Fig. 15.5) and also had a greater dispersion ΔAPD (109 vs. 62 ms, +76%).

Hyperkalemia, hypo-/hypercalcaemia and hypo-/hypernatremia did not alter the Einthoven ECG signals. As an example the Einthoven II signal under hypoercalcaemia conditions is shown in Figure 15.5. The offset from the pre-HD signal was smaller in the other settings. End-HD conditions and hypokalemia led to a pro-

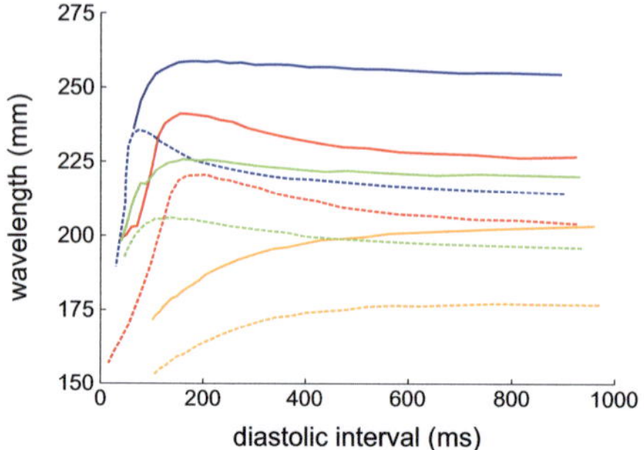

Fig. 15.4. Wavelength restitution curves before (solid curves) and just after hemodialysis (dashed curves). Hemodialysis decreased the wavelength. This allows reentries to develop and maintain in a smaller geometric area. Blue: CT, red: CAM, green: RAA, yellow: TVR. Figure adapted from [399].

longation of the P-wave (+13 and +24 ms) and a more negative amplitude of the Ta-wave. The effect was stronger under hypokalemia influence (Fig. 15.5). The pronounced Ta-wave amplitude under end-HD conditions correlates to the larger dispersion of APD, and thus repolarization, in the simulations.

15.3 Discussion

Using a multi-scale modeling approach, the effects of hemodialysis therapy on the atrial electrophysiology have been investigated to close the knowledge gap between cellular experimental findings and macroscopic observations.

Electrophysiological tissue properties were determined in a beam of atrial cells. Such coupled-cell environment usually has quantitatively different properties compared to single cell preparations (Sec. 5.2). The results for changes in transmembrane resting voltage and upstroke duration correspond to previous single cell simulations [398]. Maximum upstroke velocity was decreased in the multi-cell setup, whereas it was increased in the single-cell study. The general effects of changes in potassium, calcium and sodium on APD and ERP were comparable between both studies. The multi-cell simulation setup allowed the determination of the influence of electrolyte changes on the conduction velocity and thus also on the minimal wavelength required for a reentry circuit (WL = CV × ERP). Conduction velocity was decreased slightly by a decrease in $[Na^+]_o$, but was not significantly altered by changes in $[Ca^{2+}]_o$. Decreasing $[K^+]_o$ led to a substantial decrease in conduc-

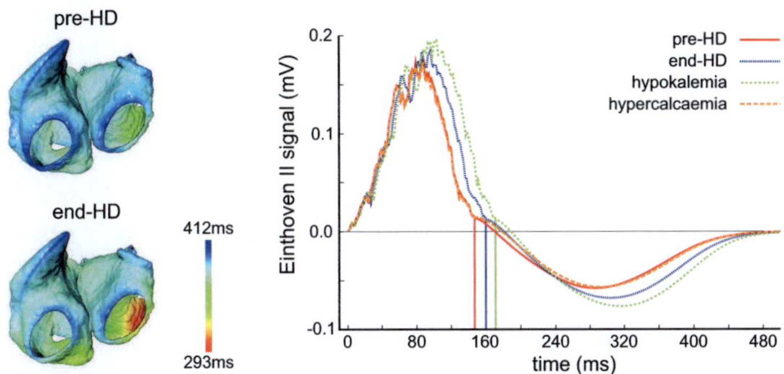

Fig. 15.5. Local APD$_{90}$ distribution under pre-HD and end-HD conditions. Einthoven II ECG signals under pre-HD, end-HD, hypokalemia and hypercalcaemia conditions.

tion velocity. The typical morphology of the restitution curves (Fig. 15.4) was not altered significantly by the changes in extracellular ion concentrations, whereas the absolute values differed.

In the sensitivity analysis, $[K^+]_o$ introduced the largest relative change. The range was determined by the measured electrolyte concentrations (Tab. 15.1). It is reasonable to assume that broader ranges for $[Ca^{2+}]_o$ and $[Na^+]_o$ would not alter the general findings, as the slopes of the $[Ca^{2+}]_o$ and $[Na^+]_o$-dependent curves were flatter than the slope of the corresponding $[K^+]_o$-dependent curves (Fig. 15.2).

The reported global slowing of conduction velocity [398] could be traced backed to the decrease of plasma potassium concentration at the end of the hemodialysis session. All right atrial cell types were affected in a similar manner and the most significant influence on the electrophysiology was a change in conduction velocity due to a change of $[K^+]_o$. ERP was decreased after hemodialysis therapy due to a combination of decreased potassium and increased calcium concentrations in the extracellular space. The changes of ERP and conduction velocity potentiate another in the wavelength, which was reduced at the end of hemodialysis therapy (Fig. 15.4). This indicates a pro-arrhythmic substrate which favors AF onset and maintenance, as reentries or rotors may establish in a smaller geometrical region or shorter pathway. This could serve as an explanation for the reported higher incidence of AF in hemodialysis patients.

The 3D bi-atrial simulations revealed an increased dispersion of APD and thus repolarization at the end of a hemodialysis session. Again, decreased extracellular potassium concentration could be identified as a cause for this. Changes in calcium and sodium did not play a significant role. These results agree with findings from Pandit et al. [447] that hyperkalemia causes an increased organization of chaotic activation patterns and vice-versa that hypokalemia promotes chaotic activation.

The simulated P-wave was longer under end-HD conditions (+13 ms). The total P-wave duration and the increase due to hemodialysis are in agreement with experimental findings (131 ± 11 ms pre-HD vs. 140 ± 12 ms end-HD) [398]. The P-wave prolongation was the result of the decreased conduction velocity in all atrial regions after hemodialysis.

The values for extracellular ion concentrations given in the original CRN model [72] differ from the measured pre-HD ion concentrations. Sodium concentration in the CRN model corresponds to the measurements (140 vs. 139.8 mM). The potassium concentrations is assumed to be higher in the original model formulation compared to the presented measurements (5.4 vs. 4.9–5.2 mM). The CRN model calcium concentration is significantly higher than the measured values (1.8 vs. 1.16–1.21 mM). The difference is most likely caused by different experimental environments. The CRN model is based on *in-vitro* experiments, which in some cases require a different cell milieu than present *in-vivo*.

The variation in model parameters was justified by the assumption that extra-cellular ion concentrations correspond directly to plasma electrolyte concentrations. This is a reasonable assumption, as hemodialysis therapy lasts several hours and changes in plasma concentrations will wash in/out into the myocardium and thereby minimize gradients between concentrations in both compartments. It remains to be proven experimentally that these concentrations correlate. The model parameterization based on blood electrolyte measurements could also allow for a personalization of the electrophysiological model in other settings (Sec. 8.2).

Hemodialysis decreases the blood volume overload and thus reduces atrial stretch and sympathetic activation. Severi et al. discussed that the observed prolongation of the P-wave after hemodialysis is most definitely caused by direct effects on the atrial electrophysiology. In contrast, acutely reduced atrial dilatation, as expected during the hemodialysis, increases the atrial conduction velocity (mechano-electrical feedback) [322]. This could partially account for the simulated P-wave prolongation, which was at the upper end of the measured prolongations.

Similar to the atrial dilatation, also an increased incidence of atrial fibrosis can be observed in hemodialysis patients. This is most likely caused by frequent changes in fluid and pressure overload. In the model, both effects were neglected, although they will influence the conduction velocity and thus P-wave duration. The conduction velocity slowing from dilatation and fibrosis will be the same at the beginning of an hemodialysis session and afterwards, as both are chronic alterations of the atrial tissue. The relative effects shown in this study therefore remain valid.

Seemann et al. [92] and Chandler et al. [239, 448] presented computational models of the electrophysiology of human sinus nodal cells. These models could be examined with the same work flow as the heterogeneous CRN model. This could in future studies provide insights into the influence of hemodialysis on cell automaticity and ectopic triggers, which are more frequent during hemodialysis.

The study results indicate that hemodialysis therapy itself could be a cause for the increased prevalence of AF in ESRD and hemodialysis patients. Very recent findings from Buiten et al. [449] support this hypothesis. They were able to show that AF episodes occur more frequently on week-days at which hemodialysis patients undergo therapy sessions.

In the future, patient-specific atrial models might enable a model-based patient-specific hemodialysis therapy control. This could be realized as an automatic control system attached to the hemodialysis hardware using ECG and blood electrolyte measurements as input. The controller could then continuously regulate the ion concentrations in the dialysate to prevent severe changes to the myocardial electrophysiology based on real-time simulations using the current parameters. Similar techniques have been shown to be useful in e.g. supporting temperature monitoring during open heart surgery [362, 450, 451]. A prerequisite for this is the reduction of the anatomical and electrophysiology model (e.g. [240] complexity and a significant speed-up of the computation, e.g. by using simplified excitation models (Sec. 5.7).

16

Conclusion

In this thesis solutions for personalized multi-scale atrial modeling were presented. These span from single cell electrophysiology modeling to ECG computation. They enabled model applications in basic and clinical research (Fig. 16.1).

For one, a method to introduce local atrial fiber orientation in patient-specific models allowed more realistic simulation of the atrial excitation sequence than before. The fiber models introduced in this thesis already found wide use in the modeling community and hopefully lead to more reliable *in-silico* studies in basic research and eventually also in clinical practice. Fiber architecture was complimented with a new model of heterogeneous atrial electrophysiology. The model provided insights into the atrial repolarization sequence and the corresponding signal in the ECG. The results support the hypothesis of the existence of action potential duration gradients in the atria. For the first time, the multi-scale simulation methodology was validated against *in-vitro* cell experiments and *in-vivo* measured P and Ta-wave from AV-block patients. Fiber orientation and heterogeneous electrophysiology enabled the initiation of atrial fibrillation in a patient-specific anatomical model without the need of unrealistic initiation protocols. This opens up a wide field of applications to understand initiation and perpetuation mechanisms of atrial fibrillation in different atrial regions. In the future, investigations of fiber- and region-related pathologies such as myocardial fibrosis can be assessed for the individual patient using these methods.

Next, a model of the heterogeneous atrial tension development was created. Together with the model of the atrial fiber architecture this enabled for the first time the biomechanical simulation of the passive and active atrial deformation during the cardiac cycle in a patient-specific whole heart model. A subsequent analysis of the effect of common ablation lesion patterns on the cardiac function revealed

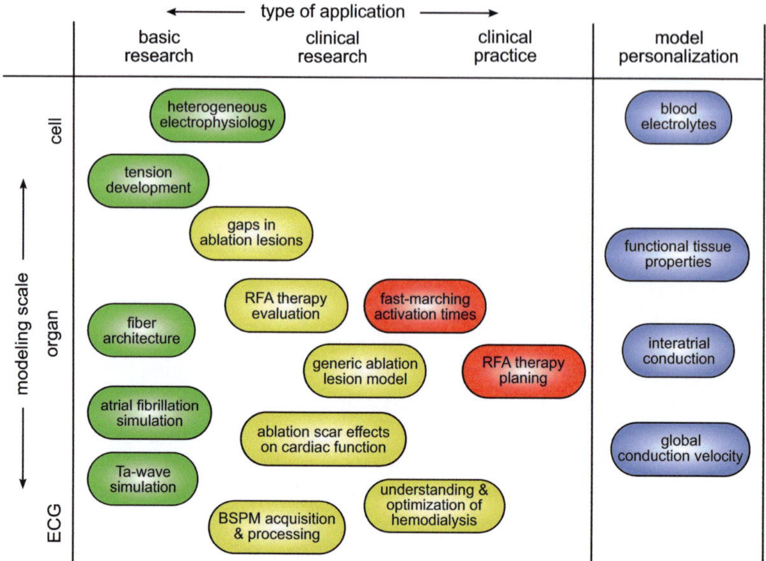

Fig. 16.1. Classification of the atrial modeling methods and simulation results presented in this thesis into a matrix of modeling scale and field of application. Additionally, the model personalization techniques introduced in this thesis are listed. Colors correspond to the field of model usage (see Fig. 1.1).

a significant decrease in atrial stroke volume after ablation which correlated to the amount of scar tissue introduced. With this methodology, ablation lesions could be optimized prior to an intervention *in-silico*, to securely eliminate atrial fibrillation while ensuring the best possible cardiac function afterwards. This could significantly improve the long-term quality of life of the patient. A coupling of the biomechanical simulations to fluid dynamic simulations may also permit the evaluation of the risk of blood clotting during atrial fibrillation and in pathological regions in the atria. This would help to understand processes which cannot be observed *in-vivo* and may contribute to the identification of new treatment opportunities.

Thirdly, an improved scheme to detect the preferential interatrial conduction route from standard ECG measurements was developed with the aid of a newly developed fast simulation framework. This allowed for high-throughput whole organ simulations in a cohort of anatomical models. This new approach of *in-silico* investigation provided more confidence in the simulation results and also showed

if significant interindividual differences were to be expected when translating the
method to clinical data. In the future, the framework could be used to gain a deeper
understanding of abnormal atrial excitation and may be used to create new ECG
acquisition and signal analysis techniques.

The multi-scale modeling approach also provided an explanation for a prolonged
P-wave duration after hemodialysis therapy in renal disease patients. Changes in
blood plasma potassium and calcium were linked to changes in cell kinetics which
subsequently caused a slowing of atrial excitation and a prolongation of the P-
wave. The effected cells also show an increased vulnerability to the development
of atrial arrhythmia caused by the therapy. This provides a new point of view on the
fact that renal disease patients have an increased incidence of AF. The simulations
hint that not solely the disease, but also the therapy might be the reason for the
arrhythmia. In the future, model-based patient monitoring during hemodialysis
therapy could adjust the dialysis solutions to the individual patient's situation, such
that the risk of arrhythmia initiation is minimized.

Besides the development of basic atrial modeling techniques, a framework to personalize and integrate anatomical, electrophysiological and excitation propagation
models was created. Statistical shape models were used to segment the variable
atrial anatomy. From the segmentation, simulation-ready models were created and
augmented with functional tissue information using LGE-MRI. Electrophysiological models were personalized based on blood plasma ion concentrations. The
global atrial conduction velocity was adjusted to match the patients' P-wave duration and interatrial conduction routes were adjusted to match local activation time
maps of the left atrium. Besides the potential application in a clinical workflow,
the framework was used to establish three cohorts of atrial and thorax models.
These can be used in multi-model simulation studies, which may provide a better
confidence in the simulation results as anatomy specific results can be ruled out.

Last, radio-frequency ablation scars were assessed on multiple modeling scales.
Microscopic simulations revealed a selective conduction of gaps in ablation lesions dependent on local fiber orientation, tissue electrophysiology and excitation
rate. The results aided the understanding of late recurrence of AF after an ablation procedure. On the whole organ scale, a framework to evaluate the risk of
pulmonary vein reconnection after ablation therapy was developed and applied to
data from three AF patients. These simulations provided insights into electrical
pathways which reconnect the pulmonary veins to the left atrium and could guide

the placement of additional ablation points. The framework can be used to plan redo-procedures for patients in whom the previous ablation intervention was not successful.

Overall, many advancements in cellular, tissue and whole atria modeling have been described in this thesis. Techniques to personalize atrial models in a smooth pipeline were presented. This allows a more realistic, more reliable and patient-specific simulation of the atrial activity. The enormous potential of the multi-scale modeling approach was shown in various applications, which already aided the in-depth understanding of pathologies and therapy outcome. Altogether, this opens a wide field of usage of personalized multi-scale atrial models for basic research and clinical applications. This will hopefully aid the more complete understanding of atrial diseases and respective therapy mechanisms. In the near future, model-based therapy evaluation and planing could shorten clinical treatment times and increase the therapy success for reduced treatment duration, therapy costs and, last but not least, the burden on the individual patient.

References

[1] H. Calkins, K. H. Kuck, R. Cappato, J. Brugada, A. J. Camm, S.-A. Chen, H. J. G. Crijns, R. J. J. Damiano, D. W. Davies, J. DiMarco, J. Edgerton, K. Ellenbogen, M. D. Ezekowitz, D. E. Haines, M. Haissaguerre, G. Hindricks, Y. Iesaka, W. Jackman, J. Jalife, P. Jais, J. Kalman, D. Keane, Y.-H. Kim, P. Kirchhof, G. Klein, H. Kottkamp, K. Kumagai, B. D. Lindsay, M. Mansour, F. E. Marchlinski, P. M. McCarthy, J. L. Mont, F. Morady, K. Nademanee, H. Nakagawa, A. Natale, S. Nattel, D. L. Packer, C. Pappone, E. Prystowsky, A. Raviele, V. Reddy, J. N. Ruskin, R. J. Shemin, H.-M. Tsao, and D. Wilber, "2012 HRS/EHRA/ECAS expert consensus statement on catheter and surgical ablation of atrial fibrillation: recommendations for patient selection, procedural techniques, patient management and follow-up, definitions, endpoints, and research trial design," *J Interv Card Electrophysiol*, vol. 33, pp. 171–257, 2012.

[2] V. Jacquemet, L. Kappenberger, and C. S. Henriquez, "Modeling atrial arrhythmias: Impact on clinical diagnosis and therapies," *IEEE Reviews in Biomedical Engineering*, vol. 1, pp. 94–114, 2008.

[3] R. C. P. Kerckhoffs, *Patient specific modeling of the cardiovascular system: technology-driven personalized medicine*. New York ; Heidelberg [u.a.] : Springer, 2010.

[4] O. Dössel, *Innovationsreport 2012 - Personalisierte Medizintechnik*, ch. Patientenmodelle, pp. 14–19. Deutsche Gesellschaft für Biomedizinische Technik (DGBMT), 2012.

[5] O. Dössel, M. W. Krueger, F. M. Weber, M. Wilhelms, and G. Seemann, "Computational modeling of the human atrial anatomy and electrophysiology," *Medical & Biological Engineering & Computing*, vol. 50, pp. 773–799, 2012.

[6] O. Dössel, M. W. Krueger, and G. Seemann, *Cardiac Mapping*, vol. 4, ch. Personalized Electrophysiological Modeling of the Human Atrium. Wiley-Blackwell, 2013.

[7] O. Ecabert and N. Smith, "euHeart: Integrated Cardiac Care using Patient-specific Cardiovascular Modeling," *SPIE Newsroom*, 2008.

[8] N. Smith, A. de Vecchi, M. McCormick, O. Camara, A. F. Frangi, H. Delingette, M. Sermesant, N. Ayache, M. W. Krueger, W. H. W. Schulze, R. Hose, I. Valverde, P. Beerbaum, C. Staicu, M. Siebes, J. Spaan, P. Hunter, J. Weese, H. Lehmann, D. Chapelle, and R. Rezavi, "euHeart: personalized and integrated cardiac care using patient-specific cardiovascular modelling," *Interface Focus*, vol. 1, pp. 349–364, 2011.

[9] S. Y. Ho and A. E. Becker, *Hurst's The Heart*, ch. Anatomy of electrophysiology. Fuster, V.; Hurst, J. W.; O'Rourke, R. A.; Walsh, R. A.; King, S. B.; Poole-Wilson, P. A., 12 ed., 2008.

[10] S. Y. Ho and D. Sanchez-Quintana, "The importance of atrial structure and fibers," *Clinical Anatomy*, vol. 22, pp. 52–63, 2009.

[11] D. Sanchez-Quintana, R. H. Anderson, J. A. Cabrera, V. Climent, R. Martin, J. Farre, and S. Y. Ho, "The terminal crest: morphological features relevant to electrophysiology," *Heart (British Cardiac Society)*, vol. 88, pp. 406–411, 2002.

[12] F. Saremi and S. Krishnan, "Cardiac conduction system: anatomic landmarks relevant to interventional electrophysiologic techniques demonstrated with 64-detector CT," *Radiographics : a Review Publication of the Radiological Society of North America, Inc*, vol. 27, pp. 1539–65, 2007.

[13] K. Wang, S. Y. Ho, D. G. Gibson, and R. H. Anderson, "Architecture of atrial musculature in humans," *British Heart Journal*, vol. 73, pp. 559–565, 1995.

[14] J. W. Papez, "Heart musculature of the atria," *American Journal of Anatomy*, vol. 27, pp. 255–286, 1920.

[15] D. Sanchez-Quintana, J. A. Cabrera, V. Climent, R. H. Anderson, and S. Y. Ho, "Sinus Node Revisited in the Era of Electroanatomical Mapping and Catheter Ablation," *Heart*, vol. 91, pp. 189–194, 2005.

[16] D. Sanchez-Quintana, J. A. Cabrera, V. Climent, J. Farre, M. C. d. Mendonca, and S. Y. Ho, "Anatomic relations between the esophagus and left atrium and relevance for ablation of atrial fibrillation," *Circulation*, vol. 112, pp. 1400–1405, 2005.

[17] C. Stöllberger, G. Ernst, E. Bonner, J. Finsterer, and J. Slany, *Left atrial appendage morphology: Comparison of transesophageal images and postmortem casts.* Springer Berlin / Heidelberg Zeitschrift für Kardiologie, 2003.

[18] A. M. Pritchett, S. J. Jacobsen, D. W. Mahoney, R. J. Rodeheffer, K. R. Bailey, and M. M. Redfield, "Left atrial volume as an index of left atrial size: a population-based study," *Journal of the American College of Cardiology*, vol. 41, pp. 1036–1043, 2003.

[19] T. S. M. Tsang, M. E. Barnes, B. J. Gersh, K. R. Bailey, and J. B. Seward, "Left atrial volume as a morpho-physiologic expression of left ventricular diastolic dysfunction and relation to cardiovascular risk burden," *The American Journal of Cardiology*, vol. 90, pp. 1284–1289, 2002.

[20] U. Schotten, H. R. Neuberger, and M. A. Allessie, "The role of atrial dilatation in the domestication of atrial fibrillation," *Progress in Biophysics and Molecular Biology*, vol. 82, pp. 151–162, 2003.

[21] J. A. Cabrera, D. Sanchez-Quintana, J. Farre, J. M. Rubio, and S. Y. Ho, "The inferior right atrial isthmus: further architectural insights for current and coming ablation technologies," *Journal of Cardiovascular Electrophysiology*, vol. 16, pp. 402–408, 2005.

[22] G. K. Feld, M. Mollerus, U. Birgersdotter-Green, O. Fujimura, T. D. Bahnson, K. Boyce, and M. Rahme, "Conduction velocity in the tricuspid valve-inferior vena cava isthmus is slower in patients with type I atrial flutter compared to those without a history of atrial flutter," *Journal of Cardiovascular Electrophysiology*, vol. 8, pp. 1338–1348, 1997.

[23] F. H. M. Wittkampf, M. F. van Oosterhout, P. Loh, R. Derksen, E.-j. Vonken, P. J. Slootweg, and S. Y. Ho, "Where to draw the mitral isthmus line in catheter ablation of atrial fibrillation: histological analysis," *European Heart Journal*, vol. 26, pp. 689–695, 2005.

[24] P. G. Platonov, V. Ivanov, S. Y. Ho, and L. Mitrofanova, "Left atrial posterior wall thickness in patients with and without atrial fibrillation: data from 298 consecutive autopsies," *Journal of Cardiovascular Electrophysiology*, vol. 19, pp. 689–692, 2008.

[25] J. M. T. de Bakker and T. Opthof, "Is the apico-basal gradient larger than the transmural gradient?," *Journal of Cardiovascular Pharmacology*, vol. 39, pp. 328–331, 2002.

[26] S. Y. Ho, R. H. Anderson, and D. Sánchez-Quintana, "Atrial structure and fibres: morphologic based of atrial conduction," *Cardiovascular Research*, vol. 54, pp. 325–336, 2002.

[27] R. Lemery, L. Soucie, B. Martin, A. S. L. Tang, M. Green, and J. Healey, "Human study of biatrial electrical coupling: determinants of endocardial septal activation and conduction over interatrial connections," *Circulation*, vol. 110, pp. 2083–2089, 2004.

[28] R. H. Anderson and A. C. Cook, "The structure and components of the atrial chambers," *Europace*, vol. 9 Suppl 6, pp. vi3–9, 2007.

[29] M. D. Lesh, J. M. Kalman, J. E. Olgin, and W. S. Ellis, "The role of atrial anatomy in clinical atrial arrhythmias," *Journal of Electrocardiology*, vol. 29 Suppl, pp. 101–113, 1996.

[30] M. Hocini, P. Loh, S. Y. Ho, D. Sanchez-Quintana, B. Thibault, J. M. de Bakker, and M. J. Janse, "Anisotropic conduction in the triangle of koch of mammalian hearts: electrophysiologic and anatomic correlations," *Journal of the American College of Cardiology*, vol. 31, pp. 629–636, 1998.

[31] T. R. Betts, S. Y. Ho, D. Sanchez-Quintana, P. R. Roberts, R. H. Anderson, and J. M. Morgan, "Three-dimensional mapping of right atrial activation during sinus rhythm and its relationship to endocardial archi-tecture," *Journal of Cardiovascular Electrophysiology*, vol. 13, pp. 1152–1159, 2002.

[32] V. Markides, R. J. Schilling, S. Y. Ho, A. W. C. Chow, D. W. Davies, and N. S. Peters, "Characterization of left atrial activation in the intact human heart," *Circulation*, vol. 107, pp. 733–739, 2003.

[33] P. G. Platonov, "Interatrial conduction in the mechanisms of atrial fibrillation: from anatomy to cardiac signals and new treatment modalities," *Europace : European Pacing, Arrhythmias, and Cardiac Electrophysiology : Journal of the Working Groups on Cardiac Pacing, Arrhythmias, and Cardiac Cellular Electrophysiology of the European Society of Cardiology*, vol. 9 Suppl 6, pp. vi10–6, 2007.

[34] R. Lemery, G. Guiraudon, and J. P. Veinot, "Anatomic description of Bachmann's bundle and its relation to the atrial septum," *The American Journal of Cardiology*, vol. 91, pp. 1482–5, A8, 2003.

[35] J. P. Veinot and R. Lemery, "Innovations in cardiovascular pathology: anatomic and electrophysiologic deter-minants associated with ablation of atrial arrhythmias," *Cardiovascular Pathology : the Official Journal of the Society for Cardiovascular Pathology*, vol. 14, pp. 204–213, 2005.

[36] F. Saremi, S. Channual, S. Krishnan, S. V. Gurudevan, J. Narula, and A. Abolhoda, "Bachmann Bundle and its arterial supply: imaging with multidetector CT–implications for interatrial conduction abnormalities and arrhythmias," *Radiology*, vol. 248, pp. 447–457, 2008.

[37] P. G. Platonov, L. Mitrofanova, V. Ivanov, and S. Y. Ho, "Substrates for intra-atrial and interatrial conduction in the atrial septum: anatomical study on 84 human hearts," *Heart Rhythm : the Official Journal of the Heart Rhythm Society*, vol. 5, pp. 1189–1195, 2008.

[38] S. Y. Ho, D. Sanchez-Quintana, J. A. Cabrera, and R. H. Anderson, "Anatomy of the left atrium: implications for radiofrequency ablation of atrial fibrillation," *J Cardiovasc Electrophysiol*, vol. 10, pp. 1525–1533, 1999.

[39] S.-I. Sakamoto, T. Nitta, Y. Ishii, Y. Miyagi, H. Ohmori, and K. Shimizu, "Interatrial electrical connections: the precise location and preferential conduction," *Journal of Cardiovascular Electrophysiology*, vol. 16, pp. 1077–1086, 2005.

[40] E. M. Marom, J. E. Herndon, Y. H. Kim, and H. P. McAdams, "Variations in pulmonary venous drainage to the left atrium: implications for radiofrequency ablation," *Radiology*, vol. 230, pp. 824–829, 2004.

[41] H. Calkins, S. Y. Ho, J. Angel Cabrera, P. Della Bella, J. Farre, J. Kautzner, and P. Tchou, *Atrial Fibrillation Ablation*, ch. Anatomy of the left atrium and pulmonary veins, pp. 1–10. Blackwell Publishing Ltd, 2008.

[42] H. M. Tsao, W. C. Yu, H. C. Cheng, M. H. Wu, C. T. Tai, W. S. Lin, Y. A. Ding, M. S. Chang, and S. A. Chen, "Pulmonary vein dilation in patients with atrial fibrillation: detection by magnetic resonance imaging," *Journal of Cardiovascular Electrophysiology*, vol. 12, pp. 809–813, 2001.

[43] R. Kato, L. Lickfett, G. Meininger, T. Dickfeld, R. Wu, G. Juang, P. Angkeow, J. LaCorte, D. Bluemke, R. Berger, H. R. Halperin, and H. Calkins, "Pulmonary vein anatomy in patients undergoing catheter ablation of atrial fibrillation: lessons learned by use of magnetic resonance imaging," *Circulation*, vol. 107, pp. 2004–2010, 2003.

[44] M. Mansour, G. Holmvang, D. Sosnovik, R. Migrino, S. Abbara, J. Ruskin, and D. Keane, "Assessment of pulmonary vein anatomic variability by magnetic resonance imaging: implications for catheter ablation techniques for atrial fibrillation," *Journal of Cardiovascular Electrophysiology*, vol. 15, pp. 387–393, 2004.

[45] H. Mlcochova, J. Tintera, V. Porod, P. Peichl, R. Cihak, and J. Kautzner, "Magnetic resonance angiography of pulmonary veins: implications for catheter ablation of atrial fibrillation," *Pacing and Clinical Electrophysiology : PACE*, vol. 28, pp. 1073–1080, 2005.

[46] M. R. M. Jongbloed, H. J. Lamb, J. J. Bax, J. D. Schuijf, A. de Roos, E. E. van der Wall, and M. J. Schalij, "Noninvasive visualization of the cardiac venous system using multislice computed tomography," *Journal of the American College of Cardiology*, vol. 45, pp. 749–753, 2005.

[47] K. Kaseno, H. Tada, K. Koyama, M. Jingu, S. Hiramatsu, M. Yokokawa, K. Goto, S. Naito, S. Oshima, and K. Taniguchi, "Prevalence and characterization of pulmonary vein variants in patients with atrial fibrillation determined using 3-dimensional computed tomography," *The American Journal of Cardiology*, vol. 101, pp. 1638–1642, 2008.

[48] R. Hanna, H. Barschdorf, T. Klinder, F. M. Weber, M. W. Krueger, O. Dössel, and C. Lorenz, "A hybrid method for automatic anatomical variant detection and segmentation," in *Functional Imaging and Modeling of the Heart 2011, Lecture Notes in Computer Science*, vol. 6666, pp. 333–340, 2011.

[49] R. J. Hassink, H. T. Aretz, J. Ruskin, and D. Keane, "Morphology of atrial myocardium in human pulmonary veins: a postmortem analysis in patients with and without atrial fibrillation," *Journal of the American College of Cardiology*, vol. 42, pp. 1108–1114, 2003.

[50] W. S. Lin, V. S. Prakash, C. T. Tai, M. H. Hsieh, C. F. Tsai, W. C. Yu, Y. K. Lin, Y. A. Ding, M. S. Chang, and S. A. Chen, "Pulmonary vein morphology in patients with paroxysmal atrial fibrillation initiated by ectopic beats originating from the pulmonary veins: implications for catheter ablation," *Circulation*, vol. 101, pp. 1274–1281, 2000.

[51] A. Keith and M. W. Flack, "The form and nature of the muscular connections between the primary divisions of the vertebrate heart," *J. Anat. Physiol.*, vol. 41, pp. 172–189, 1907.

[52] C. V. Desimone, A. Noheria, N. Lachman, W. D. Edwards, A. S. Gami, J. J. Maleszewski, P. A. Friedman, T. M. Munger, S. C. Hammill, D. L. Packer, and S. J. Asirvatham, "Myocardium of the Superior Vena Cava, Coronary Sinus, Vein of Marshall, and the Pulmonary Vein Ostia: Gross Anatomic Studies in 620 Hearts," *Journal of Cardiovascular Electrophysiology*, 2012.

[53] S. A. Chen, M. H. Hsieh, C. T. Tai, C. F. Tsai, V. S. Prakash, W. C. Yu, T. L. Hsu, Y. A. Ding, and M. S. Chang, "Initiation of atrial fibrillation by ectopic beats originating from the pulmonary veins: electrophysiological characteristics, pharmacological responses, and effects of radiofrequency ablation," *Circulation*, vol. 100, pp. 1879–1886, 1999.

[54] T. Saito, K. Waki, and A. E. Becker, "Left atrial myocardial extension onto pulmonary veins in humans: anatomic observations relevant for atrial arrhythmias," *J Cardiovasc Electrophysiol*, vol. 11, pp. 888–894, 2000.

[55] S. Y. Ho, J. A. Cabrera, V. H. Tran, J. Farre, R. H. Anderson, and D. Sanchez-Quintana, "Architecture of the pulmonary veins: relevance to radiofrequency ablation," *Heart*, vol. 86, pp. 265–270, 2001.

[56] M. Haissaguerre, P. Jais, D. C. Shah, T. Arentz, D. Kalusche, A. Takahashi, S. Garrigue, M. Hocini, J. T. Peng, and J. Clementy, "Catheter ablation of chronic atrial fibrillation targeting the reinitiating triggers," *Journal of Cardiovascular Electrophysiology*, vol. 11, pp. 2–10, 2000.

[57] M. Hocini, S. Y. Ho, T. Kawara, A. C. Linnenbank, M. Potse, D. Shah, P. Jais, M. J. Janse, M. Haissaguerre, and J. M. De Bakker, "Electrical conduction in canine pulmonary veins: electrophysiological and anatomic correlation," *Circulation*, vol. 105, pp. 2442–2448, 2002.

[58] A. Y. Tan, H. Li, S. Wachsmann-Hogiu, L. S. Chen, P.-S. Chen, and M. C. Fishbein, "Autonomic innervation and segmental muscular disconnections at the human pulmonary vein-atrial junction: implications for catheter ablation of atrial-pulmonary vein junction," *Journal of the American College of Cardiology*, vol. 48, pp. 132–143, 2006.

[59] H. Nathan and M. Eliakim, "The junction between the left atrium and the pulmonary veins. An anatomic study of human hearts," *Circulation*, vol. 34, pp. 412–422, 1966.

[60] P. Neher, H. Barschdorf, S. Dries, F. M. Weber, M. W. Krueger, O. Dössel, and C. Lorenz, "Automatic segmentation of cardiac CTs - personalized atrial models augmented with electrophysiological structures," in *Functional Imaging and Modeling of the Heart 2011, Lecture Notes in Computer Science*, vol. 6666, pp. 80–87, 2011.

[61] K. Otomo, K. Uno, H. Fujiwara, M. Isobe, and Y. Iesaka, "Local unipolar and bipolar electrogram criteria for evaluating the transmurality of atrial ablation lesions at different catheter orientations relative to the endocardial surface," *Heart Rhythm : the Official Journal of the Heart Rhythm Society*, vol. 7, pp. 1291–1300, 2010.

[62] B. Hall, V. Jeevanantham, R. Simon, J. Filippone, G. Vorobiof, and J. Daubert, "Variation in left atrial transmural wall thickness at sites commonly targeted for ablation of atrial fibrillation," *Journal of Interventional Cardiac Electrophysiology*, vol. 17, pp. 127–132, 2006.

[63] T. Deneke, K. Khargi, K.-M. Muller, B. Lemke, A. Mugge, A. Laczkovics, A. E. Becker, and P. H. Grewe, "Histopathology of intraoperatively induced linear radiofrequency ablation lesions in patients with chronic atrial fibrillation," *European Heart Journal*, vol. 26, pp. 1797–1803, 2005.

[64] J. L. Coffey, M. Cristy, and G. G. Warner, "Specific absorbed fractions for photon sources uniformly distributed in the heart chambers and heart wall of a heterogeneous phantom," *Journal of Nuclear Medicine : Official Publication, Society of Nuclear Medicine*, vol. 22, pp. 65–71, 1981.

[65] F. W. Sunderman, *Normal values in clinical medicine*, ch. Anatomical normals. WB Saunders Co, 1949.

[66] J. A. Cabrera, S. Y. Ho, V. Climent, and D. Sanchez-Quintana, "The architecture of the left lateral atrial wall: a particular anatomic region with implications for ablation of atrial fibrillation," *European Heart Journal*, vol. 29, pp. 356–362, 2008.

[67] F. O. Campos, T. Wiener, A. J. Prassl, H. Ahammer, G. Plank, R. Weber Dos Santos, D. Sanchez-Quintana, and E. Hofer, "A 2D-computer model of atrial tissue based on histographs describes the electro-anatomical impact of microstructure on endocardiac potentials and electric near-fields," *Conf Proc IEEE Eng Med Biol Soc*, vol. 1, pp. 2541–2544, 2010.

[68] J. Zhao, M. L. Trew, I. J. Legrice, B. H. Smaill, and A. J. Pullan, "A tissue-specific model of reentry in the right atrial appendage," *Journal of Cardiovascular Electrophysiology*, vol. 20, pp. 675–684, 2009.

[69] O. V. Aslanidi, M. R. Boyett, and H. Zhang, "Left to right atrial electrophysiological differences: substrate for a dominant reentrant source during atrial fibrillation," *Lecture Notes in Computer Science*, vol. 5528, pp. 154–161, 2009.

[70] F. M. Weber, D. U. J. Keller, S. Bauer, G. Seemann, C. Lorenz, and O. Dössel, "Predicting tissue conductivity influences on body surface potentials-an efficient approach based on principal component analysis," *IEEE Transactions on Biomedical Engineering*, vol. 58, pp. 265–273, 2011.

[71] D. U. J. Keller, F. M. Weber, G. Seemann, and O. Dössel, "Ranking the influence of tissue conductivities on forward-calculated ECGs," *IEEE Transactions on Biomedical Engineering*, vol. 57, pp. 1568–1576, 2010.

[72] M. Courtemanche, R. J. Ramirez, and S. Nattel, "Ionic mechanisms underlying human atrial action potential properties: Insights from a mathematical model," *Am. J. Physiol.*, vol. 275, pp. H301–H321, 1998.

[73] G. Seemann, P. Carrillo Bustamante, S. Ponto, M. Wilhelms, E. P. Scholz, and O. Dössel, "Atrial fibrillation-based electrical remodeling in a computer model of the human atrium," in *Proc. Computing in Cardiology*, vol. 37, pp. 417–420, 2010.

[74] U. Schotten, M. Greiser, D. Benke, K. Buerkel, B. Ehrenteidt, C. Stellbrink, J. F. Vazquez-Jimenez, F. Schoendube, P. Hanrath, and M. Allessie, "Atrial fibrillation-induced atrial contractile dysfunction: a tachycardiomyopathy of a different sort," *Cardiovascular Research*, vol. 53, pp. 192–201, 2002.

[75] U. Schotten, S. Verheule, P. Kirchhof, and A. Goette, "Pathophysiological mechanisms of atrial fibrillation: a translational appraisal," *Physiological Reviews*, vol. 91, pp. 265–325, 2011.

[76] G. Seemann, *Modeling of electrophysiology and tension development in the human heart*. PhD thesis, Universitätsverlag Karlsruhe, Institut für Biomedizinische Technik, 2005.

[77] A. Hansson, M. Holm, P. Blomstrom, R. Johansson, C. Luhrs, J. Brandt, and S. Olsson, "Right atrial free wall conduction velocity and degree of anisotropy in patients with stable sinus rhythm studied during open heart surgery," *Eur. Heart. J.*, vol. 19, pp. 293–300, 1998.

[78] S.-F. Lin and J. P. Wikswo, "Panoramic optical imaging of electrical propagation in isolated heart," *Journal of Biomedical Optics*, vol. 4, pp. 200–207, 1999.

[79] M. Spach, P. C. Dolber, and J. F. Heidlage, "Influence of the passive anisotropic properties on directional differences in propagation following modification of the sodium conductance in human atrial muscle: a model of reentry based on anisotropic discontinuous propagation," *Circulation Research*, vol. 62, pp. 811–832, 1988.

[80] H. Dimitri, M. Ng, A. G. Brooks, P. Kuklik, M. K. Stiles, D. H. Lau, N. Antic, A. Thornton, D. A. Saint, D. McEvoy, R. Antic, J. M. Kalman, and P. Sanders, "Atrial remodeling in obstructive sleep apnea: Implications for atrial fibrillation," *Heart Rhythm : the Official Journal of the Heart Rhythm Society*, vol. (In Press Uncorrected Proof), 2011.

[81] P. Kanagaratnam, S. Rothery, P. Patel, N. J. Severs, and N. S. Peters, "Relative expression of immunolocalized connexins 40 and 43 correlates with human atrial conduction properties," *Journal of the American College of Cardiology*, vol. 39, pp. 116–123, 2002.

[82] R. J. Schilling, N. S. Peters, J. Goldberger, A. H. Kadish, and D. W. Davies, "Characterization of the anatomy and conduction velocities of the human right atrial flutter circuit determined by noncontact mapping," *Journal of the American College of Cardiology*, vol. 38, pp. 385–393, 2001.

[83] D. C. Shah, P. Jais, M. Haissaguerre, S. Chouairi, A. Takahashi, M. Hocini, S. Garrigue, and J. Clementy, "Three-dimensional mapping of the common atrial flutter circuit in the right atrium," *Circulation*, vol. 96, pp. 3904–3912, 1997.

[84] F. M. Weber, A. Luik, C. Schilling, G. Seemann, M. W. Krueger, C. Lorenz, C. Schmitt, and O. Dössel, "Conduction velocity restitution of the human atrium–an efficient measurement protocol for clinical electrophysiological studies," *IEEE Transactions on Biomedical Engineering*, vol. 58, pp. 2648–2655, 2011.

[85] J. P. Boineau, T. E. Canavan, R. B. Schuessler, M. E. Cain, P. B. Corr, and J. L. Cox, "Demonstration of a widely distributed atrial pacemaker complex in the human heart," *Circulation*, vol. 77, pp. 1221–1237, 1988.

[86] D. Durrer, R. T. van Dam, G. E. Freud, M. J. Janse, F. L. Meijler, and R. C. Arzbaecher, "Total excitation of the isolated human heart," *Circulation*, vol. 41, pp. 899–912, 1970.

[87] Y. Wang and Y. Rudy, "Electrocardiographic imaging of normal human atrial repolarization," *Heart Rhythm*, vol. 6, pp. 582–583, 2009.

[88] J. E. Saffitz, H. L. Kanter, K. G. Green, T. K. Tolley, and E. C. Beyer, "Tissue-specific determinants of anisotropic conduction velocity in canine atrial and ventricular myocardium," *Circulation Research*, vol. 74, pp. 1065–1070, 1994.

[89] R. De Ponti, S. Y. Ho, J. A. Salerno-Uriarte, M. Tritto, and G. Spadacini, "Electroanatomic analysis of sinus impulse propagation in normal human atria," *Journal of Cardiovascular Electrophysiology*, vol. 13, pp. 1–10, 2002.

[90] P. Kojodjojo, P. Kanagaratnam, V. Markides, D. W. Davies, and N. Peters, "Age-related changes in human left and right atrial conduction," *Journal of Cardiovascular Electrophysiology*, vol. 17, pp. 120–127, 2006.

[91] D. M. Harrild and C. S. Henriquez, "A computer model of normal conduction in the human atria," *Circ. Res.*, vol. 87, pp. e25–e36, 2000.

[92] G. Seemann, C. Höper, F. B. Sachse, O. Dössel, A. V. Holden, and H. Zhang, "Heterogeneous three-dimensional anatomical and electrophysiological model of human atria," *Phil. Trans. Roy. Soc. A*, vol. 364, pp. 1465–1481, 2006.

[93] C. Tobon, C. Ruiz, J. F. Rodriguez, F. Hornero, J. M. Ferrero, and J. Saiz, "Vulnerability for reentry in a three dimensional model of human atria: a simulation study," *Conf Proc IEEE Eng Med Biol Soc*, vol. 2010, pp. 224–227, 2010.

[94] G. Plank, A. J. Prassl, J. I. Wang, G. Seemann, D. Scherr, D. Sanchez-Quintana, H. Calkins, and N. A. Trayanova, "Atrial fibrosis promotes the transistion of pulmonary vein ectopy into reentrant arrhathmias," in *Heart Rhythm*, 2008.

[95] O. V. Aslanidi, M. Al-Owais, A. P. Benson, M. Colman, C. J. Garratt, S. H. Gilbert, J. P. Greenwood, A. V. Holden, S. Kharche, E. Kinnell, E. Pervolaraki, S. Plein, J. Stott, and H. Zhang, "Virtual tissue engineering of the human atrium: Modelling pharmacological actions on atrial arrhythmogenesis," *European Journal of Pharmaceutical Sciences*, 2011.

[96] J. Zhao, T. D. Butters, H. Zhang, A. J. Pullan, I. J. LeGrice, G. B. Sands, and B. H. Smaill, "An image-based model of atrial muscular architecture: Effects of structural anisotropy on electrical activation," *Circulation: Arrhythmia and Electrophysiology*, vol. 5, pp. 361–370, 2012.

[97] V. Fuster, L. E. Ryden, D. S. Cannom, H. J. Crijns, A. B. Curtis, K. A. Ellenbogen, J. L. Halperin, J.-Y. Le Heuzey, G. N. Kay, J. E. Lowe, S. B. Olsson, E. N. Prystowsky, J. L. Tamargo, S. Wann, S. C. J. Smith,

A. K. Jacobs, C. D. Adams, J. L. Anderson, E. M. Antman, S. A. Hunt, R. Nishimura, J. P. Ornato, R. L. Page, B. Riegel, S. G. Priori, J.-J. Blanc, A. Budaj, A. J. Camm, V. Dean, J. W. Deckers, C. Despres, K. Dickstein, J. Lekakis, K. McGregor, M. Metra, J. Morais, A. Osterspey, J. L. Zamorano, and Heart Rhythm Society, "ACC/AHA/ESC 2006 Guidelines for the Management of Patients with Atrial Fibrillation: a report of the American College of Cardiology/American Heart Association Task Force on Practice Guidelines and the European Society of Cardiology Committee for Practice Guidelines (Writing Committee to Revise the 2001 Guidelines for the Management of Patients With Atrial Fibrillation): developed in collaboration with the European Heart Rhythm Association and the Heart Rhythm Society," 2006.

[98] T. Narumiya, T. Sakamaki, Y. Sato, and K. Kanmatsuse, "Relationship between left atrial appendage function and left atrial thrombus in patients with nonvalvular chronic atrial fibrillation and atrial flutter," *Circulation Journal*, vol. 67, pp. 68–72, 2003.

[99] M. C. Wijffels, C. J. Kirchhof, R. Dorland, and M. A. Allessie, "Atrial fibrillation begets atrial fibrillation. A study in awake chronically instrumented goats," *Circulation*, vol. 92, pp. 1954–1968, 1995.

[100] M. Haissaguerre, P. Jaïs, D. C. Shah, A. Takahashi, M. Hocini, G. Quiniou, S. Garrigue, A. Le Mouroux, P. Le Métayer, and J. Clémenty, "Spontaneous initiation of atrial fibrillation by ectopic beats originating in the pulmonary veins," *New England Journal of Medicine*, vol. 339, pp. 659–666, 1998.

[101] A. S. Go, E. M. Hylek, K. A. Phillips, Y. Chang, L. E. Henault, J. V. Selby, and D. E. Singer, "Prevalence of diagnosed atrial fibrillation in adults: national implications for rhythm management and stroke prevention: the AnTicoagulation and Risk Factors in Atrial Fibrillation (ATRIA) Study," *JAMA*, vol. 285, pp. 2370–2375, 2001.

[102] C. D. Furberg, B. M. Psaty, T. A. Manolio, J. M. Gardin, V. E. Smith, and P. M. Rautaharju, "Prevalence of atrial fibrillation in elderly subjects (the Cardiovascular Health Study)," *The American Journal of Cardiology*, vol. 74, pp. 236–241, 1994.

[103] Y. Miyasaka, M. E. Barnes, B. J. Gersh, S. S. Cha, K. R. Bailey, W. P. Abhayaratna, J. B. Seward, and T. S. M. Tsang, "Secular trends in incidence of atrial fibrillation in Olmsted County, Minnesota, 1980 to 2000, and implications on the projections for future prevalence," *Circulation*, vol. 114, pp. 119–125, 2006.

[104] M. Allessie, J. Ausma, and U. Schotten, "Electrical, contractile and structural remodeling during atrial fibrillation," *Cardiovascular Research*, vol. 54, pp. 230–246, 2002.

[105] D. J. Wilber, C. Pappone, P. Neuzil, A. De Paola, F. Marchlinski, A. Natale, L. Macle, E. G. Daoud, H. Calkins, B. Hall, V. Reddy, G. Augello, M. R. Reynolds, C. Vinekar, C. Y. Liu, S. M. Berry, and D. A. Berry, "Comparison of antiarrhythmic drug therapy and radiofrequency catheter ablation in patients with paroxysmal atrial fibrillation: a randomized controlled trial," *JAMA*, vol. 303, pp. 333–340, 2010.

[106] D. McBride, A. M. Mattenklotz, S. N. Willich, and B. Bruggenjurgen, "The costs of care in atrial fibrillation and the effect of treatment modalities in Germany," *Value in Health*, vol. 12, pp. 293–301, 2009.

[107] J. Kautzner, V. Bulkova, G. Hindricks, N. Maniadakis, P. Della Bella, P. Jais, and K.-H. Kuck, "Atrial fibrillation ablation: a cost or an investment?," *Europace*, vol. 13 Suppl 2, pp. ii39–43, 2011.

[108] P. Vardas, A. Auricchio, and J. L. Merino, *The EHRA White Book 2011 - The Current Status of Cardiac Electrophysiology in ESC Member Countries*. 2011.

[109] J. D. Fisher, M. A. Spinelli, D. Mookherjee, A. K. Krumerman, and E. C. Palma, "Atrial fibrillation ablation: reaching the mainstream," *PACE*, vol. 29, pp. 523–537, 2006.

[110] R. Ranjan, R. Kato, M. M. Zviman, T. M. Dickfeld, A. Roguin, R. D. Berger, G. F. Tomaselli, and H. R. Halperin, "Gaps in the Ablation Line as a Potential Cause of Recovery from Electrical Isolation and their Visualization using MRI," *Circulation: Arrhythmia and Electrophysiology*, 2011.

[111] F. Ouyang, M. Antz, S. Ernst, H. Hachiya, H. Mavrakis, F. T. Deger, A. Schaumann, J. Chun, P. Falk, D. Hennig, X. Liu, D. Bansch, and K.-H. Kuck, "Recovered pulmonary vein conduction as a dominant factor for recurrent atrial tachyarrhythmias after complete circular isolation of the pulmonary veins: lessons from double Lasso technique," *Circulation*, vol. 111, pp. 127–135, 2005.

[112] S. Nath, C. r. Lynch, J. G. Whayne, and D. E. Haines, "Cellular electrophysiological effects of hyperthermia on isolated guinea pig papillary muscle. Implications for catheter ablation," *Circulation*, vol. 88, pp. 1826–1831, 1993.

[113] S. Nath, J. P. DiMarco, and D. E. Haines, "Basic aspects of radiofrequency catheter ablation," *Journal of Cardiovascular Electrophysiology*, vol. 5, pp. 863–876, 1994.

[114] M. Wood, S. Goldberg, M. Lau, A. Goel, D. Alexander, F. Han, and S. Feinstein, "Direct Measurement of the Lethal Isotherm for Radiofrequency Ablation of Myocardial Tissue," *Circulation Arrhythmia and Electrophysiology*, 2011.

[115] H. Cao, V. R. Vorperian, S. Tungjitkusolmun, J. Z. Tsai, D. Haemmerich, Y. B. Choy, and J. G. Webster, "Flow effect on lesion formation in RF cardiac catheter ablation," *IEEE Transactions on bio-Medical Engineering*, vol. 48, pp. 425–433, 2001.

[116] P. Kovoor, M. P. J. Daly, J. Pouliopoulos, K. Byth, B. I. Dewsnap, V. E. Eipper, T. Yung, J. F. B. Uther, and D. L. Ross, "Comparison of radiofrequency ablation in normal versus scarred myocardium," *Journal of Cardiovascular Electrophysiology*, vol. 17, pp. 80–86, 2006.

[117] F. Holmqvist, D. Husser, J. M. Tapanainen, J. Carlson, R. Jurkko, Y. Xia, R. Havmoller, O. Kongstad, L. Toivonen, S. B. Olsson, and P. G. Platonov, "Interatrial conduction can be accurately determined using standard 12-lead electrocardiography: validation of P-wave morphology using electroanatomic mapping in man," *Heart Rhythm*, vol. 5, pp. 413–418, 2008.

[118] D. E. Haines, *Catheter Ablation of Cardiac Arrhythmias: Basic Concepts and Clinical Applications*, vol. Third Edition, ch. Biophysics and pathophysiology of lesion formation by transcatheter radiofrequency ablation, pp. 20–34. Blackwell Futura, 2008.

[119] G. Ndrepepa and H. Estner, vol. Catheter Ablation of Cardiac Arrhythmias, ch. Ablation of cardiac arrhythmias ? energy sources and mechanisms of lesion formation, pp. 35–53. Steinkopff, 2006.

[120] O. J. Eick, "Factors influencing lesion formation during radiofrequency catheter ablation," *Indian Pacing and Electrophysiology Journal*, vol. 3, pp. 117–128, 2003.

[121] D. E. Haines, "The biophysics of radiofrequency catheter ablation in the heart: the importance of temperature monitoring," *Pacing and Clinical Electrophysiology : PACE*, vol. 16, pp. 586–591, 1993.

[122] F. H. Wittkampf, R. N. Hauer, and E. O. Robles de Medina, "Control of radiofrequency lesion size by power regulation," *Circulation*, vol. 80, pp. 962–968, 1989.

[123] D. E. Haines, "Determinants of Lesion Size During Radiofrequency Catheter Ablation: The Role of Electrode-Tissue Contact Pressure and Duration of Energy Delivery," *Journal of Cardiovascular Electrophysiology*, vol. 2, pp. 509–515, 1991.

[124] A. Thiagalingam, A. D'Avila, L. Foley, J. L. Guerrero, H. Lambert, G. Leo, J. N. Ruskin, and V. Y. Reddy, "Importance of catheter contact force during irrigated radiofrequency ablation: evaluation in a porcine ex vivo model using a force-sensing catheter," *Journal of Cardiovascular Electrophysiology*, vol. 21, pp. 806–811, 2010.

[125] S. Tungjitkusolmun, E. J. Woo, H. Cao, J. Z. Tsai, V. R. Vorperian, and J. G. Webster, "Thermal–electrical finite element modelling for radio frequency cardiac ablation: effects of changes in myocardial properties," *Medical & Biological Engineering & Computing*, vol. 38, pp. 562–568, 2000.

[126] M. A. Wood and I. A. Fuller, "Acute and chronic electrophysiologic changes surrounding radiofrequency lesions," *Journal of Cardiovascular Electrophysiology*, vol. 13, pp. 56–61, 2002.

[127] Y. Z. Ge, P. Z. Shao, J. Goldberger, and A. Kadish, "Cellular electrophysiological changes induced in vitro by radiofrequency current: comparison with electrical ablation," *PACE*, vol. 18, pp. 323–333, 1995.

[128] C.-C. Wu, R. W. Fasciano, H. Calkins, and L. Tung, "Sequential change in action potential of rabbit epicardium during and following radiofrequency ablation," *Journal of Cardiovascular Electrophysiology*, vol. 10, pp. 1252–1261, 1999.

[129] S. P. Thomas, D. J. R. Guy, A. C. Boyd, V. E. Eipper, D. L. Ross, and R. B. Chard, "Comparison of epicardial and endocardial linear ablation using handheld probes," *The Annals of Thoracic Surgery*, vol. 75, pp. 543–548, 2003.

[130] J. M. T. de Bakker and H. M. V. van Rijen, "Continuous and discontinuous propagation in heart muscle," *Journal of Cardiovascular Electrophysiology*, vol. 17, pp. 567–573, 2006.

[131] D. Haemmerich, "Biophysics of radiofrequency ablation," *Critical Reviews in Biomedical Engineering*, vol. 38, pp. 53–63, 2010.

[132] I. D. McRury and D. E. Haines, "Ablation for the treatment of arrhythmias," *Proc. of the IEEE*, vol. 84, pp. 404–416, 1996.

[133] E. J. Berjano, "Theoretical modeling for radiofrequency ablation: state-of-the-art and challenges for the future," *Biomedical Engineering Online*, vol. 5, p. 24, 2006.

[134] S. Nath, J. G. Whayne, S. Kaul, N. C. Goodman, A. R. Jayaweera, and D. E. Haines, "Effects of radiofrequency catheter ablation on regional myocardial blood flow. Possible mechanism for late electrophysiological outcome," *Circulation*, vol. 89, pp. 2667–2672, 1994.

[135] G. Hindricks, W. Haverkamp, H. Gulker, U. Rissel, T. Budde, K. D. Richter, M. Borggrefe, and G. Breithardt, "Radiofrequency coagulation of ventricular myocardium: improved prediction of lesion size by monitoring catheter tip temperature," *European Heart Journal*, vol. 10, pp. 972–984, 1989.

[136] J. P. Saul, J. E. Hulse, J. Papagiannis, R. Van Praagh, and E. P. Walsh, "Late enlargement of radiofrequency lesions in infant lambs. Implications for ablation procedures in small children," *Circulation*, vol. 90, pp. 492–499, 1994.

[137] T. J. van Brakel, G. Bolotin, K. J. Salleng, L. W. Nifong, M. A. Allessie, W. R. J. Chitwood, and J. G. Maessen, "Evaluation of epicardial microwave ablation lesions: histology versus electrophysiology," *The Annals of Thoracic Surgery*, vol. 78, pp. 1397–402, 2004.

[138] T. J. Badger, R. S. Oakes, M. Daccarett, N. S. Burgon, N. Akoum, E. N. Fish, J. J. E. Blauer, S. N. Rao, Y. Adjei-Poku, E. G. Kholmovski, S. Vijayakumar, E. V. R. Di Bella, R. S. MacLeod, and N. F. Marrouche, "Temporal left atrial lesion formation after ablation of atrial fibrillation," *Heart Rhythm*, vol. 6, pp. 161–168, 2009.

[139] H. P. Schuster and H. J. Trappe, *EKG-Kurs für Isabel.* 2005.

[140] P. E. Dilaveris, E. J. Gialafos, S. K. Sideris, A. M. Theopistou, G. K. Andrikopoulos, M. Kyriakidis, J. E. Gialafos, and P. K. Toutouzas, "Simple electrocardiographic markers for the prediction of paroxysmal idiopathic atrial fibrillation," *American Heart Journal*, vol. 135, pp. 733–738, 1998.

[141] S. A. Guidera and J. S. Steinberg, "The signal-averaged P wave duration: a rapid and noninvasive marker of risk of atrial fibrillation," *Journal of the American College of Cardiology*, vol. 21, pp. 1645–1651, 1993.

[142] A. De Sisti, J. F. Leclercq, M. Stiubei, P. Fiorello, F. Halimi, and P. Attuel, "P wave duration and morphology predict atrial fibrillation recurrence in patients with sinus node dysfunction and atrial-based pacemaker," *Pacing and Clinical Electrophysiology : PACE*, vol. 25, pp. 1546–1554, 2002.

[143] L. Clavier, J. M. Boucher, R. Lepage, J. J. Blanc, and J. C. Cornily, "Automatic P-wave analysis of patients prone to atrial fibrillation," *Medical & Biological Engineering & Computing*, vol. 40, pp. 63–71, 2002.

[144] A. SippensGroenewegen, H. A. Peeters, E. R. Jessurun, A. C. Linnenbank, E. O. Robles de Medina, M. D. Lesh, and N. M. van Hemel, "Body surface mapping during pacing at multiple sites in the human atrium: P-wave morphology of ectopic right atrial activation," *Circulation*, vol. 97, pp. 369–380, 1998.

[145] D. M. Mirvis, "Body surface distribution of electrical potential during atrial depolarization and repolarization," *Circulation*, vol. 62, pp. 167–173, 1980.

[146] J. Lian, G. Li, J. Cheng, B. Avitall, and B. He, "Body surface Laplacian mapping of atrial depolarization in healthy human subjects," *Medical & Biological Engineering & Computing*, vol. 40, pp. 650–659, 2002.

[147] H. B. Sprague and P. D. White, "Clinical Observations On The T Wave Of The Auricle Appearing In The Human Electrocardiogram," *The Journal of Clinical Investigation*, vol. 1, pp. 389–402, 1925.

[148] D. I. Abramson, N. M. Fenichel, and C. Shookhoff, "A study of electrical activity in the auricles," *American Heart Journal*, vol. 15, pp. 471–481, 1938.

[149] H. Hayashi, M. Okajima, and K. Yamada, "Atrial T (Ta) loop in patients with A-V block: a trial to differentiate normal and abnoral groups," *American Heart Journal*, vol. 91, pp. 492–500, 1976.

[150] N. M. Debbas, S. H. Jackson, D. de Jonghe, A. Robert, and A. J. Camm, "Human atrial repolarization: effects of sinus rate, pacing and drugs on the surface electrocardiogram," *Journal of the American College of Cardiology*, vol. 33, pp. 358–365, 1999.

[151] F. Holmqvist, J. Carlson, and P. G. Platonov, "Detailed ECG analysis of atrial repolarization in humans," *Annals of Noninvasive Electrocardiology*, vol. 14, pp. 13–18, 2009.

[152] H. Hayashi, M. Okajima, and K. Yamada, "Atrial T(Ta) wave and atrial gradient in patients with A-V block," *American Heart Journal*, vol. 91, pp. 689–698, 1976.

[153] G. E. Dower, H. B. Machado, and J. A. Osborne, "On deriving the electrocardiogram from vectorcardiographic leads," *Clinical Cardiology*, vol. 3, pp. 87–95, 1980.

[154] J. Carlson, R. Havmoller, A. Herreros, P. Platonov, R. Johansson, and B. Olsson, "Can orthogonal lead indicators of propensity to atrial fibrillation be accurately assessed from the 12-lead ECG?," *Europace*, vol. 7 Suppl 2, pp. 39–48, 2005.

[155] P. V. Bayly, B. H. KenKnight, J. M. Rogers, E. E. Johnson, R. E. Ideker, and W. M. Smith, "Spatial organization, predictability, and determinism in ventricular fibrillation," *Chaos*, vol. 8, pp. 103–115, 1998.

[156] A. R. Barnette, P. V. Bayly, S. Zhang, G. P. Walcott, R. E. Ideker, and W. M. Smith, "Estimation of 3-D conduction velocity vector fields from cardiac mapping data," *IEEE Transactions on Biomedical Engineering*, vol. 47, pp. 1027–1035, 2000.

[157] P. Kojodjojo, N. S. Peters, D. W. Davies, and P. Kanagaratnam, "Characterization of the electroanatomical substrate in human atrial fibrillation: the relationship between changes in atrial volume, refractoriness, wavefront propagation velocities, and AF burden," *Journal of Cardiovascular Electrophysiology*, vol. 18, pp. 269–275, 2007.

[158] M. Mase and F. Ravelli, "Automatic reconstruction of activation and velocity maps from electro-anatomic data by radial basis functions," *Conf Proc IEEE Eng Med Biol Soc*, vol. 2010, pp. 2608–2611, 2010.

[159] M. Mase, M. Del Greco, M. Marini, and F. Ravelli, "Velocity Field Analysis of Activation Maps in Atrial Fibrillation - A Simulation Study," in *IFMBE Proceedings World Congress on Medical Physics and Biomedical Engineering*, vol. 25/4, pp. 1014–1017, 2009.

[160] F. M. Weber, C. Schilling, G. Seemann, A. Luik, C. Schmitt, C. Lorenz, and O. Dössel, "Wave-direction and conduction-velocity analysis from intracardiac electrograms–a single-shot technique," *IEEE Transactions on Biomedical Engineering*, vol. 57, pp. 2394–2401, 2010.

[161] F. B. Sachse, R. Frech, C. D. Werner, and O. Dössel, "A Model Based Approach to Assignment of Myocardial Fibre Orientation," in *Proc. Computers in Cardiology*, vol. 26, (Hannover), pp. 145–148, 1999.

[162] C. D. Werner, F. B. Sachse, and O. Dössel, "Electrical excitation propagation in the human heart," *Int. J. Bioelectromagnetism*, vol. 2, 2000.

[163] E. J. Vigmond, R. Ruckdeschel, and N. Trayanova, "Reentry in a morphologically realistic atrial model," *Journal of Cardiovascular Electrophysiology*, vol. 12, pp. 1046–1054, 2001.

[164] C. W. Zemlin, H. Herzel, S. Y. Ho, and A. Panfilow, *Computer Simulation and Experimental Assessment of Cardiac Electrophysiology*, ch. A realistic and efficient model of excitation propagation in the human atria, pp. 29–34. Futura, 2001.

[165] V. Jacquemet, N. Virag, Z. Ihara, L. Dang, O. Blanc, S. Zozor, J.-M. Vesin, L. Kappenberger, and C. Henriquez, "Study of unipolar electrogram morphology in a computer model of atrial fibrillation," *Journal of Cardiovascular Electrophysiology*, vol. 14, pp. S172–9, 2003.

[166] J.-M. Peyrat, M. Sermesant, X. Pennec, H. Delingette, C. Xu, E. R. McVeigh, and N. Ayache, "A computational framework for the statistical analysis of cardiac diffusion tensors: application to a small database of canine hearts," *IEEE Transactions on Medical Imaging*, vol. 26, pp. 1500–1514, 2007.

[167] D. D. Streeter, H. M. Spotnitz, D. P. Patel, R. J. J, and E. H. Sonnenblick, "Fiber orientation in the canine left ventricle during diastole and systole," *Circ. Res.*, vol. 24, pp. 339–347, 1969.

[168] B. D. F. Hermosillo, "Semi-automatic enhancement of atrial models to include atrial architecture and patient specific data: for biophysical simulations," in *Proc. Computers in Cardiology*, 2008.

[169] M. W. Krueger, F. M. Weber, G. Seemann, and O. Dössel, "Semi-automatic segmentation of sinus node, Bachmann's Bundle and Terminal Crest for patient specific atrial models," in *World Congress on Medical Physics and Biomedical Engineering. IFMBE Proceedings*, vol. 25/4, pp. 673–676, Springer Heidelberg, 2009.

[170] A. V. Shahidi and P. Savard, "A finite element model for radiofrequency ablation of the myocardium," *IEEE Transactions on Biomedical Engineering*, vol. 41, pp. 963–968, 1994.

[171] S. Labonte, "A computer simulation of radio-frequency ablation of the endocardium," *IEEE Transactions on bio-Medical Engineering*, vol. 41, pp. 883–890, 1994.

[172] D. Panescu, J. G. Webster, W. J. Tompkins, and R. A. Stratbucker, "Optimization of cardiac defibrillation by three-dimensional finite element modeling of the human thorax," *IEEE Trans Biomed Eng*, vol. 42, pp. 185–192, 1995.

[173] P. Panescu, S. D. Fleischmann, J. G. Whayne, D. K. Swanson, M. S. Mirotznik, I. McRury, and D. E. Haines, "Radiofrequency multielectrode catheter ablation in the atrium," *Physics in Medicine and Biology*, vol. 44, pp. 899–915, 1999.

[174] M. K. Jain and P. D. Wolf, "Temperature-controlled and constant-power radio-frequency ablation: what affects lesion growth?," *IEEE Transactions on bio-Medical Engineering*, vol. 46, pp. 1405–1412, 1999.

[175] A. K. Jain, R. P. W. Duin, and J. Mao, "Statistical pattern recognition: a review," *Pattern Analysis and Machine Intelligence, IEEE Transactions on*, vol. 22, pp. 4–37, 2000.

[176] M. K. Jain and P. D. Wolf, "A three-dimensional finite element model of radiofrequency ablation with blood flow and its experimental validation," *Annals of Biomedical Engineering*, vol. 28, pp. 1075–1084, 2000.

[177] N. Shimko, P. Savard, and K. Shah, "Radio frequency perforation of cardiac tissue: modelling and experimental results," *Medical & Biological Engineering & Computing*, vol. 38, pp. 575–582, 2000.

[178] S. Tungjitkusolmun, V. R. Vorperian, N. Bhavaraju, H. Cao, J. Z. Tsai, and J. G. Webster, "Guidelines for predicting lesion size at common endocardial locations during radio-frequency ablation," *IEEE Transactions on bio-Medical Engineering*, vol. 48, pp. 194–201, 2001.

[179] S. A. Solazzo, Z. Liu, S. M. Lobo, M. Ahmed, A. U. Hines-Peralta, R. E. Lenkinski, and S. N. Goldberg, "Radiofrequency ablation: importance of background tissue electrical conductivity–an agar phantom and computer modeling study," *Radiology*, vol. 236, pp. 495–502, 2005.

[180] J. A. L. Molina, M. J. Rivera, M. Trujillo, and E. J. Berjano, "Effect of the thermal wave in radiofrequency ablation modeling: an analytical study," *Physics in Medicine and Biology*, vol. 53, pp. 1447–1462, 2008.

[181] T. J. Bunch, J. D. Day, and D. L. Packer, "Insights into energy delivery to myocardial tissue during radiofrequency ablation through application of the first law of thermodynamics," *Journal of Cardiovascular Electrophysiology*, vol. 20, pp. 461–465, 2009.

[182] J. Alba, M. Trujillo, R. Blasco, and E. J. Berjano, "Computer Modeling to Study the Dynamic Response of the Temperature Control Loop in RF Cardiac Ablation," in *XII Mediterranean Conference on Medical and Biological Engineering and Computing 2010* (R. Magjarevic, P. D. Bamidis, and N. Pallikarakis, eds.), vol. 29 of *IFMBE Proceedings*, pp. 725–728, Springer Berlin Heidelberg, 2010.

[183] E. J. Berbari and C. Vasquez, "Cardiac late potential signals and sources," *Journal of Electrocardiology*, vol. 43, pp. 530–534, 2010.

[184] W. S. Ellis, A. SippensGroenewegen, and M. D. Lesh, "The effect of linear lesions on atrial defibrillation threshold and spontaneous termination. A computer modeling study," *PACE*, vol. 20, p. 1145, 1997.

[185] L. Dang, N. Virag, Z. Ihara, V. Jacquemet, J. M. Vesin, J. Schlaepfer, P. Ruchat, and L. Kappenberger, "Evaluation of ablation patterns using a biophysical model of atrial fibrillation," *Annals of Biomedical Engineering*, vol. 33, pp. 465–474, 2005.

[186] M. Rotter, L. Dang, V. Jacquemet, N. Virag, L. Kappenberger, and M. Haissaguerre, "Impact of varying ablation patterns in a simulation model of persistent atrial fibrillation," *PACE*, vol. 30, pp. 314–321, 2007.

[187] M. Reumann, J. Bohnert, G. Seemann, B. Osswald, and O. Dössel, "Preventive ablation strategies in a biophysical model of atrial fibrillation based on realistic anatomical data," *IEEE Transactions on Biomedical Engineering*, vol. 55, pp. 399–406, 2008.

[188] M. Lorange and R. M. Gulrajani, "A computer heart model incorporating anisotropic propagation. I. Model construction and simulation of normal activation," *Journal of Electrocardiology*, vol. 26, pp. 245–261, 1993.

[189] R. A. Gray and J. Jalife, "Ventricular fibrillation and atrial fibrillation are two different beasts," *Chaos*, vol. 8, pp. 65–78, 1998.

[190] R. A. FitzHugh, "Impulses and physiological states in theoretical models of nerve membran," *Biophysical Journal*, vol. 1, pp. 445–466, 1961.

[191] J. Freudenberg, T. Schiemann, U. Tiede, and K. H. Hohne, "Simulation of cardiac excitation patterns in a three-dimensional anatomical heart atlas," *Computers in Biology and Medicine*, vol. 30, pp. 191–205, 2000.

[192] R. J. Ramirez, S. Nattel, and M. Courtemanche, "Mathematical analysis of canine atrial action potentials: rate, regional factors, and electrical remodeling," *Am J Physiol Heart Circ Physiol*, vol. 279, pp. H1767–H1785, 2000.

[193] O. Blanc, N. Virag, J.-M. Vesin, and L. Kappenberger, "A computer model of human atria with reasonable computation load and realistic anatomical properties," *IEEE Transactions on Biomedical Engineering*, vol. 48, pp. 1229–1237, 2001.

[194] G. W. Beeler and H. Reuter, "Reconstruction of the action potential of ventricular myocardial fibers," *Journal of Physiology*, vol. 268, pp. 177–210, 1977.

[195] C. H. Luo and Y. Rudy, "A model of the ventricular cardiac action potential. Depolarization, repolarization, and their interaction," *Circulation Research*, vol. 68, pp. 1501–1526, 1991.

[196] N. Virag, V. Jacquemet, C. S. Henriquez, S. Zozor, O. Blanc, J. M. Vesin, E. Pruvot, and L. Kappenberger, "Study of atrial arrhythmias in a computer model based on magnetic resonance images of human atria," *Chaos*, vol. 12, pp. 754–763, 2002.

[197] P. M. van Dam and A. van Oosterom, "Atrial excitation assuming uniform propagation," *Journal of Cardiovascular Electrophysiology*, vol. 14, pp. S166–71, 2003.

[198] V. Jacquemet, A. van Oosterom, J.-M. Vesin, and L. Kappenberger, "Analysis of electrocardiograms during atrial fibrillation. A biophysical model approach," *IEEE Eng Med Biol Mag*, vol. 25, pp. 79–88, 2006.

[199] P. Ruchat, L. Dang, J. Schlaepfer, N. Virag, L. K. von Segesser, and L. Kappenberger, "Use of a biophysical model of atrial fibrillation in the interpretation of the outcome of surgical ablation procedures," *Eur J Cardiothorac Surg*, vol. 32, pp. 90–95, 2007.

[200] P. Ruchat, L. Dang, N. Virag, J. Schlaepfer, L. K. von Segesser, and L. Kappenberger, "A biophysical model of atrial fibrillation to define the appropriate ablation pattern in modified maze," *Eur J Cardiothorac Surg*, vol. 31, pp. 65–69, 2007.

[201] L. Uldry, V. Jacquemet, N. Virag, L. Kappenberger, and J.-M. Vesin, "Estimating the time scale and anatomical location of atrial fibrillation spontaneous termination in a biophysical model," *Medical & Biological Engineering & Computing*, vol. 50, pp. 155–163, 2012.

[202] M. Ridler, D. M. McQueen, C. S. Peskin, and E. Vigmond, "Action potential duration gradient protects the right atrium from fibrillating," in *Conf Proc IEEE Eng Med Biol Soc*, vol. 1, pp. 3978–3981, 2006.

[203] S. Dokos, S. L. Cloherty, and N. H. Lovell, "Computational model of atrial electrical activation and propagation," in *Conf Proc IEEE Eng Med Biol Soc*, vol. 2007, pp. 908–911, 2007.

[204] J. M. Rogers and A. D. McCulloch, "A collocation–Galerkin finite element model of cardiac action potential propagation," *IEEE Transactions on Biomedical Engineering*, vol. 41, pp. 743–757, 1994.

[205] N. H. L. Kuijpers, R. J. Rijken, H. M. M. ten Eikelder, and P. A. J. Hilbers, "Vulnerability to Atrial Fibrillation under Stretch Can Be Explained by Stretch-Activated Channels," *Computers in Cardiology*, vol. 34, pp. 237–240, 2007.

[206] Y. Gong, F. Xie, K. M. Stein, A. Garfinkel, C. A. Culianu, B. B. Lerman, and D. J. Christini, "Mechanism underlying initiation of paroxysmal atrial flutter/atrial fibrillation by ectopic foci: a simulation study," *Circulation*, vol. 115, pp. 2094–2102, 2007.

[207] C. Tobón, C. Ruiz, E. Heidenreich, F. Hornero, and J. Sáiz, "Effect of the ectopic beats location on vulnerability to reentries in a three-dimensional realistic model of human atria," in *Computers in Cardiology*, vol. 36, pp. 449–452, 2009.

[208] C. Tobon, C. Ruiz, J. F. Rodriguez, F. Hornero, J. M. Ferrero, and J. Saiz, "A biophysical model of atrial fibrillation to simulate the Maze III ablation pattern," *Proc. Computing in Cardiology*, vol. 37, pp. 621–624, 2010.

[209] C. Tobón, C. Ruiz, J. Sáiz, E. Heidenreich, and F. Hornero, "Reentrant mechanisms triggered by ectopic activity in a three-dimensional realistic model of human atrium. a computer simulation study," in *Computers in Cardiology*, vol. 35, pp. 629–632, 2008.

[210] L. Wieser, C. N. Nowak, B. Tilg, and G. Fischer, "Mother rotor anchoring in branching tissue with heterogeneous membrane properties," *Biomedizinische Technik / Biomedical Engineering*, vol. 53, pp. 25–35, 2008.

[211] L. Wieser, H. E. Richter, G. Plank, B. Pfeifer, B. Tilg, C. N. Nowak, and G. Fischer, "A finite element formulation for atrial tissue monolayer," *Methods of Information in Medicine*, vol. 47, pp. 131–139, 2008.

[212] B. Tilg, G. Fischer, R. Modre, F. Hanser, B. Messnarz, M. Schocke, C. Kremser, T. Berger, F. Hintringer, and F. X. Roithinger, "Model-based imaging of cardiac electrical excitation in humans," *IEEE Transactions on Medical Imaging*, vol. 21, pp. 1031–1039, 2002.

[213] S. Kharche, G. Seemann, L. Margetts, J. Leng, A. V. Holden, and H. Zhang, "Simulation of clinical electrophysiology in 3D human atria: a high-performance computing and high-performance visualization application," *Concurrency and Computation: Practice and Experience*, vol. 20, pp. 1317–1328, 2008.

[214] M. W. Krueger, K. Rhode, F. M. Weber, D. U. J. Keller, D. Caulfield, G. Seemann, B. R. Knowles, R. Razavi, and O. Dössel, "Patient-Specific Volumetric Atrial Models with Electrophysiological Components: A Comparison of Simulations and Measurements," in *Biomedizinische Technik / Biomedical Engineering*, vol. 55(s1), pp. 54–57, 2010.

[215] A. A. Abed, T. Guo, N. H. Lovell, and S. Dokos, "An anatomically realistic 3d model of atrial propagation based on experimentally recorded action potentials," *Conference Proceedings : ... Annual International Conference of the IEEE Engineering in Medicine and Biology Society. IEEE Engineering in Medicine and Biology Society. Conference*, vol. 2010, pp. 243–246, 2010.

[216] M. W. Krueger, V. Schmidt, C. Tobón, F. M. Weber, C. Lorenz, D. U. J. Keller, H. Barschdorf, M. Burdumy, P. Neher, G. Plank, K. Rhode, G. Seemann, D. Sanchez-Quintana, J. Saiz, R. Razavi, and O. Dössel, "Modeling atrial fiber orientation in patient-specific geometries: a semi-automatic rule-based approach," in *Functional Imaging and Modeling of the Heart 2011, Lecture Notes in Computer Science* (L. Axel and D. Metaxas, eds.), vol. 6666, pp. 223–232, 2011.

[217] W. Lu, X. Zhu, W. Chen, and D. Wei, "A computer model based on real anatomy for electrophysiology study," *Advances in Engineering Software*, vol. 42, pp. 463–476, 2011.

[218] O. V. Aslanidi, M. A. Colman, J. Stott, H. Dobrzynski, M. R. Boyett, A. V. Holden, and H. Zhang, "3D virtual human atria: A computational platform for studying clinical atrial fibrillation," *Progress in Biophysics and Molecular Biology*, vol. 107, pp. 156–168, 2011.

[219] M. A. Colman, O. V. Aslanidi, J. Stott, A. V. Holden, and H. Zhang, "Correlation between P-wave morphology and origin of atrial focal tachycardia–insights from realistic models of the human atria and torso," *IEEE Transactions on bio-Medical Engineering*, vol. 58, pp. 2952–2955, 2011.

[220] A. Cristoforetti, M. Masè, and F. Ravelli, "A computationally efficient patient-specific simulation model for the investigation of arrhythmia mechanisms," in *JECG (Proc. ICE 2010)*, vol. 44, p. e61, 2010.

[221] M.-E. Ridler, M. Lee, D. McQueen, C. Peskin, and E. Vigmond, "Arrhythmogenic consequences of action potential duration gradients in the atria," *The Canadian Journal of Cardiology*, vol. 27, pp. 112–119, 2011.

[222] J. Kneller, R. Zou, E. J. Vigmond, Z. Wang, L. J. Leon, and S. Nattel, "Cholinergic atrial fibrillation in a computer model of a two-dimensional sheet of canine atrial cells with realistic ionic properties," *Circulation Research*, vol. 90, p. 73, 2002.

[223] M. Burdumy, A. Luik, P. Neher, R. Hanna, M. W. Krueger, C. Schilling, H. Barschdorf, C. Lorenz, G. Seemann, C. Schmitt, O. Dössel, and F. M. Weber, "Comparing measured and simulated wave directions in the left atrium - a workflow for model personalization and validation," *Biomed Tech (Berl)*, vol. 57, pp. 79–87, 2012.

[224] F. Fenton and A. Karma, "Vortex dynamics in three-dimensional continuous myocardium with fiber rotation: Filament instability and fibrillation," *Chaos*, vol. 8, pp. 20–27, 1998.

[225] M. W. Krueger, G. Seemann, K. Rhode, D. U. J. Keller, C. Schilling, A. Arujuna, J. Gill, M. D. O'Neill, R. Razavi, and O. Dössel, "Personalization of Atrial Anatomy and Elelectophysiology as a Basis for Clinical Modeling of Radio-Frequency-Ablation of Atrial Fibrillation," *IEEE Transactions on Medical Imaging (epub ahead of print)*, vol. 31, 2012.

[226] O. Aslanidi, T. Nikolaidou, J. Zhao, B. Smaill, S. Gilbert, J. Jarvis, R. Stephenson, J. Hancox, M. Boyett, and H. Zhang, "Application of Micro-Computed Tomography with Iodine Staining to Cardiac Imaging, Segmentation and Computational Model Development," *Medical Imaging, IEEE Transactions on*, vol. Pp, p. 1, 2012.

[227] A. Nygren, C. Fiset, L. Firek, J. W. Clark, D. S. Lindblad, R. B. Clark, and W. R. Giles, "Mathematical model of a adult human atrial cell. The role of K+ currents in repolarization," *Circulation Research*, vol. 82, pp. 63–81, 1998.

[228] C. J. Kafer, "Internodal pathways in the human atria: a model study," *Computers and Biomedical Research*, vol. 24, pp. 549–563, 1991.

[229] P. M. van Dam and A. van Oosterom, "Volume conductor effects involved in the genesis of the P wave," *Europace*, vol. 7 Suppl 2, pp. 30–38, 2005.

[230] A. van Oosterom and V. Jacquemet, "Genesis of the P wave: atrial signals as generated by the equivalent double layer source model," *Europace*, vol. 7 Suppl 2, pp. 21–29, 2005.

[231] V. Jacquemet, M. Lemay, A. van Oosterom, and L. Kappenberger, "The equivalent dipole used to characterize atrial fibrillation," in *Proc. Computers in Cardiology*, vol. 33, (17-20 Sept., Valencia), pp. 149–152, IEEE, 2006.

[232] A. L. Hodgkin and A. F. Huxley, "A quantitative description of membrane current and its application to conduction and excitation in nerve," *Journal of Physiology*, vol. 117, pp. 500–544, 1952.

[233] D. Noble, "A modification of the Hodgkin-Huxley equations applicable to Purkinje fibre action and pace-maker potentials," *Journal of Physiology*, vol. 160, pp. 317–352, 1962.

[234] J. Feng, L. Yue, Z. Wang, and S. Nattel, "Ionic mechanisms of regional action potential heterogeneity in the canine right atrium," *Circ. Res.*, vol. 83, pp. 541–551, 1998.

[235] D. Li, L. Zhang, J. Kneller, and S. Nattel, "Potential ionic mechanism for repolarization differences between canine right and left atrium," *Circ. Res.*, vol. 88, pp. 1168–1175, 2001.

[236] M. M. Maleckar, J. L. Greenstein, W. R. Giles, and N. A. Trayanova, "Electrotonic coupling between human atrial myocytes and fibroblasts alters myocyte excitability and repolarization," *Biophysical Journal*, vol. 97, pp. 2179–2190, 2009.

[237] J. T. Koivumaeki, T. Korhonen, and P. Tavi, "Impact of sarcoplasmic reticulum calcium release on calcium dynamics and action potential morphology in human atrial myocytes: a computational study," *PLoS Computational Biology*, vol. 7, p. e1001067, 2011.

[238] E. Grandi, S. V. Pandit, N. Voigt, A. J. Workman, D. Dobrev, J. Jalife, and D. M. Bers, "Human Atrial Action Potential and Ca2+ Model: Sinus Rhythm and Chronic Atrial Fibrillation," *Circulation Research*, vol. 109, pp. 1055–1066, 2011.

[239] N. J. Chandler, I. D. Greener, J. O. Tellez, S. Inada, H. Musa, P. Molenaar, D. Difrancesco, M. Baruscotti, R. Longhi, R. H. Anderson, R. Billeter, V. Sharma, D. C. Sigg, M. R. Boyett, and H. Dobrzynski, "Molecular architecture of the human sinus node: insights into the function of the cardiac pacemaker," *Circulation*, vol. 119, pp. 1562–1575, 2009.

[240] F. M. Weber, S. Lurz, D. U. J. Keller, D. L. Weiss, G. Seemann, C. Lorenz, and O. Dössel, "Adaptation of a minimal four-state cell model for reproducing atrial excitation properties," in *Proc. Computers in Cardiology*, (Bologna, 14-17 Sept. 2008), pp. 61–64, IEEE, 2008.

[241] R. F. Bosch, X. Zeng, J. B. Gramer, K. Popovic, C. Mewis, and V. Kühlkamp, "Ionic mechanisms of electrical remodeling in human atrial fibrillation," *Cardiovascular Research*, vol. 44, pp. 121–131, 1999.

[242] H. M. van der Velden, J. Ausma, M. B. Rook, A. J. Hellemons, T. A. van Veen, M. A. Allessie, and H. J. Jongsma, "Gap junctional remodeling in relation to stabilization of atrial fibrillation in the goat," *Cardiovascular Research*, vol. 46, pp. 476–486, 2000.

[243] M. Wilhelm, W. Kirste, S. Kuly, K. Amann, W. Neuhuber, M. Weyand, W. G. Daniel, and C. Garlichs, "Atrial distribution of connexin 40 and 43 in patients with intermittent, persistent, and postoperative atrial fibrillation," *Heart, Lung and Circulation*, vol. 15, pp. 30–37, 2006.

[244] D. R. V. Wagoner, A. L. Pond, and P. M. McCarthy, "Outward K+ current densities and Kv1.5 expression are reduced in chronic human atrial fibrillation," *Circ. Res.*, vol. 80, pp. 772–781, 1997.

[245] D. Dobrev, E. Graf, E. Wettwer, H. M. Himmel, O. Hala, C. Doerfel, T. Christ, S. Schuler, and U. Ravens, "Molecular basis of downregulation of G-protein-coupled inward rectifying K+ current (IK, ACh) in chronic human atrial fibrillation: Decrease in GIRK4 mRNA correlates with reduced IK, ACh and muscarinic receptor-mediated shortening of action potentials," *Circulation*, vol. 104, pp. 2551–2557, 2001.

[246] P. Carrillo, "Correlation between pharmacological binding properties of Ikur inhibitors and antiarrhythmic efficiency in a virtual model of chronic atrial fibrillation," Diplomarbeit, Institut für Biomedizinische Technik, Karlsruher Institut für Technologie (KIT), Karlsruhe, 2009.

[247] R. Caballero, M. G. de la Fuente, R. Gomez, A. Barana, I. Amoros, P. Dolz-Gaiton, L. Osuna, J. Almendral, F. Atienza, F. Fernandez-Aviles, A. Pita, J. Rodriguez-Roda, A. Pinto, J. Tamargo, and E. Delpon, "In humans, chronic atrial fibrillation decreases the transient outward current and ultrarapid component of the delayed rectifier current differentially on each atria and increases the slow component of the delayed rectifier current in both," *Journal of the American College of Cardiology*, vol. 55, pp. 2346–2354, 2010.

[248] A. J. Workman, K. A. Kane, and A. C. Rankin, "The contribution of ionic currents to changes in refractoriness of human atrial myocytes associated with chronic atrial fibrillation," *Cardiovascular Research*, vol. 52, pp. 226–235, 2001.

[249] N. Voigt, A. Trausch, M. Knaut, K. Matschke, A. Varro, D. R. Van Wagoner, S. Nattel, U. Ravens, and D. Dobrev, "Left-to-right atrial inward rectifier potassium current gradients in patients with paroxysmal versus chronic atrial fibrillation," *Circulation. Arrhythmia and Electrophysiology*, vol. 3, pp. 472–480, 2010.

[250] D. Li, P. Melnyk, J. Feng, Z. Wang, K. Petrecca, A. Shrier, and S. Nattel, "Effects of experimental heart failure on atrial cellular and ionic electrophysiology," *Circulation*, vol. 101, pp. 2631–2638, 2000.

[251] H. Zhang, C. J. Garratt, J. Zhu, and A. V. Holden, "Role of up-regulation of IK1 in action potential shortening associated with atrial fibrillation in humans," *Cardiovascular Research*, vol. 66, pp. 493–502, 2005.

[252] C. S. Henriquez, "Simulating the electrical behavior of cardiac tissue using the bidomain model," *Critical Reviews ins Biomedical Engineering*, vol. 21, pp. 1–77, 1993.

[253] R. H. Clayton, O. Bernus, E. M. Cherry, H. Dierckx, F. H. Fenton, L. Mirabella, A. V. Panfilov, F. B. Sachse, G. Seemann, and H. Zhang, "Models of cardiac tissue electrophysiology: progress, challenges and open questions.," *Progress in Biophysics and Molecular Biology*, vol. 104, pp. 22–48, 2011.

[254] F. B. Sachse, A. P. Moreno, G. Seemann, and J. A. Abildskov, "A model of electrical conduction in cardiac tissue including fibroblasts," *Annals of Biomedical Engineering*, vol. 37, pp. 874–889, 2009.

[255] G. K. Moe, W. C. Rheinboldt, and J. A. Abildskov, "A computer model of atrial fibrillation," *American Heart Journal*, vol. 67, pp. 200–220, 1964.

[256] D. D. Correa de Sa, N. Thompson, J. Stinnett-Donnelly, P. Znojkiewicz, N. Habel, J. G. Muller, J. H. Bates, J. S. Buzas, and P. S. Spector, "Electrogram fractionation: the relationship between spatio-temporal variation of tissue excitation and electrode spatial resolution," *Circulation. Arrhythmia and Electrophysiology*, 2011.

[257] M. Reumann, J. Bohnert, B. Osswald, S. Hagl, and O. Dössel, "Multiple wavelets, rotors and snakes in atrial fibrillation - a computer simulation study," *Journal of Electrocardiology*, vol. 40, pp. 328–334, 2007.

[258] J. A. Sethian, *Level Set Methods and Fast Marching Methods*. Cambridge University Press, 1999.

[259] J. A. Baerentzen, "On the implementation of fast marching methods for 3D lattices," tech. rep., Lyngby, Denmark, 2001.

[260] E. Konukoglu, J. Relan, U. Cilingir, B. H. Menze, P. Chinchapatnam, A. Jadidi, H. Cochet, M. Hocini, H. Delingette, P. Jais, M. Haissaguerre, N. Ayache, and M. Sermesant, "Efficient probabilistic model personalization integrating uncertainty on data and parameters: Application to Eikonal-Diffusion models in cardiac electrophysiology," *Progress in Biophysics and Molecular Biology*, vol. 107, pp. 134–146, 2011.

[261] H. Dierckx, O. Bernus, and H. Verschelde, "Accurate eikonal-curvature relation for wave fronts in locally anisotropic reaction-diffusion systems," *Physical Review Letters*, vol. 107, p. 108101, 2011.

[262] V. Jacquemet, "An eikonal-diffusion solver and its application to the interpolation and the simulation of reentrant cardiac activations," *Computer Methods and Programs in Biomedicine*, 2011.

[263] V. Jacquemet, "An eikonal approach for the initiation of reentrant cardiac propagation in reaction-diffusion models," *IEEE Transactions on bio-Medical Engineering*, vol. 57, pp. 2090–2098, 2010.

[264] M. Sermesant, E. Konukoglu, H. Delingette, Y. Coudiere, P. Chinchapatnam, K. Rhode, R. Razavi, and N. Ayache, *Functional Imaging and Modeling of the Heart*, vol. 4466, ch. An Anisotropic Multi-front Fast Marching Method for Real-Time Simulation of Cardiac Electrophysiology, pp. 160–169. Springer Berlin / Heidelberg, 2007.

[265] M. Wallman, N. P. Smith, and B. Rodriguez, "A comparative study of graph-based, eikonal, and monodomain simulations for the estimation of cardiac activation times," *IEEE Transactions on bio-Medical Engineering*, vol. 59, pp. 1739–1748, 2012.

[266] A. J. Pullan, M. L. Buist, and L. K. Cheng, *Mathematically Modelling the Electrical Activity of the Heart: From Cell to Body Surface and Back Again*. Singapore: World Scientific, 2005.

[267] D. Keller, *Multiscale modeling of the ventricles: From cellular electrophysiology to body surface electrocardiograms*. PhD thesis, Karlsruhe, KIT Scientific Publishing, 2011.

[268] F. M. Weber, *Personalizing simulations of the human atria: Intracardiac measurements, tissue conductivities, and cellular electrophysiology*. PhD thesis, Karlsruhe, KIT Scientific Publishing, 2011.

[269] R. S. Oakes, T. J. Badger, E. G. Kholmovski, N. Akoum, N. S. Burgon, E. N. Fish, J. J. E. Blauer, S. N. Rao, E. V. R. DiBella, N. M. Segerson, M. Daccarett, J. Windfelder, C. J. McGann, D. Parker, R. S. MacLeod, and N. F. Marrouche, "Detection and quantification of left atrial structural remodeling with delayed-enhancement magnetic resonance imaging in patients with atrial fibrillation," *Circulation*, vol. 119, pp. 1758–1767, 2009.

[270] B. R. Knowles, D. Caulfield, M. Cooklin, C. A. Rinaldi, J. Gill, J. Bostock, R. Razavi, T. Schaeffter, and K. S. Rhode, "3-D visualization of acute RF ablation lesions using MRI for the simultaneous determination of the patterns of necrosis and edema," *IEEE Transactions on bio-Medical Engineering*, vol. 57, pp. 1467–1475, 2010.

[271] R. Karim, A. Arujuna, A. Brazier, J. Gill, C. A. Rinaldi, M. O?Neill, R. Razavi, T. Schaeffter, D. Rueckert, and K. S. Rhode, "Automatic Segmentation of Left Atrial Scar from Delayed-Enhancement Magnetic Resonance Imaging," *Functional Imaging and Modeling of the Heart 2011*, vol. 6666, pp. 63–70, 2011.

[272] N. Akoum, M. Daccarett, C. McGann, N. Segerson, G. Vergara, S. Kuppahally, T. Badger, N. Burgon, T. Haslam, E. Kholmovski, R. Macleod, and N. Marrouche, "Atrial fibrosis helps select the appropriate patient and strategy in catheter ablation of atrial fibrillation: a DE-MRI guided approach," *Journal of Cardiovascular Electrophysiology*, vol. 22, pp. 16–22, 2011.

[273] R. Karim, A. Arujuna, A. Brazier, J. Gill, C. Rinaldi, M. Cooklin, M. O?Neill, R. Razavi, T. Schaeffter, D. Rueckert, and K. Rhode, *Statistical Atlases and Computational Models of the Heart. Imaging and Modelling Challenges*, vol. 7085, ch. Validation of a Novel Method for the Automatic Segmentation of Left Atrial Scar from Delayed-Enhancement Magnetic Resonance, pp. 254–262. Springer Berlin / Heidelberg, 2012.

[274] R. Karim, H. Cliffe, R. Razavi, T. Schaeffter, K. Rhode, J. Cates, D. Perry, E. Kholmovski, R. Ranjan, R. MacLeod, and D. Peters, "Delayed-enhancement MRI Segmentation Challenge," in *cDEMRIS 2012*, 2012.

[275] A. Khawaja and O. Dössel, "Predicting the QRS complex and detecting small changes using principal component analysis," *Biomed Tech (Berl)*, vol. 52, pp. 11–17, 2007.

[276] K. S. Rhode, D. L. G. Hill, P. J. Edwards, J. Hipwell, D. Rueckert, G. Sanchez-Ortiz, S. Hegde, V. Rahunathan, and R. Razavi, "Registration and tracking to integrate X-ray and MR images in an XMR facility," *IEEE Transactions on Medical Imaging*, vol. 22, pp. 1369–1378, 2003.

[277] A. Thorpe, Z. Chen, W. H. W. Schulze, M. W. Krueger, J. Realn, H. Delingette, R. Razavi, C. Rinaldi, and K. Rhode, "Enhanced workflow for BSPM and ECGi using X-ray based Electrode Localisation," in *Bioengineering12*, 2012.

[278] Z. Zhang, "Determining the Epipolar Geometry and its Uncertainty: A Review," *International Journal of Computer Vision*, vol. 27, pp. 161–195, 1998.

[279] P. S. Cuculich, J. Zhang, Y. Wang, K. A. Desouza, R. Vijayakumar, P. K. Woodard, and Y. Rudy, "The electrophysiological cardiac ventricular substrate in patients after myocardial infarction: noninvasive characterization with electrocardiographic imaging," *Journal of the American College of Cardiology*, vol. 58, pp. 1893–1902, 2011.

[280] Y. Wang, P. S. Cuculich, J. Zhang, K. A. Desouza, R. Vijayakumar, J. Chen, M. N. Faddis, B. D. Lindsay, T. W. Smith, and Y. Rudy, "Noninvasive electroanatomic mapping of human ventricular arrhythmias with electrocardiographic imaging," *Science Translational Medicine*, vol. 3, p. 98ra84, 2011.

[281] A. Khawaja, S. Sanyal, and O. Dössel, "A wavelet-based technique for baseline wander correction in ECG and multi-channel ECG," in *IFMBE Proceedings*, vol. 9, pp. 291–292, 2005.

[282] A. Khawaja, *Automatic ECG analysis using principal component analysis and wavelet transformation*. PhD thesis, Karlsruhe, Universitätsverlag, 2006.

[283] M. E. Nygards and L. Sornmo, "Delineation of the QRS complex using the envelope of the e.c.g," *Medical & Biological Engineering & Computing*, vol. 21, pp. 538–547, 1983.

[284] P. Dilaveris, V. Batchvarov, J. Gialafos, and M. Malik, "Comparison of different methods for manual P wave duration measurement in 12-lead electrocardiograms," *PACE*, vol. 22, pp. 1532–1538, 1999.

[285] O. Skipa, *Linear inverse problem of electrocardiography: epicardial potentials and transmembrane voltages*. PhD thesis, Helmes Verlag Karlsruhe, 2004.

[286] M. Wilhelms, O. Dössel, and G. Seemann, "In silico investigation of electrically silent acute cardiac ischemia in the human ventricles," *IEEE Transactions on Biomedical Engineering*, vol. 58, pp. 2961–2964, 2011.

[287] D. T. Rudolph, W. H. W. Schulze, D. Potyagaylo, M. W. Krueger, and O. Dössel, "Reconstruction of atrial excitation conduction velocities and implementation into the inverse problem of electrocardiography," in *Biomedizinische Technik / Biomedical Engineering*, vol. 57, pp. 179–182, 2012.

[288] K. H. W. J. t. Tusscher, D. Noble, P. J. Noble, and A. V. Panfilov, "A model for human ventricular tissue," *Am. J. Physiol.*, vol. 286, pp. H1573–H1589, 2004.

[289] T. Shibasaki, "Conductance and kinetics delayed rectifier potassium channels in nodal cells of the rabbit heart," *J. Physiol.*, vol. 387, pp. 227–250, 1987.

[290] H. Gelband, H. L. Bush, M. R. Rosen, R. J. Myerburg, and B. F. Hoffman, "Electrophysiologic properties of isolated preparations of human atrial myocardium," *Circulation Research*, vol. 30, pp. 293–300, 1972.

[291] M. S. Spach, P. C. Dolber, and P. A. W. Anderson, "Multiple regional differences in cellular properties that regulate repolarization and contraction in the right atrium of adult and newborn dogs," *Circulation Research*, vol. 65, pp. 1594–1611, 1989.

[292] G. Schram, M. Pourrier, P. Melnyk, and S. Nattel, "Differential distribution of cardiac ion channel expression as a basis for regional specialization in electrical function," *Circ Res*, vol. 90, pp. 939–950, 2002.

[293] S. R. Kuo and N. A. Trayanova, "Action potential morphology heterogeneity in the atrium and its effect on atrial reentry: a two-dimensional and quasi-three-dimensional study," *Philosophical Transactions. Series A, Mathematical, Physical, and Engineering Sciences*, vol. 364, pp. 1349–1366, 2006.

[294] M. M. Maleckar, J. L. Greenstein, N. A. Trayanova, and W. R. Giles, "Mathematical simulations of ligand-gated and cell-type specific effects on the action potential of human atrium," *Progress in Biophysics and Molecular Biology*, vol. 98, pp. 161–170, 2008.

[295] M. W. Krueger, A. Dorn, D. U. J. Keller, F. Holmqvist, J. Carlson, P. G. Platonov, K. S. Rhode, R. Razavi, G. Seemann, and O. Dössel, "In-Silico Modeling of Atrial Repolarization in Normal and Atrial Fibrillation Remodeled State," *in revision*, 2012.

[296] A. Burashnikov, S. Mannava, and C. Antzelevitch, "Transmembrane action potential heterogeneity in the canine isolated arterially perfused right atrium: effect of IKr and IKur/Ito block," *American Journal of Physiology. Heart and Circulatory Physiology*, vol. 286, pp. H2393–400, 2004.

[297] T.-J. Cha, J. R. Ehrlich, L. Zhang, D. Chartier, T. K. Leung, and S. Nattel, "Atrial tachycardia remodeling of pulmonary vein cardiomyocytes: comparison with left atrium and potential relation to arrhythmogenesis," *Circulation*, vol. 111, pp. 728–735, 2005.

[298] J. R. Ehrlich, T.-J. Cha, L. Zhang, D. Chartier, P. Melnyk, S. H. Hohnloser, and S. Nattel, "Cellular electrophysiology of canine pulmonary vein cardiomyocytes: action potential and ionic current properties," *The Journal of Physiology*, vol. 551, pp. 801–813, 2003.

[299] D. Gong, Y. Zhang, B. Cai, Q. Meng, S. Jiang, X. Li, L. Shan, Y. Liu, G. Qiao, Y. Lu, and B. Yang, "Characterization and comparison of Na+, K+ and Ca2+ currents between myocytes from human atrial right appendage and atrial septum," *Cell Physiol Biochem*, vol. 21, pp. 385–394, 2008.

[300] P. Melnyk, J. R. Ehrlich, M. Pourrier, L. Villeneuve, T.-J. Cha, and S. Nattel, "Comparison of ion channel distribution and expression in cardiomyocytes of canine pulmonary veins versus left atrium," *Cardiovascular Research*, vol. 65, pp. 104–116, 2005.

[301] A. Qi, J. A. Yeung-Lai-Wah, J. Xiao, and C. R. Kerr, "Regional differences in rabbit atrial repolarization: importance of transient outward current," *The American Journal of Physiology*, vol. 266, pp. H643–9, 1994.

[302] Z. Wang, J. Feng, H. Shi, A. Pond, J. M. Nerbonne, and S. Nattel, "Potential molecular basis of different physiological properties of the transient outward K+ current in rabbit and human atrial myocytes," *Circulation Research*, vol. 84, pp. 551–561, 1999.

[303] R. W. Childers, J. Merideth, and G. K. Moe, "Supernormality in Bachmann's bundle. An in vitro and in vivo study in the dog," *Circulation Research*, vol. 22, pp. 363–370, 1968.

[304] S. Rajamani, C. L. Anderson, C. R. Valdivia, L. L. Eckhardt, J. D. Foell, G. A. Robertson, T. J. Kamp, J. C. Makielski, B. D. Anson, and C. T. January, "Specific serine proteases selectively damage KCNH2 (hERG1) potassium channels and I(Kr)," *American Journal of Physiology. Heart and Circulatory Physiology*, vol. 290, pp. H1278–88, 2006.

[305] L. X. Yue, J. Feng, G. R. Li, and S. Nattel, "Transient outward and delayed rectifier currents in canine atrium: Properties and role of isolation methods," *Am J Physiol Heart Circ Physiol*, vol. 270, pp. H2157–H1568, 1996.

[306] M. R. Boyett, H. Honjo, M. Yamamoto, M. R. Nikmaram, R. Niwa, and I. Kodama, "Downward gradient in action potential duration along conduction path in and around the sinoatrial node," *Am. J. Physiol.*, vol. 276, p. 686, 1999.

[307] Z. Li, E. Hertervig, O. Kongstad, M. Holm, E. Grins, S. B. Olsson, and S. Yuan, "Global repolarization sequence of the right atrium: monphasic action potential mapping in health pigs," *PACE*, vol. 26, pp. 1803–1808, 2003.

[308] G. Seemann, D. L. Weiß, F. B. Sachse, and O. Dössel, "Electrophysiology and Tension Development in a Transmural Heterogeneous Model of the Visible Female Left Ventricle," in *Lecture Notes in Computer Science* (A. Franji, P. Radeva, A. Santos, and M. Hernandez, eds.), vol. 3504, pp. 172–182, 2005.

[309] J. Sainte-Marie, D. Chapelle, R. Cimrman, and M. Sorine, "Modeling and estimation of the cardiac electromechanical activity," *Computers & Structures*, vol. 84, pp. 1743–1759, 2006.

[310] U. Schotten, J. Ausma, C. Stellbrink, I. Sabatschus, M. Vogel, D. Frechen, F. Schoendube, P. Hanrath, and M. A. Allessie, "Cellular mechanisms of depressed atrial contractility in patients with chronic atrial fibrillation," *Circulation*, vol. 103, pp. 691–698, 2001.

[311] F. B. Sachse, K. Glänzel, and G. Seemann, "Modeling of protein interactions involved in cardiac tension development," *Int. J. Bifurcation and Chaos*, vol. 13, pp. 3561–3578, 2003.

[312] F. B. Sachse, G. Seemann, K. Chaisaowong, and M. B. Mohr, "Modeling of electro-mechanics of human cardiac myocytes: Parameterization with numerical minimization techniques," in *Proc. IEEE EMBS*, pp. 2810–2813, 2003.

[313] H. D. White and E. W. Taylor, "Energetics and mechanism of actomyosin adenosine triphosphatase," *Biochemistry*, vol. 15, pp. 5818–5826, 1976.

[314] M. Brune, J. L. Hunter, J. E. Corrie, and M. R. Webb, "Direct, real-time measurement of rapid inorganic phosphate release using a novel fluorescent probe and its application to actomyosin subfragment 1 ATPase," *Biochemistry*, vol. 33, pp. 8262–8271, 1994.

[315] A. M. Gordon, M. Regnier, and E. Homsher, "Skeletal and cardiac muscle contractile activation: tropomyosin "rocks and rolls"," *News Physiol. Sci.*, vol. 16, pp. 49–55, 2001.

[316] G. Seemann, S. Lurz, D. U. J. Keller, D. L. Weiss, E. P. Scholz, and O. Dössel, "Adaption of Mathematical Ion Channel Models to measured data using the Particle Swarm Optimization," in *IFMBE Proceedings*, vol. 22, pp. 2507?–2510, 2008.

[317] S. R. Jernigan, G. D. Buckner, J. W. Eischen, and D. R. Cormier, "Finite element modeling of the left atrium to facilitate the design of an endoscopic atrial retractor," *Journal of Biomechanical Engineering*, vol. 129, pp. 825–837, 2007.

[318] E. S. Di Martino, C. Bellini, and D. S. Schwartzman, "In vivo porcine left atrial wall stress: Computational model," *Journal of Biomechanics*, vol. 44, pp. 2589–2594, 2011.

[319] E. S. Di Martino, C. Bellini, and D. S. Schwartzman, "In vivo porcine left atrial wall stress: Effect of ventricular tachypacing on spatial and temporal stress distribution," *Journal of Biomechanics*, vol. 44, pp. 2755–2760, 2011.

[320] U. Ravens, "Mechano-electric feedback and arrhythmias," *Prog. Biophys. Mol. Biol.*, vol. 82, pp. 255–266, 2003.

[321] M. R. Franz and F. Bode, "Mechano-electrical feedback underlying arrhythmias: The atrial fibrillation case," *Prog. Biophys. Mol. Biol.*, vol. 82, pp. 163–174, 2003.

[322] F. Ravelli, "Mechano-electric feedback and atrial fibrillation," *Progress in Biophysics and Molecular Biology*, vol. 82, pp. 137–149, 2003.

[323] N. Kuijpers, H. T. Eikelder, P. Bovendeerd, S. Verheule, T. Arts, and P. Hilbers, "Mechano-electric feedback leads to conduction slowing and block in acutely dilated atria: a modeling study of cardiac electromechanics," *Am J Physiol Heart Circ Physiol.*, vol. 292, pp. H2832–H2853, 2007.

[324] N. Kuijpers, H. ten Eikelder, and S. Verheule, "Atrial anatomy influences onset and termination of atrial fibrillation: a computer model study," *Lecture Notes in Computer Science*, vol. 5528, pp. 285–294, 2009.

[325] N. H. L. Kuijpers, M. Potse, P. M. van Dam, H. M. M. ten Eikelder, S. Verheule, F. W. Prinzen, and U. Schotten, "Mechanoelectrical coupling enhances initiation and affects perpetuation of atrial fibrillation during acute atrial dilation," *Heart Rhythm*, vol. 8, pp. 429–436, 2011.

[326] D. F. Scollan, A. Holmes, J. Zhang, and R. L. Winslow, "Reconstruction of cardiac ventricular geometry and fiber orientation using magnetic resonance imaging," *Annals of Biomedical Engineering*, vol. 28, pp. 934–944, 2000.

[327] D. Rohmer, A. Sitek, and G. T. Gullberg, "Reconstruction and visualization of fiber and laminar structure in the normal human heart from ex vivo diffusion tensor magnetic resonance imaging (DTMRI) data," *Investigative Radiology*, vol. 42, pp. 777–789, 2007.

[328] N. Toussaint, M. Sermesant, C. T. Stoeck, S. Kozerke, and P. G. Batchelor, "In vivo human 3D cardiac fibre architecture: reconstruction using curvilinear interpolation of diffusion tensor images," *Med Image Comput Comput Assist Interv (MICCAI)*, vol. 13, pp. 418–425, 2010.

[329] H. Lombaert, J.-M. Peyrat, P. Croisille, S. Rapacchi, L. Fanton, P. Clarysse, H. Delingette, and N. Ayache, *Functional Imaging and Modeling of the Heart 2011, LNCS*, vol. 6666 of *Lecture Notes in Computer Science*, ch. Statistical Analysis of the Human Cardiac Fiber Architecture from DT-MRI, pp. 171–179. Springer Berlin / Heidelberg, 2011.

[330] H. Lombaert, J. Peyrat, P. Croisille, S. Rapacchi, L. Fanton, F. Cheriet, P. Clarysse, I. Magnin, H. Delingette, and N. Ayache, "Human Atlas of the Cardiac Fiber Architecture: Study on a Healthy Population," *IEEE Transactions on Medical Imaging*, 2012.

[331] D. D. J. Streeter and D. L. Bassett, "An engineering analysis of myocardial fibre orientation in pig," *The Anatomical Record*, vol. 155, pp. 503–511, 1966.

[332] P. Helm, M. F. Beg, M. I. Miller, and R. L. Winslow, "Measuring and mapping cardiac fiber and laminar architecture using diffusion tensor MR imaging," *Annals of the New York Academy of Sciences*, vol. 1047, pp. 296–307, 2005.

[333] S. Y. Wan and W. E. Higgins, "Symmetric Region Growing," *IEEE Transactions on Image Processing*, vol. 12, pp. 1007–1015, 2003.

[334] I. J. LeGrice, B. H. Smaill, L. Z. Chai, S. G. Edgar, J. B. Gavin, and P. J. Hunter, "Laminar structure of the heart: ventricular myocyte arrangement and connective tissue architecture in the dog," *Am. J. Physiol.*, vol. 269, pp. H571–H582, 1995.

[335] G. Seemann, D. L. Weiss, E. P. Scholz, F. B. Sachse, and O. Dössel, "Effects of a KCNA5 Mutation in a Human Atrial Myocyte: A Computational Study," in *Biophys. J (Annual Meeting Abstracts)*, 2008.

[336] A. Frangi, W. Niessen, K. Vincken, and M. Viergever, "Multiscale vessel enhancement filtering," in *Medical Image Computing and Computer-Assisted Interventation (MICCAI 1998)* (W. Wells, A. Colchester, and S. Delp, eds.), vol. 1496, pp. 130–137, Springer Berlin / Heidelberg, 1998.

[337] J. Zhao, M. W. Krueger, G. Seemann, S. Meng, H. Zhang, O. Dössel, I. J. LeGrice, and B. H. Smaill, "Myofiber orientation and electrical activation in human and sheep atrial models," in *34th Annual International IEEE EMBS Conference*, 2012.

[338] J. Weese, J. Peters, I. Waechter, R. Kneser, H. Lehmann, O. Ecabert, H. Barschdorf, F. M. Weber, O. Doessel, and C. Lorenz, "The generation of patient-specific heart models for diagnosis and interventions," in *Lecture Notes in Computer Science*, vol. 6364, pp. 25–35, 2010.

[339] H. Nickisch, H. Barschdorf, F. M. Weber, M. W. Krueger, and O. Dössel, "From Image to Personalized Cardiac Simulation: Encoding Biophysical Structures into a Model-Based Segmentation Framework," in *Statistical Atlases and Computational Models of the Heart 2012*, 2012.

[340] M. W. Krueger, W. H. W. Schulze, K. Rhode, R. Razavi, G. Seemann, and O. Dössel, "Towards Personalized Clinical in-silico Modeling of Atrial Anatomy and Electrophysiology," *Medical & Biological Engineering & Computing (epub ahead of print)*, 2012.

[341] M. J. Earley and R. J. Schilling, "Catheter and surgical ablation of atrial fibrillation," *Heart*, vol. 92, pp. 266–274, 2006.

[342] J. E. Marine, J. Dong, and H. Calkins, "Catheter ablation therapy for atrial fibrillation," *Progress in Cardiovascular Diseases*, vol. 48, pp. 178–192, 2005.

[343] G. Seemann, F. B. Sachse, C. D. Werner, and O. Dössel, "Simulation of surgical interventions: Atrial radio frequency ablation with a haptic interface," in *Proc. CARS 2002*, pp. 49–54, 2002.

[344] H. Kottkamp, G. Hindricks, D. Hammel, R. Autschbach, J. Mergenthaler, M. Borggrefe, G. Breithardt, F. W. Mohr, and H. H. Scheld, "Intraoperative radiofrequency ablation of chronic atrial fibrillation: a left atrial curative approach by elimination of anatomic "anchor" reentrant circuits," *Journal of Cardiovascular Electrophysiology*, vol. 10, pp. 772–780, 1999.

[345] S.-A. Chen and C.-T. Tai, "Catheter ablation of atrial fibrillation originating from the non-pulmonary vein foci," *Journal of Cardiovascular Electrophysiology*, vol. 16, pp. 229–232, 2005.

[346] C. Pappone, G. Oreto, F. Lamberti, G. Vicedomini, M. L. Loricchio, S. Shpun, M. Rillo, M. P. Calabro, A. Conversano, S. A. Ben-Haim, R. Cappato, and S. Chierchia, "Catheter ablation of paroxysmal atrial fibrillation using a 3D mapping system," *Circulation*, vol. 100, pp. 1203–1208, 1999.

[347] S. Benussi, C. Pappone, S. Nascimbene, G. Oreto, A. Caldarola, P. L. Stefano, V. Casati, and O. Alfieri, "A simple way to treat chronic atrial fibrillation during mitral valve surgery: the epicardial radiofrequency approach," *Eur J Cardiothorac Surg*, vol. 17, pp. 524–529, 2000.

[348] C. Pappone, F. Manguso, G. Vicedomini, F. Gugliotta, O. Santinelli, A. Ferro, S. Gulletta, S. Sala, N. Sora, G. Paglino, G. Augello, E. Agricola, A. Zangrillo, O. Alfieri, and V. Santinelli, "Prevention of iatrogenic atrial tachycardia after ablation of atrial fibrillation: a prospective randomized study comparing circumferential pulmonary vein ablation with a modified approach," *Circulation*, vol. 110, pp. 3036–3042, 2004.

[349] J. Melo, P. Adragao, J. Neves, M. M. Ferreira, M. M. Pinto, M. J. Rebocho, L. Parreira, and T. Ramos, "Surgery for atrial fibrillation using radiofrequency catheter ablation: assessment of results at one year," *Eur J Cardiothorac Surg*, vol. 15, pp. 851–4; discussion 855, 1999.

[350] I. Deisenhofer, H. Estner, and A. Pustowoit, vol. Catheter Ablation of Cardiac Arrhythmias, ch. Catheter ablation of atrial fibrillation, pp. 211–246. Steinkopff.

[351] A. T. Winfree, *Computational Biology of the Heart*, ch. Rotors, fibrillation, and dimensionality. Wiley, 1997.

[352] O. Camara, M. Sermesant, P. Lamata, L. Wang, M. Pop, J. Relan, M. D. Craene, H. Delingette, H. Liu, S. Niederer, A. Pashaei, G. Plank, D. Romero, R. Sebastian, K. C. L. Wong, H. Zhang, N. Ayache, A. F. Frangi, P. Shi, N. P. Smith, and G. A. Wright, "Inter-model consistency and complementarity: Learning from ex-vivo imaging and electrophysiological data towards an integrated understanding of cardiac physiology," *Progress in Biophysics and Molecular Biology*, vol. 107, pp. 122–133, 2011.

[353] F. B. Sachse, M. Glas, M. Müller, and K. Meyer-Waarden, "Segmentation and Tissue-Classification of the Visible Man Dataset Using the Computertomographic Scans and the Thin-Section Photos," in *Proc. First Users Conference of the National Library of Medicines Visible Human Project*, 1996.

[354] C. Werner, F. Sachse, and O. Dössel, "Applications of the visible human male dataset in electrocardiology: simulation of the electrical excitation propagation," in *Proceedings of 2nd User Conference of the National Liberary of Medicines Visible Human Project*, pp. 69–70, 1998.

[355] N. S. Couto, S. Barros, F. B. Sachse, C. D. Werner, and G. Seemann, "Segmentation and tissue-classification of the visible female dataset - non cardiac thoracic organs," in *Biomedizinische Technik*, vol. 46, pp. 512–513, 2001.

[356] S. Barros, N. S. Couto, F. B. Sachse, C. D. Werner, and G. Seemann, "Segmentation and tissue-classification of the visible female dataset - thoracic muscles, bones and blood vessels," in *Biomedizinische Technik*, vol. 46-1, pp. 510–511, 2001.

[357] C. Lorenz and J. von Berg, "A comprehensive shape model of the heart," *Medical Image Analysis*, vol. 10, pp. 657–670, 2006.

[358] O. Ecabert, J. Peters, H. Schramm, C. Lorenz, J. von Berg, M. J. Walker, M. Vembar, M. E. Olszewski, K. Subramanyan, G. Lavi, and J. Weese, "Automatic model-based segmentation of the heart in CT images," *IEEE Transactions on Medical Imaging*, vol. 27, pp. 1189–1201, 2008.

[359] O. Ecabert, J. Peters, M. J. Walker, T. Ivanc, C. Lorenz, J. von Berg, J. Lessick, M. Vembar, and J. Weese, "Segmentation of the heart and great vessels in CT images using a model-based adaptation framework," *Medical Image Analysis*, vol. 15, pp. 863–876, 2011.

[360] R. Hanna, "Model-based segmentation of the left atrium for electrophysiological simulations," Diplomarbeit, Institut für Biomedizinische Technik, Karlsruher Institut für Technologie (KIT), Karlsruhe, 2009.

[361] P. Neher, "Model-based segmentation of the right atrium for electrophysiological simulations," Diplomarbeit, Institut für Biomedizinische Technik, Karlsruher Institut für Technologie (KIT), Karlsruhe, 2010.

[362] M. W. Krueger, M. Schwarz, C. Heilmann, and O. Doessel, "A real-time haemodynamics-based temperature model for hypothermic patients," in *IV International Conference on Computational Bioengineering, Book of Abstracts*, 2009.

[363] D. Kutra, "Automatic Model-based Segmentation of the Left Atrium in Cardiac MRI-Scans," Diplomarbeit, Institute of Biomedical Engineering, Karlsruhe Institute of Technology (KIT), Karlsruhe, Hamburg, 2011.

[364] D. Kutra, A. Saalbach, H. Lehmann, A. Groth, S. Dries, M. W. Krueger, O. Dössel, and J. Weese, "Automatic multi-model-based segmentation of the left atrium in cardiac MRI scans," in *MICCAI 2012, Lecture Notes in Computer Science*, vol. 7510, 2012.

[365] A. Teimourian, "Simulation der Vorhoferregung in einer neuen Softwareumgebung," Bachelor Thesis, Institut für Biomedizinische Technik, Karlsruher Institut für Technologie (KIT), Karlsruhe, 2011.

[366] C. Sánchez, M. W. Krueger, G. Seemann, O. Dössel, E. Pueyo, and B. Rodríguez, "Ionic Modulation of Atrial Fibrillation Dynamics in a Human 3D Atrial Model," in *Computing in Cardiology*, 2012.

[367] M. O. Bernabeu, R. Bordas, P. Pathmanathan, J. Pitt-Francis, J. Cooper, A. Garny, D. J. Gavaghan, B. Rodriguez, J. A. Southern, and J. P. Whiteley, "CHASTE: incorporating a novel multi-scale spatial and temporal algorithm into a large-scale open source library," *Philosophical Transactions. Series A, Mathematical, Physical, and Engineering Sciences*, vol. 367, pp. 1907–1930, 2009.

[368] D. Gianni, S. McKeever, T. Yu, R. Britten, H. Delingette, A. Frangi, P. Hunter, and N. Smith, "Sharing and reusing cardiovascular anatomical models over the Web: a step towards the implementation of the virtual physiological human project," *Philosophical Transactions. Series A, Mathematical, Physical, and Engineering Sciences*, vol. 368, pp. 3039–3056, 2010.

[369] W. Dzeakou, "Evaluation des Einflusses von Ablationsnarben auf die Erregungsausbreitung in patientenspezifischen Vorhofmodellen," Bachelor Thesis, Institut für Biomedizinische Technik, Karlsruher Institut für Technologie (KIT), Karlsruhe, 2011.

[370] R. Ionasec, I. Voigt, V. Mihalef, S. Grbic, D. Vitanovski, Y. Wang, Y. Zheng, J. Hornegger, N. Navab, B. Georgescu, and D. Comaniciu, "Patient-Specific Modeling of the Heart: Applications to Cardiovascular Disease Management," in *Statistical Atlases; Computational Models of the Heart 2010, Lecture Notes in Computer Science* (O. Camara, M. Pop, K. Rhode, M. Sermesant, N. Smith, and A. Young, eds.), vol. 6364, pp. 14–24, Springer Berlin / Heidelberg, 2010.

[371] V. Mihalef, R. I. Ionasec, P. Sharma, B. Georgescu, I. Voigt, M. Suehling, and D. Comaniciu, "Patient-specific modelling of whole heart anatomy, dynamics and haemodynamics from four-dimensional cardiac CT images," *Interface Focus*, vol. 1, pp. 286–296, 2011.

[372] R. Manzke, C. Meyer, O. Ecabert, J. Peters, N. J. Noordhoek, A. Thiagalingam, V. Y. Reddy, R. C. Chan, and J. Weese, "Automatic segmentation of rotational x-ray images for anatomic intra-procedural surface generation in atrial fibrillation ablation procedures," *IEEE Transactions on Medical Imaging*, vol. 29, pp. 260–272, 2010.

[373] I. Voigt, T. Mansi, V. Mihalef, R. Ionasec, A. Calleja, E. Mengue, P. Sharma, H. Houle, B. Georgescu, J. Hornegger, and D. Comaniciu, *LNCS (FIMH 2011)*, vol. 6666 of *Lecture Notes in Computer Science*, ch. Patient-Specific Model of Left Heart Anatomy, Dynamics and Hemodynamics from 4D TEE: A First Validation Study, pp. 341–349. Springer Berlin / Heidelberg, 2011.

[374] Y. Zheng, T. Wang, M. John, S. K. Zhou, J. Boese, and D. Comaniciu, "Multi-part left atrium modeling and segmentation in C-arm CT volumes for atrial fibrillation ablation," *Med Image Comput Comput Assist Interv (MICCAI)*, vol. 14, pp. 487–495, 2011.

[375] P. Zerfass, F. B. Sachse, C. D. Werner, and O. Dössel, "Deformation of surface nets for interactive segmentation of tomographic data," in *Biomedizinische Technik*, vol. 45-1, pp. 483–484, 2000.

[376] "CGAL, Computational Geometry Algorithms Library." http://www.cgal.org.

[377] B. N. Delaunay, "Sur la sphere vide," *Izv. Akad. Nauk SSSR, Otdelenie Matematicheskii i Estestvennyka Nauk*, vol. 6, pp. 793–800, 1934.

[378] T. Heimann and H.-P. Meinzer, "Statistical shape models for 3D medical image segmentation: a review," *Medical Image Analysis*, vol. 13, pp. 543–563, 2009.

[379] H. Homann, P. Bornert, H. Eggers, K. Nehrke, O. Dössel, and I. Graesslin, "Toward individualized SAR models and in vivo validation," *Magnetic Resonance in Medicine*, vol. 66, pp. 1767–1776, 2011.

[380] H. Homann, *SAR Prediction and SAR Management for Parallel Transmit MRI*. PhD thesis, Karlsruhe, KIT Scientific Publishing, 2012.

[381] N. Toussaint, T. Mansi, H. Delingette, N. Ayache, and M. Sermesant, "An Integrated Platform for Dynamic Cardiac Simulation and Image Processing: Application to Personalised Tetralogy of Fallot Simulation.," in *Proc. Eurographics Workshop on Visual Computing for Biomedicine (VCBM)*, 2008.

[382] Asclepios Team, Inria, Sophia-Antipolis, "Cardioviz3d." http://www-sop.inria.fr/asclepios/software/CardioViz3D.

[383] Scientific Computing and Imaging Institute (SCI), "Seg3d: Volumetric image segmentation and visualization." http://www.seg3d.org.

[384] "3dslicer." http://www.slicer.org/.

[385] D. U. J. Keller, O. Dössel, and G. Seemann, "Evaluation of rule-based approaches for the incorporation of skeletal muscle fiber orientation in patient-specific anatomies," in *Proceedings Computers in Cardiology*, vol. 36, pp. 181–184, 2009.

[386] D. U. J. Keller, O. Jarrousse, T. Fritz, S. Ley, O. Dössel, and G. Seemann, "Impact of physiological ventricular deformation on the morphology of the T-wave: a hybrid, static-dynamic approach," *IEEE Trans Biomed Eng.*, vol. 58, pp. 2109–2119, 2011.

[387] D. U. J. Keller, D. L. Weiss, O. Dössel, and G. Seemann, "Influence of I(Ks) heterogeneities on the genesis of the T-wave: a computational evaluation," *IEEE Trans. Biomed. Engineering*, vol. 59, pp. 311–322, 2012.

[388] M. W. Krueger, V. Schmidt, F. M. Weber, D. U. J. Keller, G. Seemann, and O. Dössel, "Influence of atrial anisotropy and conduction on the body surface ECG in a variety of patient specific models," in *Cardiac Physiome Conference 2011*, (Merton College, Oxford), p. 53, Smith, N., 2011.

[389] K. S. McDowell, F. Vadakkumpadan, R. Blake, J. Blauer, G. Plank, R. S. Macleod, and N. A. Trayanova, "Methodology for patient-specific modeling of atrial fibrosis as a substrate for atrial fibrillation," *Journal of Electrocardiology*, 2012.

[390] W. H. W. Schulze, M. W. Krueger, Y. Jiang, K. Rhode, F. M. Weber, D. Caulfield, B. R. Knowles, R. Razavi, and O. Dössel, "Localization of the Atrial Excitation Origin by Reconstruction of Time-Integrated Transmembrane Voltages," in *Biomedizinische Technik/Biomedical Engineering*, vol. 55, pp. 103–106, 2010.

[391] W. H. W. Schulze, M. W. Krueger, K. Rhode, R. Razavi, and O. Dössel, "Non-invasive imaging of activation times in the atria - can excitation patterns be reconstructed?," in *Biomedizinische Technik / Biomedical Engineering (Proc. BMT 2011)*, vol. 56, 2011.

[392] A. A. Young and A. F. Frangi, "Computational cardiac atlases: from patient to population and back," *Experimental Physiology*, vol. 94, pp. 578–596, 2009.

[393] S. Ordas, E. Oubel, R. Sebastian, and A. Frangi, "Computational Anatomy Atlas of the Heart," in *Proc. 5th Internat. Symp. Image and Signal Processing and Analysis*, pp. 338–342, 2007.

[394] M. Backhaus, R. Britten, J. Do Chung, B. Cowan, C. Fonseca, P. Medrano-Gracia, W. Tao, and A. Young, "The Cardiac Atlas Project: Development of a Framework Integrating Cardiac Images and Models," in *Statistical Atlases and Computational Models of the Heart* (O. Camara and Pop, eds.), vol. 6364, pp. 54–64, 2010.

[395] H. Koch, R. D. Bousseljot, O. Kosch, C. Jahnke, I. Paetsch, E. Fleck, and B. Schnackenburg, "A reference dataset for verifying numerical electrophysiological heart models," *Biomedical Engineering Online*, vol. 10, p. 11, 2011.

[396] S. A. Niederer, E. Kerfoot, A. P. Benson, M. O. Bernabeu, O. Bernus, C. Bradley, E. M. Cherry, R. Clayton, F. H. Fenton, A. Garny, E. Heidenreich, S. Land, M. Maleckar, P. Pathmanathan, G. Plank, J. F. Rodriguez, I. Roy, F. B. Sachse, G. Seemann, O. Skavhaug, and N. P. Smith, "Verification of cardiac tissue electrophysiology simulators using an N-version benchmark," *Philosophical Transactions of the Royal Society A: Mathematical, Physical and Engineering Sciences*, vol. 369, pp. 4331–4351, 2011.

[397] T. Ashihara, R. Haraguchi, K. Nakazawa, T. Namba, T. Ikeda, Y. Nakazawa, T. Ozawa, M. Ito, M. Horie, and N. A. Trayanova, "The role of fibroblasts in complex fractionated electrograms during persistent/permanent atrial fibrillation: implications for electrogram-based catheter ablation," *Circulation Research*, vol. 110, pp. 275–284, 2012.

[398] S. Severi, D. Pogliani, G. Fantini, P. Fabbrini, M. R. Viganò, E. Galbiati, G. Bonforte, A. Vincenti, A. Stella, and S. Genovesi, "Alterations of atrial electrophysiology induced by electrolyte variations: combined computational and P-wave analysis," *Europace*, 2010.

[399] M. W. Krueger, S. Severi, K. Rhode, S. Genovesi, F. M. Weber, A. Vincenti, P. Fabbrini, G. Seemann, R. Razavi, and O. Dössel, "Alterations of atrial electrophysiology related to hemodialysis session: insights from a multi-scale computer model," *Journal of Electrocardiology*, vol. 44, pp. 176–183, 2011.

[400] G. Seemann, P. Carillo, D. L. Weiss, M. W. Krueger, O. Dössel, and E. P. Scholz, "Investigating Arrhythmogenic Effects of the hERG Mutation N588K in Virtual Human Atria," in *Lecture Notes in Computer Science*, vol. 5528, pp. 144–153, Ayache, N.; Delingette, H.; Sermesant, M., 2009.

[401] N. Konrad, "Methoden für die Integration von Ionenstrommessdaten in Modelle kardialer Elektrophysiologie," Bachelor Thesis, Institute of Biomedical Engineering, Karlsruhe Institute of Technology (KIT), Karlsruhe, 2012.

[402] M. Wilhelms, J. Schmid, M. J. Krause, N. Konrad, J. Maier, E. P. Scholz, V. Heuveline, O. Dössel, and G. Seemann, "Calibration of Human Cardiac Ion Current Models to Patch Clamp Measurement Data," in *Computing in Cardiology*, vol. 39, (Kraków, Poland), p. 62, 2012.

[403] F. Holmqvist, P. G. Platonov, R. Havmoller, and J. Carlson, "Signal-averaged P wave analysis for delineation of interatrial conduction - further validation of the method," *BMC Cardiovascular Disorders*, vol. 7, p. 29, 2007.

[404] F. Holmqvist, J. Carlson, J. E. P. Waktare, and P. G. Platonov, "Noninvasive evidence of shortened atrial refractoriness during sinus rhythm in patients with paroxysmal atrial fibrillation," *Pacing and Clinical Electrophysiology*, vol. 32, pp. 302–307, 2009.

[405] D. Deng and L. Xia, "Study the effect of tissue heterogeneity and anisotropy in atrial fibrillation based on a human atrial model," in *Proc. Computing in Cardiology*, vol. 37, (Belfast, 26-29 Sept.), pp. 433–436, 2010.

[406] L. Uldry, N. Virag, V. Jacquemet, J.-M. Vesin, and L. Kappenberger, "Spontaneous termination of atrial fibrillation: study of the effect of atrial geometry in a biophysical model," *Conf Proc IEEE Eng Med Biol Soc*, vol. 2009, pp. 4504–4507, 2009.

[407] L. Uldry, N. Virag, V. Jacquemet, J.-M. Vesin, and L. Kappenberger, "Optimizing local capture of atrial fibrillation by rapid pacing: study of the influence of tissue dynamics," *Annals of Biomedical Engineering*, vol. 38, pp. 3664–3673, 2010.

[408] N. M. S. de Groot, R. P. M. Houben, J. L. Smeets, E. Boersma, U. Schotten, M. J. Schalij, H. Crijns, and M. A. Allessie, "Electropathological substrate of longstanding persistent atrial fibrillation in patients with structural heart disease: epicardial breakthrough," *Circulation*, vol. 122, pp. 1674–1682, 2010.

[409] M. A. Allessie, N. M. S. de Groot, R. P. M. Houben, U. Schotten, E. Boersma, J. L. Smeets, and H. J. Crijns, "Electropathological substrate of long-standing persistent atrial fibrillation in patients with structural heart disease: longitudinal dissociation," *Circulation. Arrhythmia and Electrophysiology*, vol. 3, pp. 606–615, 2010.

[410] S. Verheule, E. Tuyls, A. van Hunnik, M. Kuiper, U. Schotten, and M. Allessie, "Fibrillatory conduction in the atrial free walls of goats in persistent and permanent atrial fibrillation," *Circulation. Arrhythmia and Electrophysiology*, vol. 3, pp. 590–599, 2010.

[411] J. Eckstein, B. Maesen, D. Linz, S. Zeemering, A. van Hunnik, S. Verheule, M. Allessie, and U. Schotten, "Time course and mechanisms of endo-epicardial electrical dissociation during atrial fibrillation in the goat," *Cardiovascular Research*, vol. 89, pp. 816–824, 2011.

[412] C. Tobón, J. Sáiz, F. Hornero, C. Ruiz, V. Hernándes, and G. Molto, "Contribution of electrophysiological remodelling to generation of anatomical re-entries around the right pulmonary veins in human atrium: a simulation study," in *Computers in Cardiology*, vol. 33, pp. 773–776, 2006.

[413] G. Seemann, F. B. Sachse, M. Karl, D. L. Weiss, V. Heuveline, and O. Dössel, "Framework for modular, flexible and efficient solving the cardiac bidomain equation using PETSc," *Mathematics in Industry*, vol. 15, pp. 363–369, 2010.

[414] A. Bollmann, D. Husser, M. Stridh, L. Soernmo, M. Majic, H. U. Klein, and S. B. Olsson, "Frequency measures obtained from the surface electrocardiogram in atrial fibrillation research and clinical decision-making," *Journal of Cardiovascular Electrophysiology*, vol. 14, pp. S154–61, 2003.

[415] J. Bohnert, M. Reumann, T. Faber, and O. Dössel, "Investigation of curative ablation techniques for atrial fibrillation in a computer model," in *Gemeinsame Jahrestagung der Deutschen, der Österreichischen und der Schweizerischen Gesellschaft für Biomedizinische Technik*, 2006.

[416] M. Wilhelms, G. Seemann, M. Weiser, and O. Dössel, "Benchmarking Solvers of the Monodomain Equation in Cardiac Electrophysiological Modeling," in *Biomedizinische Technik / Biomedical Engineering*, vol. 55, pp. 99–102, 2010.

[417] L. Thomas, "Topologisch veränderbare dreidimensionale Aktive Konturen in der medizinischen Bildverarbeitung," Diplomarbeit, Universität Karlsruhe (TH), Institut für Biomedizinische Technik, 2000.

[418] F. J. Perez, M. A. Wood, and C. M. Schubert, "Effects of gap geometry on conduction through discontinuous radiofrequency lesions," *Circulation*, vol. 113, pp. 1723–1729, 2006.

[419] C. Cabo, A. M. Pertsov, W. T. Baxter, J. M. Davidenko, R. A. Gray, and J. Jalife, "Wave-front curvature as a cause of slow conduction and block in isolated cardiac muscle," *Circulation Research*, vol. 75, pp. 1014–1028, 1994.

[420] L. Uldry, N. Virag, J.-M. Vesin, and L. Kappenberger, *Patient-Specific Modeling of the Cardiovascular System*, ch. Studies of Therapeutic Strategies for Atrial Fibrillation Based on a Biophysical Model of the Human Atria, pp. 63–79. Springer New York, 2010.

[421] P. C. Dolber and M. S. Spach, "Structure of canine Bachmann's bundle related to propagation of excitation," *The American Journal of Physiology*, vol. 257, pp. H1446–57, 1989.

[422] T. Koura, M. Hara, S. Takeuchi, K. Ota, Y. Okada, S. Miyoshi, A. Watanabe, K. Shiraiwa, H. Mitamura, I. Kodama, and S. Ogawa, "Anisotropic conduction properties in canine atria analyzed by high-resolution optical mapping: preferential direction of conduction block changes from longitudinal to transverse with increasing age," *Circulation*, vol. 105, pp. 2092–2098, 2002.

[423] F. Atienza, D. Calvo, J. Almendral, S. Zlochiver, K. R. Grzeda, N. Martinez-Alzamora, E. Gonzalez-Torrecilla, A. Arenal, F. Fernandez-Aviles, and O. Berenfeld, "Mechanisms of fractionated electrograms formation in the posterior left atrium during paroxysmal atrial fibrillation in humans," *Journal of the American College of Cardiology*, vol. 57, pp. 1081–1092, 2011.

[424] V. Jacquemet and C. S. Henriquez, "Genesis of complex fractionated atrial electrograms in zones of slow conduction: a computer model of microfibrosis," *Heart Rhythm*, vol. 6, pp. 803–810, 2009.

[425] T. Fritz and O. Dössel, "Simulating the beating heart within the pericardium using finite element analysis," in *Biomedical Engineering / Biomedizinische Technik*, vol. 57, p. 330, 2012.

[426] K. H. W. J. ten Tusscher and A. V. Panfilov, "Alternans and spiral breakup in a human ventricular tissue model," *American Journal of Physiology. Heart and Circulatory Physiology*, vol. 291, pp. H1088–100, 2006.

[427] D. U. J. Keller, R. Kalayciyan, O. Dössel, and G. Seemann, "Fast creation of endocardial stimulation profiles for the realistic simulation of body surface ECGs," in *IFMBE Proceedings World Congress on Medical Physics and Biomedical Engineering*, vol. 25/4, pp. 145–148, 2009.

[428] J. M. Guccione, A. D. McCulloch, and L. K. Waldman, "Passive material properties of intact ventricular myocardium determined from a cylindrical model," *J. Biomechanical Engineering*, vol. 113, pp. 42–55, 1991.

[429] J. M. Guccione, K. D. Costa, and A. D. McCulloch, "Finite element stress analysis of left ventricular mechanics in the beating dog heart," *J. Biomechanics*, vol. 28, pp. 1167–1177, 1995.

[430] Y. Takahashi, M. D. O'Neill, M. Hocini, P. Reant, A. Jonsson, P. Jais, P. Sanders, T. Rostock, M. Rotter, F. Sacher, S. Laffite, R. Roudaut, J. Clementy, and M. Haissaguerre, "Effects of stepwise ablation of chronic atrial fibrillation on atrial electrical and mechanical properties," 2007.

[431] K. B. Gupta, M. B. Ratcliffe, M. A. Fallert, L. H. J. Edmunds, and D. K. Bogen, "Changes in passive mechanical stiffness of myocardial tissue with aneurysm formation," *Circulation*, vol. 89, pp. 2315–2326, 1994.

[432] D. Tang, C. Yang, T. Geva, and P. J. Del Nido, "Patient-specific MRI-based 3D FSI RV/LV/patch models for pulmonary valve replacement surgery and patch optimization," *Journal of Biomechanical Engineering*, vol. 130, p. 041010, 2008.

[433] C. Yang, D. Tang, T. Geva, R. Rathod, H. Yamauchi, V. Gooty, A. Tang, G. Gaudette, K. L. Billiar, M. H. Kural, and P. J. Del Nido, "Using contracting band to improve right ventricle ejection fraction for patients with repaired tetralogy of Fallot, a modeling study using patient-specific CMR-based 2-layer anisotropic models of human right and left ventricles," *The Journal of Thoracic and Cardiovascular Surgery*, 2012.

[434] L. E. Hudsmith, S. E. Petersen, J. M. Francis, M. D. Robson, and S. Neubauer, "Normal human left and right ventricular and left atrial dimensions using steady state free precession magnetic resonance imaging," *Journal of Cardiovascular Magnetic Resonance*, vol. 7, pp. 775–782, 2005.

[435] P. Marino, A. M. Prioli, G. Destro, I. LoSchiavo, G. Golia, and P. Zardini, "The left atrial volume curve can be assessed from pulmonary vein and mitral valve velocity tracings," *American Heart Journal*, vol. 127, pp. 886–898, 1994.

[436] J. Richter, "Modellierung der Kraftentwicklung und Kontraktion des menschlichen Vorhofs," Diplomarbeit, Institut für Biomedizinische Technik, Karlsruhe, 2012.

[437] R. S. Rivlin and A. G. Thomas, "Rupture of rubber. I. Characteristic energy for tearing," *Journal of Polymer Science*, vol. 10, pp. 291–318, 1953.

[438] Y. Matsuda, Y. Toma, H. Ogawa, M. Matsuzaki, K. Katayama, T. Fujii, F. Yoshino, K. Moritani, T. Kumada, and R. Kusukawa, "Importance of left atrial function in patients with myocardial infarction," *Circulation*, vol. 67, pp. 566–571, 1983.

[439] T. Poutanen, A. Ikonen, P. Vainio, E. Jokinen, and T. Tikanoja, "Left atrial volume assessed by transthoracic three dimensional echocardiography and magnetic resonance imaging: dynamic changes during the heart cycle in children," *Heart*, vol. 83, pp. 537–542, 2000.

[440] D. M. Bers, "Cardiac excitation-contraction coupling," *Nature*, vol. 415, pp. 198–205, 2002.

[441] S. Genovesi, D. Pogliani, A. Faini, M. G. Valsecchi, A. Riva, F. Stefani, I. Acquistapace, A. Stella, G. Bonforte, A. DeVecchi, V. DeCristofaro, G. Buccianti, and A. Vincenti, "Prevalence of atrial fibrillation and associated

factors in a population of long-term hemodialysis patients," *American Journal of Kidney Diseases*, vol. 46, pp. 897–902, 2005.

[442] S. Genovesi, A. Vincenti, E. Rossi, D. Pogliani, I. Acquistapace, A. Stella, and M. G. Valsecchi, "Atrial fibrillation and morbidity and mortality in a cohort of long-term hemodialysis patients," *American Journal of Kidney Diseases*, vol. 51, pp. 255–262, 2008.

[443] E. Vazquez, C. Sanchez-Perales, F. Garcia-Garcia, P. Castellano, M.-J. Garcia-Cortes, A. Liebana, and C. Lozano, "Atrial fibrillation in incident dialysis patients," *Kidney International*, vol. 76, pp. 324–330, 2009.

[444] P. Korantzopoulos, S. Kokkoris, T. Liu, I. Protopsaltis, G. Li, and J. A. Goudevenos, "Atrial fibrillation in end-stage renal disease," *PACE*, vol. 30, pp. 1391–1397, 2007.

[445] M. W. Krueger, S. Severi, G. Seemann, F. M. Weber, S. Genovesi, A. Vincenti, P. Fabbrini, and O. Dössel, "Alterations of atrial electrophysiology after hemodialysis session: a simulation study in volumetric atrial tissues," in *Journal of Electrocardiology (Proc. ICE 2010)*, vol. 44, p. e2, 2011.

[446] S. Genovesi, C. Dossi, M. R. Vigano, E. Galbiati, F. Prolo, A. Stella, and M. Stramba-Badiale, "Electrolyte concentration during haemodialysis and QT interval prolongation in uraemic patients," *Europace*, vol. 10, pp. 771–777, 2008.

[447] S. V. Pandit, M. Warren, S. Mironov, E. G. Tolkacheva, J. Kalifa, O. Berenfeld, and J. Jalife, "Mechanisms underlying the antifibrillatory action of hyperkalemia in Guinea pig hearts," *Biophysical Journal*, vol. 98, pp. 2091–2101, 2010.

[448] N. Chandler, O. Aslanidi, D. Buckley, S. Inada, S. Birchall, A. Atkinson, D. Kirk, O. Monfredi, P. Molenaar, R. Anderson, V. Sharma, D. Sigg, H. Zhang, M. Boyett, and H. Dobrzynski, "Computer three-dimensional anatomical reconstruction of the human sinus node and a novel paranodal area," *Anatomical Record (Hoboken, N.J. : 2007)*, vol. 294, pp. 970–979, 2011.

[449] M. S. Buiten, M. K. De Bie, J. Rotmans, L. Van Erven, A. J. Rabelink, M. J. Schalij, and J. W. Jukema, "Is the high prevelance of atrial fibrillation in end stage renal disease related to the hemodialysis procedure itself?," in *European Heart Journal*, vol. 33, p. 374, 2012.

[450] M. Schwarz, C. Heilmann, M. W. Krueger, and U. Kiencke, "Model based monitoring of hypothermic patients," *Metrology & Measurement Systems*, vol. XVI, pp. 443–455, 2009.

[451] M. Schwarz, M. W. Krueger, H.-. J. Busch, C. Benk, and C. Heilmann, "Model-Based Assessment of Tissue Perfusion and Temperature in Deep Hypothermic Patients," *IEEE Transactions on Biomedical Engineering*, vol. 57, pp. 1577–1586, 2010.

List of Publications and Supervised Thesis

Journal Articles

- **M. W. Krueger**, A. Dorn, D. U. J. Keller, F. Holmqvist, J. Carlson, P. G. Platonov, K. S. Rhode, R. Razavi, G. Seemann, and O. Dössel, *In-Silico Modeling of Atrial Repolarization in Normal and Atrial Fibrillation Remodeled State*, in revision, 2012

- **M. W. Krueger**, W. H. W. Schulze, K. Rhode, R. Razavi, G. Seemann, and O. Dössel, *Towards Personalized Clinical in-silico Modeling of Atrial Anatomy and Electrophysiology*, Medical & Biological Engineering & Computing (in press), 2012

- **M. W. Krueger**, G. Seemann, K. Rhode, D. U. J. Keller, C. Schilling, A. Arujuna, J. Gill, M. D. O'Neill, R. Razavi, and O. Dössel. *Personalization of atrial anatomy and elelectophysiology as a basis for clinical modeling of radio-frequency-ablation of atrial fibrillation.* IEEE Transactions on Medical Imaging (epub ahead of print), 31, 2012

- **M. W. Krueger**, S. Severi, K. Rhode, S. Genovesi, F. M. Weber, A. Vincenti, P. Fabbrini, G. Seemann, R. Razavi, and O. Dössel. *Alterations of atrial electrophysiology related to hemodialysis session: insights from a multiscale computer model.* Journal of Electrocardiology, 44:176–183, 2011.

- O. Dössel, **M. W. Krueger**, F. M. Weber, M. Wilhelms, and G. Seemann. *Computational modeling of the human atrial anatomy and electrophysiology.* Medical & Biological Engineering & Computing, 50:773–799, 2012

- M. Burdumy, A. Luik, P. Neher, R. Hanna, **M. W. Krueger**, C. Schilling, H. Barschdorf, C. Lorenz, G. Seemann, C. Schmitt, O. Dössel, and F. M. Weber. *Comparing measured and simulated wave directions in the left atrium – a workflow for model personalization and validation.* Biomed Tech (Berl), 57:79–87, 2012

- F. M. Weber, A. Luik, C. Schilling, G. Seemann, **M. W. Krueger**, C. Lorenz, C. Schmitt, and O. Dössel. *Conduction velocity restitution of the human atrium – an efficient measurement protocol for clinical electrophysiological studies.* IEEE Transactions on Biomedical Engineering, 58:2648–2655, 2011

- N. Smith, A. de Vecchi, M. McCormick, O. Camara, A. F. Frangi, H. Delingette, M. Sermesant, N. Ayache, **M. W. Krueger**, W. H. W. Schulze, R. Hose, I. Valverde, P. Beerbaum, C. Staicu, M. Siebes, J. Spaan, P. Hunter, J. Weese, H. Lehmann, D. Chapelle, and R. Rezavi. *euHeart: personalized and integrated cardiac care using patient-specific cardiovascular modelling.* Interface Focus, 1(3):349–364, 2011.

- G. Seemann, D. U. J. Keller, **M. W. Krueger**, F. M. Weber, M. Wilhelms, and O. Dössel. *Electrophysiological modeling for cardiology: methods and potential applications.* Information Technology, 52:242–249, 2010.

- M. Schwarz, M. W. Krueger, H.-J. Busch, C. Benk, and C. Heilmann. *Model-based assessment of tissue perfusion and temperature in deep hypothermic patients.* IEEE Transactions on Biomedical Engineering, 57:1577–1586, 2010.

- M. Schwarz, C. Heilmann, **M. W. Krueger**, and U. Kiencke. *Model based monitoring of hypothermic patients.* Metrology & Measurement Systems, XVI:443–455, 2009

Book Chapters

- G. Seemann, **M. W. Krueger**, and M. Wilhelms, Health Academy, vol. 16, ch. *Elektrophysiologische Modellierung und Virtualisierung für die Kardiologie – Methoden und potenzielle Anwendungen.* 2013

- O. Dössel, **M. W. Krueger**, and G. Seemann, Cardiac Mapping, vol. 4, ch. *Personalized Electrophysiological Modeling of the Human Atrium.* Wiley-Blackwell, 2013

Refereed Conference Articles

- **M. W. Krueger**, V. Schmidt, C. Tobón, F. M. Weber, C. Lorenz, D. U. J. Keller, H. Barschdorf, M. Burdumy, P. Neher, G. Plank, K. Rhode, G. Seemann, D. Sanchez-Quintana, J. Saiz, R. Razavi, and O. Dössel, *Modeling atrial fiber orientation in patient-specific geometries: a semi-automatic rule-based approach*, Functional Imaging and Modeling of the Heart 2011, Lecture Notes in Computer Science (L. Axel and D. Metaxas, eds.), vol. 6666, pp. 223–232, 2011

- **M. W. Krueger**, K. Rhode, F. M. Weber, D. U. J. Keller, D. Caulfield, G. Seemann, B. R. Knowles, R. Razavi, and O. Dössel, *Patient-Specific Volumetric Atrial Models with Electrophysiological Components: A Comparison of Simulations and Measurements*, Biomedizinische Technik / Biomedical Engineering, vol. 55(s1), pp. 54–57, 2010

- **M. W. Krueger**, F. M. Weber, G. Seemann, and O. Dössel, *Semi-automatic segmentation of sinus node, Bachmann's Bundle and Terminal Crest for patient specific atrial models*, World Congress on Medical Physics and Biomedical Engineering. IFMBE Proceedings, vol. 25/4, pp. 673–676, Springer Heidelberg, 2009

- **M. W. Krueger**, D. L. Weiss, and O. Doessel, *Intraventricular outweighs transmural dispersion of repolarization after epicardial pacing in a virtual human left ventricle*, Biomedical Engineering / Biomedizinische Technik, vol. 52(s1), 2007

- **M. W. Krueger**, D. L. Weiss, G. Seemann, and O. Dössel, *Die Begrenztheit theoretischer Modelle der menschlichen Biologie und ihr großer Nutzen für das Verständnis des Körpers*, Challenges and Limitations of Models in Science and Theology, 2007

- C. Sánchez, **M. W. Krueger**, G. Seemann, O. Dössel, E. Pueyo, and B. Rodríguez, *Ionic Modulation of Atrial Fibrillation Dynamics in a Human 3D Atrial Model*, Computing in Cardiology, 2012

- J. Zhao, **M. W. Krueger**, G. Seemann, S. Meng, H. Zhang, O. Dössel, I. J. LeGrice, and B. H. Smaill, *Myofiber orientation and electrical activation in human and sheep atrial models*, 34th Annual International IEEE EMBS Conference, 2012

- A. Dorn, **M. W. Krueger**, G. Seemann, and O. Dössel, *Modelling of heterogeneous human atrial electrophysiology*, Biomedical Engineering / Biomedizinische Technik, vol. 57(s1), 2012

- D. T. Rudolph, W. H. W. Schulze, D. Potyagaylo, **M. W. Krueger**, and O. Dössel, *Reconstruction of atrial excitation conduction velocities and implementation into the inverse problem of electrocardiography*, Biomedizinische Technik / Biomedical Engineering, vol. 57(s1), pp. 179–182, 2012

- D. Kutra, A. Saalbach, H. Lehmann, A. Groth, S. Dries, **M. W. Krueger**, O. Dössel, and J. Weese, *Automatic multi-model-based segmentation of the left atrium in cardiac MRI scans*, 15th International Conference on Medical Image Computing and Computer Assisted Intervention (MICCAI), 2012

- H. Nickisch, H. Barschdorf, F. M. Weber, **M. W. Krueger**, and O. Dössel, *From Image to Personalized Cardiac Simulation: Encoding Biophysical Structures into a Model-Based Segmentation Framework*, Statistical Atlases and Computational Models of the Heart (STACOM) 2012, 2012

- O. Dössel, **M. W. Krueger**, F. M. Weber, C. Schilling, H. W. H. Schulze, and G. Seemann, *A framework for personalization of computational models of the human atria*, Proc EMBS, vol. 2011, pp. 4324–4328, 2011

- S. Ponto, C. Schilling, **M. W. Krueger**, F. M. Weber, S. Seemann, and O. Dössel, *Influence of endocardial catheter contact on properties of the atrial signal and comparison with simulated electrograms*, Biomedizinische Technik / Biomedical Engineering, vol. 56(s1), 2011

- P. Neher, H. Barschdorf, S. Dries, F. M. Weber, **M. W. Krueger**, O. Dössel, and C. Lorenz, *Automatic segmentation of cardiac CTs – personalized atrial models augmented with electrophysiological structures*, Functional Imaging and Modeling of the Heart 2011, Lecture Notes in Computer Science, vol. 6666, pp. 80–87, 2011

- R. Hanna, H. Barschdorf, T. Klinder, F. M. Weber, **M. W. Krueger**, O. Dössel, and C. Lorenz, *A hybrid method for automatic anatomical variant detection and segmentation*, Functional Imaging and Modeling of the Heart 2011, Lecture Notes in Computer Science, vol. 6666, pp. 333–340, 2011

- W. H. W. Schulze, **M. W. Krueger**, Y. Jiang, K. Rhode, F. M. Weber, D. Caulfield, B. R. Knowles, R. Razavi, and O. Dössel, *Localization of the Atrial Excitation Origin by Reconstruction of Time-Integrated Transmembrane Voltages*, Biomedizinische Technik/Biomedical Engineering, vol. 55(s1), pp. 103–106, 2010

- M. Burdumy, F. M. Weber, A. Luik, R. Hanna, **M. W. Krueger**, C. Schilling, H. Barschdorf, C. Lorenz, G. Seemann, C. Schmitt, and O. Dössel, *Comparing Measured and Simulated Incidence Directions in the Left Atrium – A Workflow for Model Personalization and Validation*, Biomedizinische Technik / Biomedical Engineering, vol. 55(s1), pp. 50–53, 2010

- G. Seemann, P. Carillo, D. L. Weiss, **M. W. Krueger**, O. Dössel, and E. P. Scholz, *Investigating Arrhythmogenic Effects of the hERG Mutation N588K in Virtual Human Atria*, Functional Imaging and Modeling of the Heart 2009, Lecture Notes in Computer Science, vol. 5528, pp. 144–153, 2009

Refereed Conference Abstracts

- **M. W. Krueger**, G. Seemann, and O. Dössel, *Towards personalized biophysical models of atrial anatomy and electrophysiology in clinical environments*, Biomedical Engineering / Biomedizinische Technik, vol. 57(s1), 2012

- **M. W. Krueger**, F. M. Weber, G. Seemann, and O. Doessel, *Personalizing Anatomical and Electrophysiological Models of the Human Atria*, Biomedizinische Technik / Biomedical Engineering, vol. 56(s1), 2011

- **M. W. Krueger**, S. Severi, G. Seemann, F. M. Weber, S. Genovesi, A. Vincenti, P. Fabbrini, and O. Dössel, *Alterations of atrial electrophysiology after hemodialysis session: a simulation study in volumetric atrial tissues*, Journal of Electrocardiology (Proc. ICE 2010), vol. 44, p. e2, 2011

- **M. W. Krueger**, M. Schwarz, C. Heilmann, and O. Doessel, *A real-time haemodynamics- based temperature model for hypothermic patients*, IV International Conference on Computational Bioengineering (ICCB), 2009

- A. Thorpe, Z. Chen, W. H. W. Schulze, **M. W. Krueger**, J. Realn, H. Delingette, R. Razavi, C. Rinaldi, and K. Rhode, *Enhanced workflow for BSPM and ECGi using X-ray based Electrode Localisation*, Bioengineering12, 2012

- K. Rhode, Y. Ma, J. Housden, R. Karim, C. A. Rinaldi, M. Cooklin, J. Gill, M. O'Neill, T. Schaeffter, J. Relan, M. Sermesant, H. Delingette, N. Ayache, **M. W. Krueger**, W. Schulze, G. Seemann, O. Dössel, and R. Razavi, *Clinical applications of image fusion for electrophysiology procedures*, ISBI 2012, (Barcelona), 2012

- W. H. W. Schulze, **M. W. Krueger**, K. Rhode, R. Razavi, and O. Dössel, *Noninvasive imaging of activation times in the atria – can excitation patterns be reconstructed?*, Biomedizinische Technik / Biomedical Engineering, vol. 56(s1), 2011

- W. H. W. Schulze, **M. W. Krueger**, K. Rhode, R. Razavi, and O. Dössel, *Critical times based activation time imaging*, Proc. 38th International Congress on Electrocardiology (ICE), 2011

- F. M. Weber, C. Schilling, A. Luik, **M. W. Krueger**, G. Seemann, C. Lorenz, C. Schmitt, and O. Dössel, *Quantitative determination of wave direction and conduction velocity in the human atrium from intracardiac electrograms*, Journal of Electrocardiology (Proc. ICE 2010), vol. 44, p. e14, 2011

Conference Presentations

- **M. W. Krueger**, V. Schmidt, F. M. Weber, D. U. J. Keller, G. Seemann, and O. Dössel, *Influence of atrial anisotropy and conduction on the body surface ECG in a variety of patient specific models*, Cardiac Physiome Conference 2011, (Merton College, Oxford), p. 53, 2011

- **M. W. Krueger**, V. Schmidt, D. U. J. Keller, T. Fritz, G. Seemann, and O. Dössel, *Comparison of Methods for Visualization of 3D Myocardial Fiber Structure in Printed Images*, KIT PhD Symposium 2010, p. 80, 2010

- **M. W. Krueger**, F. M. Weber, G. Seemann, and O. Dössel, *Influence of myocardial structures on electrophysiologic simulations in patient specific atrial models*, The Cardiac Physiome: Multi-scale and Multi-physics Mathematical Modelling Applied to the Heart, Isaac Newton Institute, Cambridge, 2009

- **M. W. Krueger**, F. M. Weber, O. Jarrousse, D. U. J. Keller, G. Seemann, and O. Dössel, *Anatomical, electrophysiological and mechanical modeling of the heart*, Chaste Users' & Developers' Workshop, 2009

- O. Dössel, **M. W. Krueger**, W. H. W. Schulze, F. M. Weber, and G. Seemann, *From atrial imaging and atrial mapping to atrial modelling*, Cardiac Physiome Conference 2011, (Merton College, Oxford), p. 58, 2011

- F. M. Weber, G. Seemann, C. Schilling, **M. W. Krueger**, and O. Dössel, *Towards patient-specific simulations of atrial fibrillation*, The Cardiac Physiome: Multi-scale and Multi-physics Mathematical Modelling Applied to the Heart, Isaac Newton Institute, Cambridge, 2009

Invited Talks

- **M. W. Krueger**, *Towards Personalized Clinical in-silico Modeling of Atrial Anatomy and Electrophysiology*, Cardiac Delayed Enhancement Magnetic Resonance Image Segmentation (cDEMRIS), IEEE International Symposium on Biomedical Imaging (ISBI) 2012 (Barcelona, Spain), 2012

- **M. W. Krueger**, *Anatomical and Electrophysiological Modeling of the Atria*, Philips Research (Hamburg, Germany), 2012

- **M. W. Krueger**, F. M. Weber, G. Seemann, and O. Dössel, *Modeling of atrial arrhythmias for therapy planing*, CardioStim 2010 (Nice, France), 2010

Reports and Theses

- **M. W. Krueger**, J. Relan, C. Zhong, M. Sermesant, N. Ayache, G. Seemann, O. Dössel, N. Linton, C. A. Rinaldi, R. Razavi, K. Rhode, H. Delingette, and euHeart Consortium, *euHeart D6.3.3: Evaluation of RFA planning based on biophysical models*, euHeart Deliverable, 2012

- C. Schilling, and **M. W. Krueger**, "Verfassen von Ausarbeiten am IBT", Workshop Manuscript, Hochschuldidaktisches Zentrum Baden-Württemberg, 2011

- J. Relan, H. Delingette, M. Sermesant, N. Ayache, **M. W. Krueger**, W. Schulze, G. Seemann, O. Dössel, and euHeart Consortium, *euHeart D6.3.2: Planning of Radio-Frequency Ablation for patients with Ventricular Tachycardia*, euHeart Deliverable, 2011

- **M. W. Krueger**, M. Keller, J. Relan, H. Delingette, M. Sermesant, N. Ayache, G. Seemann, O. Dössel, and euHeart Consortium, *euHeart D6.3.1: Planning of RFA for patients with AF*, euHeart Deliverable, 2011

- W. H. W. Schulze, **M. W. Krueger**, H. Delingette, K. Rhode, G. Seemann, O. Doessel, and euHeart Consortium, *euHeart D6.1.2: Update of ECGI acquisition and processing*, euHeart Deliverable, 2010

- **M. W. Krueger**, J. Relan, H. Delingette, M. Sermesant, N. Ayache, W. H. W. Schulze, G. Seemann, O. Dössel, and euHeart Consortium, *euHeart D6.2.2: Model Personalization for AF and VT simulation*, euHeart Deliverable, 2011

- **M. W. Krueger**, J. Relan, H. Delingette, M. Sermesant, N. Ayache, G. Seemann, O. Dössel, and euHeart Consortium, *euHeart D6.2.1: Computational Models for AF and VT simulation*, euHeart Deliverable, 2010

- **M. W. Krueger**, W. Schulze, K. Rhode, O. Dössel, G. Seemann, H. Delingette, and euHeart Consortium, *euHeart D6.1.1: ECGI acquisition and processing*, euHeart Deliverable, 2009

- **M. W. Krüger**, *Modellierung des Einflusses der Hämodynamik auf die Temperaturverteilung während extrakorporaler Zirkulation*, Diplomarbeit, Universität Karlsruhe (TH), Karlsruhe, 2008

- **M. W. Krüger**, *Erhöhung der transmuralen Dispersion der Repolarisation durch epikardiale Stimulation eines virtuellen linken Ventrikels*, Studienarbeit, Universität Karlsruhe, Institute of Biomedical Engineering, 2007

Supervised Student Theses

- J. Wingerter, *Modellerzeugung für die Segmentierung der linken Vorhofwand des Herzens*, Diplomarbeit (extern), Institute of Biomedical Engineering, Karlsruhe Institute of Technology (KIT), Karlsruhe, Philips Research, Hamburg, ongoing

- M. Ly, *Elektrophysiologische und anatomische Auswirkungen von Ablationstherapie*, Studienarbeit, Institute of Biomedical Engineering, Karlsruhe Institute of Technology (KIT), Karlsruhe, ongoing

- L.-M. Busch, *Modeling of the Elastomechnical Properties of Radio-Frequency Ablation Scars: Influence of Different Lesion Patterns on Cardiac Contraction*, Bachelor Thesis, Institute of Biomedical Engineering, Karlsruhe Institute of Technology (KIT), Karlsruhe, 2012

- J. Richter, *Modellierung der Kraftentwicklung und Kontraktion des menschlichen Vorhofs*, Diplomarbeit, Institute of Biomedical Engineering, Karlsruhe Institute of Technology (KIT), Karlsruhe, 2012

- B. Verma, *Analysis of clinical 3D activation time data for the personalization of human atrial models*, Master Thesis, Institute of Biomedical Engineering, Karlsruhe Institute of Technology (KIT), Karlsruhe, Karlsruhe, 2012

- A. Dorn, *Modellierung von elektrophysiologischen Heterogenitäten in den Herz-Vorhöfen*, Bachelor Thesis, Institute of Biomedical Engineering, Karlsruhe Institute of Technology (KIT), Karlsruhe, 2011

- D. Kutra, *Automatic Model-based Segmentation of the Left Atrium in Cardiac MRI- Scans*, Diplomarbeit (extern), Institute of Biomedical Engineering, Karlsruhe Institute of Technology (KIT), Karlsruhe, Philips Research, Hamburg, 2011

- M. Sirkin, *Simulation der Vorhofaktivierung zur Modellpersonalisierung*, Studienarbeit, Institute of Biomedical Engineering, Karlsruhe Institute of Technology (KIT), Karlsruhe, 2011

- A. Teimourian, *Simulation der Vorhoferregung in einer neuen Softwareumgebung*, Bachelor Thesis, Institute of Biomedical Engineering, Karlsruhe Institute of Technology (KIT), Karlsruhe, 2011

- W. Dzeakou, *Evaluation des Einflusses von Ablationsnarben auf die Erregungsausbreitung in patientenspezifischen Vorhofmodellen*, Bachelor Thesis, Institute of Biomedical Engineering, Karlsruhe Institute of Technology (KIT), Karlsruhe, 2011

- V. Schmidt, *Semi-automatische Methode zur Modellierung der Faserorientierung in patienten-spezifischen Vorhofgeometrien*, Diplomarbeit, Institute of Biomedical Engineering, Karlsruhe Institute of Technology (KIT), Karlsruhe, 2010

Awards & Grants

- Third place DGBMT student competition, BMT 2012 – 46. Jahrestagung der DGBMT:
  A. Dorn, **M. W. Krueger**, G. Seemann, and O. Dössel, *Modelling of heterogeneous human atrial electrophysiology*

- Young Investigator Award, International Congress on Electrocardiology, 2011:
  W. H. W. Schulze, **M. W. Krueger**, K. Rhode, R. Razavi, and O. Dössel, *Critical times based activation time imaging*

- Further Education Scholarship, Karlsruhe House of Young Scientists (KHYS), 2011

- Networking Scholarship, Karlsruhe House of Young Scientists (KHYS), 2010

- Special equipment acquisition grant, Universität Karlsruhe (TH), 2009

- Student Scholarship, Evangelisches Studienwerk Villigst e.V., 2005 – 2008

- Student Scholarship, e-fellows.net, 2003 – 2012

Karlsruhe Transactions on Biomedical Engineering
(ISSN 1864-5933)

Karlsruhe Institute of Technology / Institute of Biomedical Engineering (Ed.)

Die Bände sind unter www.ksp.kit.edu als PDF frei verfügbar oder als Druckausgabe bestellbar.

Karlsruhe Transactions on Biomedical Engineering
(ISSN 1864-5933)

Band 10 Sebastian Seitz
Magnetic Resonance Imaging on Patients with Implanted Cardiac Pacemakers. 2011
ISBN 978-3-86644-610-6

Band 11 Tobias Voigt
Quantitative MR Imaging of the Electric Properties and Local SAR based on Improved RF Transmit Field Mapping. 2011
ISBN 978-3-86644-598-7

Band 12 Frank Michael Weber
Personalizing Simulations of the Human Atria: Intracardiac Measurements, Tissue Conductivities, and Cellular Electrophysiology. 2011
ISBN 978-3-86644-646-5

Band 13 David Urs Josef Keller
Multiscale Modeling of the Ventricles: from Cellular Electrophysiology to Body Surface Electrocardiograms. 2011
ISBN 978-3-86644-714-1

Band 14 Oussama Jarrousse
Modified Mass-Spring System for Physically Based Deformation Modeling. 2012
ISBN 978-3-86644-742-4

Band 15 Julia Bohnert
Effects of Time-Varying Magnetic Fields in the Frequency Range 1 kHz to 100 kHz upon the Human Body: Numerical Studies and Stimulation Experiment. 2012
ISBN 978-3-86644-782-0

Band 16 Hanno Homann
SAR Prediction and SAR Management for Parallel Transmit MRI. 2012
ISBN 978-3-86644-800-1

Band 17 Christopher Schilling
Analysis of Atrial Electrograms. 2012
ISBN 978-3-86644-894-0

Band 18 Tobias Baas
ECG Based Analysis of the Ventricular Repolarisation in the Human Heart. 2012
ISBN 978-3-86644-882-7

Karlsruhe Transactions on Biomedical Engineering
(ISSN 1864-5933)

Band 19   Martin Wolfgang Krüger
**Personalized Multi-Scale Modeling of the Atria: Heterogeneities,
Fiber Architecture, Hemodialysis and Ablation Therapy.** 2012
ISBN 978-3-86644-948-0